P9-CMW-762

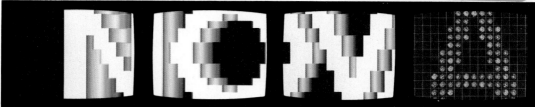

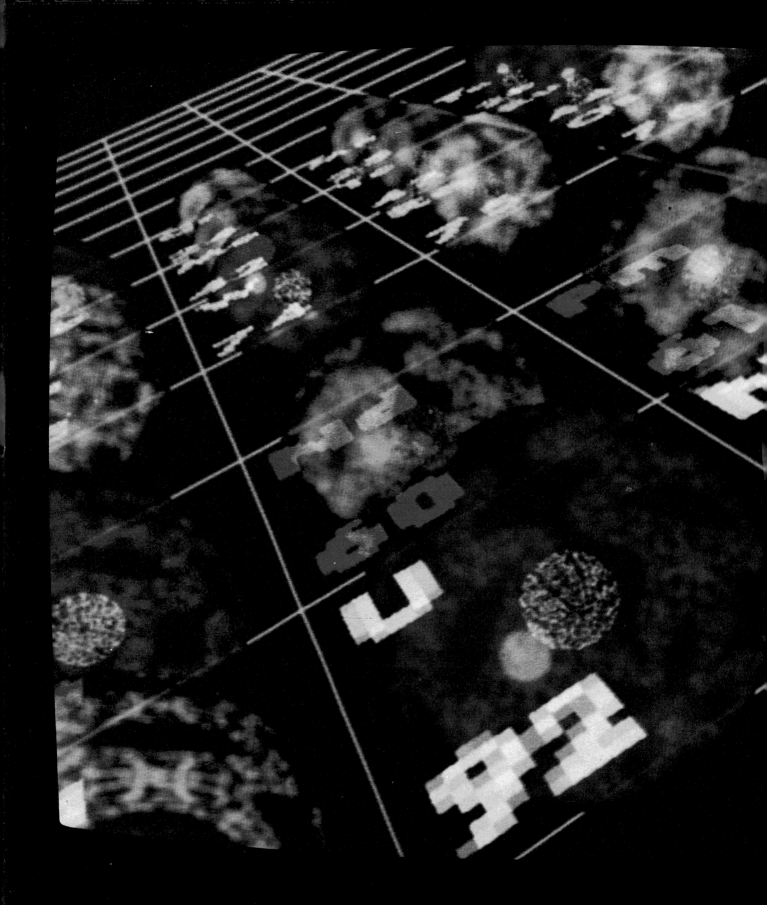

NOVA

ADVENTURES IN SCIENCE

WGBH BOSTON

ADDISON-WESLEY

PUBLISHING COMPANY

Throughout its history, "NOVA" has received funding from a broad range of agencies, foundations and corporations without whose generous support the series would not have been possible. They include: The Corporation for Public Broadcasting, The National Science Foundation, Carnegie Corporation of New York, The Arthur Vining Davis Foundations, Polaroid, Exxon, TRW Inc., and The Johnson & Johnson Family of Companies. In addition, WGBH would like to acknowledge the continuing support of "NOVA" by Public Television Stations, nationwide.

Copyright © 1983
by WGBH Educational Foundation

All rights reserved. No part of this publication may be reproduced, stored in a retrieval system, or transmitted, in any form or by any means, electronic, mechanical, photocopying, recording, or otherwise, without the prior written permission of the publisher. Printed in the United States of America. Published simultaneously in Canada. Library of Congress Catalog Card No. 82-16306.

ISBN 0-201-05358-6
ISBN 0-201-05359-4 paperback

BCDEFGHIJ-KR-85432

Library of Congress
Cataloging in Publication Data
Main entry under title:
NOVA, adventures in science.

  Includes index.
   1. Television in science education.
   2. NOVA (Television program)
   I. WGBH (Television station: Boston, Mass.)
Q196.N68 1982   500   82-16306
ISBN 0-201-05358-6
ISBN 0-201-05359-4 (pbk.)

Second printing November 1982

Science is, to many people, mysterious and incomprehensible–equations, computers, laboratories, unfamiliar language, unthinkable miracles. Some even have said that science is the new religion of the twentieth century, with scientists in priestly robes murmuring incantations. Often the processes and products of science seem so obscure that we are tempted not to pry.

Yet "NOVA" has always held that science is neither secret lore nor sacred ritual. Rather, at its heart, science is simply people asking questions. Behind every dazzling new display of technology, every miracle cure or drug, every startling feat or discovery are men and women who have searched for answers. Scientists in all disciplines–from computers and chemistry to geology and genetics–embody this unique combination of qualities–curiosity, perserverance, and a willingness to take risks, to hazard guesses rather than remain unknowing.

This book offers you glimpses of scientists and the work they do. We have tried to peer briefly but carefully into many areas to extract the nugget or two that reveal the nature of scientific inquiry. In doing so we do not expect to have defined science in its entirety; its domain is too vast for any one book to accomplish that. Rather, this book offers an eclectic sampling of some of the adventures of scientists at work. Its aim, like that of science itself, is to unravel some of the puzzles of life and to reveal a rich universe of new perspectives and possibilities.

Contents

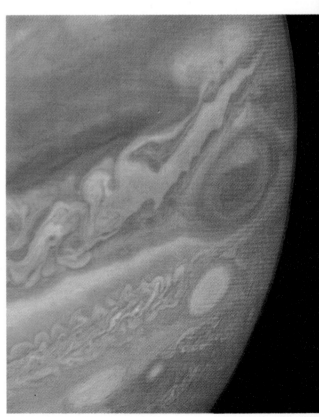

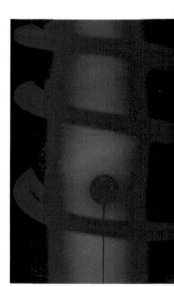

The Making of "NOVA"
John Mansfield, Executive Producer

Did you see that extraordinary 'NOVA' program last night all about dinosaurs?"

"No, I'm afraid we missed it."

"You missed a treat. Still, you must have seen their show on Mount St. Helens."

"No, we were out. It's a shame; I heard it was quite good."

That typical conversation reflects one of the most frustrating elements of television production and is one of the main reasons for this, the first "NOVA" book. Television is such a transitory medium that it's only too easy to miss an important program when it first airs or when it's repeated. Thus *NOVA*, a volume to enable you to relish the subjects of many of the "NOVAs" you've never managed to see and to give you fresh recollections of the "NOVAs" you did experience.

John Mansfield, current executive producer of "NOVA."

NOVA: Adventures in Science is the celebration of a major television achievement, for the book heralds the beginning in 1983 of the tenth season of the longest-running science documentary series in America. The "NOVA" series has already spanned a critical period in the history of virtually every branch of science, ranging from astronomy to medicine, from ecology to genetics. Judging from "NOVA's" popularity and success, television viewers seem to value the kind of depth and responsibility that have become the hallmark of our television journalism. And popular it is indeed: "NOVA" broke PBS records in 1981, when more than 20 million viewers watched "The Anatomy of a Volcano," an amazing film about the eruptions of Mount St. Helens. The "NOVA" production team risked their lives to reveal the awesome spectacle and lethal power as the mountain literally blew its top.

But "NOVA's" popularity isn't evidenced merely by our having become a national viewing habit for millions of Americans. It's also reflected in the fact that the show regularly wins major television awards. In 1982 "NOVA's" investigation into the current spate of fires and arson in the United States, entitled "Why America Burns," was awarded an Emmy. "NOVA" also won the Ohio State Award for all network series. And since 1977 "NOVA" has won no fewer than twelve major awards at the American Film Festival, five of them in 1982.

As executive producer, I'd love to boast that this success is all due to me. I'm sorry to have to admit that it's not, and that I'm just one player—and a recent one at that—in a team effort and a story that goes back to the Second World War. Hollywood had firmly established itself as the world center of feature films, and the major studios were beginning to achieve dominant status in creative quality as well as in box office success. The art of film documentary was, however, with a few major exceptions, still in its infancy. The United States, Germany, and Great Britain all attempted to set up documentary units whose prime purpose was propaganda in support of the war effort. In Britain, under brilliant producers such as John Grierson, documentary films emerged not as jingoist war propaganda but as a new and very popular art form. The techniques in writing and cinematography they developed were to be the mainstay of documentaries in Britain after the postwar rebirth of television. Every week on BBC, and then on the new commercial channel Independent Television, one-hour film documentaries without reporters or anchorpeople became part of the British way of life.

In America, documentary makers were wooed by television news and developed, on the major networks, probably the best reportage of its type on Earth. But their expertise was focused mainly on five-minute segments, and rarely on the complexities of a documentary lasting a whole hour. Then, in the early 1960s, the BBC was given a second channel. Suddenly it needed quality programs beyond opera, soap opera, Shakespeare, and public affairs.

The BBC is still the largest single broadcasting conglomerate there is, and British subjects, with characteristic lack of modesty, expect their network to turn out the best television in the world, night after night. And it thrives, although doesn't always succeed, in that ambition. When BBC 2 arrived, it was agreed among the executives on the sixth floor of BBC Television Center in Shepherd's Bush, London, that Science with a capital *S* must be given a special chance. There was little hope that it would be popular, but it was generally agreed that a dose of science television would do the country good. A few BBC entrepreneurs realized that they in turn might make good by doing just that.

Within a department called Television Outside Broadcasts, which had previously concerned itself mainly with Wimbledon, soccer, cricket, and British events such as the trooping of the colors and royal weddings, a genial ball-of-fire named Aubrey Singer was promoted to set up a new department called Outside Broadcasts, Science and Features.

One of the first series Aubrey Singer encouraged his new department to produce was "Horizon," a forty-five-minute studio-based science program with major film segments. It was anchored by Olympic runner Chris Chataway, one of the athletes who paced Roger Bannister into running the world's first four-minute mile. "Horizon" went on to recruit several prestigious BBC film documentary producers, who soon rebelled, unanimously, at the idea of their films being anything less than one hour, and all shot in the field. Fortunately, the series was well received, so "Horizon" was formalized as a series of single-subject, all-on-film documentaries, each producer having the privilege of making almost an hour of television by himself.

At about this time, on the other side of the Atlantic, Boston's fine public television station was reconsidering its policy. WGBH (its call letters derived from the site of its transmitter, Great Blue Hill) was thinking seriously about its long-term educational role in public broadcasting.

Science broadcasting had had its roots in shows such as "Spectrum" and "M.I.T. Science Reporter." But by the early 1970s, all but a few science shows were off the air. On the one hand, Don Herbert as Mr. Wizard was a family friend; on the other, Walter Cronkite seemed to know more about the *Apollo* mission than even the astronauts themselves. But in between there was little else of substance. WGBH Boston decided this educational gap needed refilling and culled its resources to launch what was to become one of the flagships of PBS.

WGBH's associate director of programming, Michael Ambrosino, had planned to take his fortieth year as sabbatical and received an American Fellowship Abroad of the Corporation for Public Broadcasting to spend a year as a working guest at the BBC. There he encountered "Horizon" and decided that, with some adaptations, it was a worthy model for America. Ambrosino returned home and spent a year developing the project and another half-year raising the money to have the series produced. He invented the title "NOVA," and returned to Britain to persuade a handful of BBC-trained producers to accompany him to the United States to help launch the program. He had also returned with several finished "Horizon" programs, and in the spring of 1974 "NOVA" was ready to debut with one of those films: "The Making of a Natural History Film," a magnificent wildlife film that revealed the unique, painstaking work of the Oxford Scientific Film Unit.

"The Making of a Natural History Film" tackled such technical problems as filming a wood-wasp laying its eggs inside a tree, the

The rising sun, "NOVA's" opening film sequence for most of its life.

Michael Ambrosino, original executive producer of "NOVA."

Cinematographer Boyd Estus takes a mug shot of the star of "NOVA" #110, "The First Signs of Washoe." *(Standing left to right)* **Dr. Roger Fouts, sound recordist Eric Neudel, and writer and producer Simon Campbell-Jones.**

hatching of a chick, and the courtship rituals of a tiny fish, appropriately called a stickleback. One epic sequence in the film, and a classic of wildlife cinematography, showed the seemingly defenseless stickleback being stealthily stalked by a predator possibly thirty times its size. The climax came as the big fish lunged forward to snap its jaws and engulf the stickleback. Poor little stickleback, we all thought. But no. A few moments later, with its eyes popping and its tail thrashing in helpless convulsions, the predator reopened its mouth to spit out its apparently unpalatable prey with an almost human expression of disgust and pain. The tiny stickleback swam away in proud defiance, presumably to live happily ever after.

Ambrosino, however, did much more than import British films and producers. He wanted "NOVA" to be a distinctly American program. He recruited American producers, directors, and writers, with the result that today "NOVA" is largely an American home-made enterprise. Moreover, whereas in the early years "NOVA" was heavily dependent on BBC productions to fill its schedule, today the BBC presents American-made "NOVAs" as part of its "Horizon" series. Similarly, American-made "NOVAs" are now shown all around the world. And in the cur-

Filming a sequence for "NOVA" #818, "Notes of a Biology Watcher: A Film with Lewis Thomas." *(Left to right)* **Lewis Thomas, production assistant Joshua Lobel, cinematographer Peter Hoving, and producer Robin Bates.**

rent "NOVA" team of twenty-two, there remain only two British expatriates.

Michael's goal in creating "NOVA" was to share good stories of the physical and natural sciences, to show examples, as he says, "of how the world works and how people interact." His personal motivation was that he was curious about how the world was made, and he translated this curiosity into every film he instigated for "NOVA." His problems, as he now relates them, were more journalistic than anything else: trying to make a good story out of an idea, creating films about subjects ranging from chimpanzees to swamp algae.

As the current executive producer of "NOVA," one of my most difficult problems is choosing the new films for each year. Sometimes I feel like a Ping-Pong ball poised on a fountain of water in a shooting range at a fair, except that the pressure comes from all sides. "NOVA" producers and the rest of the "NOVA" team bombard me with ideas for new programs. So does our advisory panel, which includes such luminaries as Professor Philip Morrison and his brilliant wife, Phylis Morrison, Professor Stephen Jay Gould, Professor Gerald Holton, Dr. Harry Woolf, Dr. Daniel Fox, Professor Arthur Kantrowitz, and several other very wise people. Lobbies also proffer suggestions—pronuke, antinuke, left, right, Moral Majority, anti–Moral Majority, antipollution, big business, and so on—to name but a very small sample.

Management of the WGBH Foundation in Boston and PBS in Washington drops ideas in our laps as well. Unfortunately, most of their ideas are very good, and my juggling act gets rather difficult. I've never had a moment when I haven't had at least fifty excellent "NOVA" ideas fighting for places in our schedule of twenty new shows per year. For while the range of potential "NOVA" subjects is almost infinite, each idea must fit neatly into the series as a whole. With only twenty new shows each year, you might neither thank us nor watch us if we brought you four programs focused solely on hazardous wastes, or six episodes about rodents. And we try to make the subjects timely and the bill of fare as lively as possible.

Once the subjects are decided, the real work begins. We're constantly asked, "How *do* you make a 'NOVA'?" And there's no easy answer. The process should be simple: research the idea (say, two months), film it (say, three weeks), edit the film (allow about ten weeks), and air it on PBS. Would that it were so simple. Every stage of that scenario is so fraught with potential mishaps that the perfectly smooth production, from our getting the original idea to your seeing the film at home, just never happens. It more often is a matter of crises and compromises.

Sometimes the key remark in the irreplaceable interview with Dr. X or Professor Y will somehow have a scratch on it, or the production team will fly to Lagos only to discover that all the equipment has gone to Moscow. In one example I can recall, filming a simple segment turned into a Keystone Kops routine.

I was directing a film on the futuristic ideas of an eminent professor of engineering. He'd invented "Able Mable: The Robot Housemaid," and although he had not succeeded in making a production-line prototype, he'd designed lots of gadgets that would be incorporated in the final model. Mabel would make the beds, do the washing, cook the meals, and so on. She was going to do all sorts of highly sophisticated things—even for superior robots—such as climb stairs and pick up fragile eggs.

We began by patiently filming one of Mabel's components as it attempted to climb

The parade of twins on Twins Day in Twinsburg, Ohio. "NOVA" teams sometimes use anything that's available to get that perfect "trucking" shot, as seen here in the filming of "NOVA" #820, "Twins." *(Left to right)* Producer and director Linda Harrar, cameraman David Westphal, and soundman Ray Cymoszinski.

a staircase. It made three steps and then suddenly every cog in its body ripped apart in a kind of chainsaw suicide. Mabel fell backwards and died, and her built-in fire-extinguishing abilities nearly asphyxiated the rest of us. As for picking up eggs (a test of sensitive dexterity), she'd have made good only in an omelet house.

Mabel was also supposed to be able to take the dog for a walk and do the shopping. As no full-scale Mabel prototype existed, we built a look-alike to complete the filming. Imagine a small refrigerator on wheels, with arms and a head (not an endearing E.T. head, but a digital display with electronic eyes and ears and a long neck). Since the mechanism that would provide Mabel's brains was still on the drawing board, we had to commission a very short actress, barely three feet tall, to "drive" the contraption. So far so good.

"Let's have a nice big dog for Mabel to walk," someone suggested. "It will give a sense of scale to the exercise, and it might be fun." My first mistake was the St. Bernard. The next mistake was not looking beyond the end of my nose: the sidewalk directly in front of the house where we were filming was hard asphalt, but ten yards beyond was

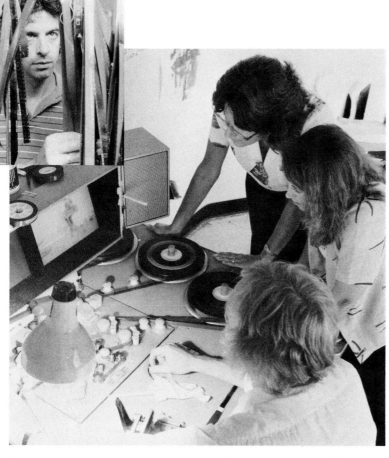

During the editing of "NOVA" #916, "Mississippi" (working title). *(Left from top to bottom)* Associate producer Betsy Anderson, producer Barbara Gullahorn-Holecek, and film editor Bill Anderson. Good sections are selected out and hung in order for possible use in the final film. Pictured in the inset is film editor Peter Neudel. Several stages of editing from rough to fine versions follow, until the film is ready to be locked in place for the exact time period required for broadcast. Pictured at right are associate producer Peter Argentine and film editor Joy Manesiotis.

fresh new tar. Our actress squeezed herself into our model of Mabel. We hung a shopping basket on one of Mabel's arms and attached the dog's leash to the other.

I'd decided to do a tracking shot, which meant that we'd put the camera on a dolly (a platform on wheels) that would move to follow the action. In true Hollywood style I called "Action!" and Mabel, with the shopping bag on one plastic arm and the St. Bernard dog attached to the other, emerged from the driveway. Our "robot" turned the corner and proceeded toward the shops. The sequence was going so well that we figured we'd probably not even have to reshoot it.

Then Mabel hit the soft tar of the new sidewalk. All movement stopped. Our actress panicked and screamed, the shopping bag dropped, and the St. Bernard bolted, pulling one of Mabel's arms out of its socket and toppling Mabel herself (with the actress inside) over. The actress climbed out, none

too pleased, and it took hours to catch the dog. Mabel herself was irreparably tired.

So much for the way Hollywood is alleged to do it. But "NOVA" is mainly a documentary series; as you can see, it rarely films on a studio set and rarely uses actors; and, as its "stars" tend to be Nobel Prize winners, spiders, bacteria, electrons, and elephants, we'd either give them cardiac arrest, fry them, or be trampled to death if we followed the clapboard routine of feature films.

"NOVA's" charge is to report both accurately and responsibly on the world of science and technology. It combines the best of investigative journalism with a sense of show biz, hoping, like any other TV program, to lure viewers back week after week. Producing "NOVA" is very much an act on the high wire, however: wobble too much in one direction, be overdidactic, and you plunge your viewers into the bog of boredom; teeter too far toward easy superficiality and glitzy gimmicks, and you invite the wrath of the academic community. What's more, this trapeze act isn't performed in a totally philanthropic climate. "NOVA" is largely funded by all the stations that make up the PBS network. All the station managers of PBS meet every year at the Station Program Cooperative (SPC) to decide what program series they want or don't want and how much they're prepared to pay for them.

To justify "NOVA's" continued existence, I'm given just fifteen minutes to make a case for their continued support of "NOVA" at a series of presentations to the PBS program managers. I show a short promotional tape, make a brief speech, and then, in a question-and-answer session, with an umpire who tells me when I've run out of time, attempt to persuade them that "NOVA" isn't just good for their viewers, but the best thing since sliced bread.

PBS station program managers know their stuff. Through every kind of fund-raising activity imaginable they are just, and only just, able to buy their share of series like "The McNeil/Lehrer Report" and so on. When it comes to "NOVA," which costs them a total of nearly $5 million a year, they're very wary. But science doesn't usually have much of a shelf life; episodes become out of date much more quickly than dramas like "Upstairs, Downstairs," and so to make a series like "NOVA" demands a critical mass of production staff. We now make ten in-house "NOVAs" a year, with five production teams working full-time at full stretch. We also buy or coproduce ten more new programs. (Coproduction means we invest in someone else's show at its inception and have editorial input from the original treatment right through "rough-cut" and "fine cut," which are trade terms for the semi-edited version, which may last seventy minutes, and the final version, which is about the right length but still needs polish in both script and picture content.) Some of the shows we buy are either too short or too long to fit the "NOVA" slot on PBS: that means additional expense in either cutting the film down or shooting addi-

tional material to bring it to our length. And would that it were as simple as that!

Perhaps in a future *NOVA* book we'll explore more of the technicalities of making the shows. In this, our first, we simply hope you'll extend your interest in "NOVA" and enjoy the breadth of programs we now try to bring you. From particle physics to pandas, from plastics to prehistoric pterodactyls, we bring you new insights into both the history and the future of science—science with a capital *S* and a small *s,* from astronomy to zoology and beyond. You name it, and "NOVA" will bring you the very latest in "ologies" and "onomies," with a little bit of style and a lot of authority. And that is the goal we've aimed for in the pages that follow. I know you'll enjoy some of the extraordinary adventures unraveled here. If they whet your appetite, I hope they'll even persuade you to bring that banned television set out of the attic so that, once a week, you can enjoy the forbidden delights of "NOVA" and at last turn to your fellow commuters and say, "Hey, hold it. I saw that 'NOVA' too! And I'm going to watch the repeat as well!"

"NOVA" staff and engineers and the elaborate electronic equipment necessary to this step in the editing process. *(Left to right)* Post-production assistant Kathy Smith, post-production supervisor Nancy Linde, and producer Theodore Bogosian.

Engineer Mary Doyle transfers the final print of a film to videotape with opening sequence, titles, subtitles, credits, and any special effects —for a final master tape ready for broadcast.

A Conversation with Philip Morrison

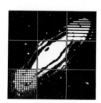

Doing science is like growing a garden," says Philip Morrison. "You sow seeds, you have to water them, you've got to tend them, and when you bring in the harvest, you have to think of next year. And not everybody gets magnificent flowers and fruits."

Sitting in his airy office at the Massachusetts Institute of Technology, Morrison grapples verbally with the underlying nature of science and scientists. A man who has worked with the most brilliant during a distinguished career as an astrophysicist, teacher, and general connoisseur of science, Morrison has thought through such general questions more thoroughly than most of his colleagues.

"A characteristic of the human mind is to formulate models of the world, so as to be able to predict what we will encounter in the next moment, the next day, the next year," Morrison begins again. "I think that science grew out of that curiosity. Everybody wants to know if it will rain tomorrow, or if I run, will I catch the bus?"

But present-day science demands much more than a well-developed curiosity. "Nowadays, you don't study things afresh. You can't go out and ask why smoke rises, for example, because an awful lot is known about that. You have to find out what's new, so that you can add something new. You have to fit into a body of knowledge that's really significant and that has many hints and helps for understanding things; therefore you don't expect to get anywhere in science if you don't have that body of knowledge which is taught in schools and included in books. There's no point trying to make a perpetual-motion machine because most people say that that's not very viable. You have to deal with what is possible—what is given to you in your time."

How do scientists recognize the areas of knowledge that are ready for plucking? "What fixes what most scientists do is that they are trained and embedded in a structure which gives them a certain feel, like citizens of the United States," Morrison continues. "If a citizen works a nine-day week and sleeps between noon and five

P.M., he's going to have a very hard time coping with life. In the same way the scientist has to deal with what's given to him—the tools, the ideas, the directions to go are given to some extent."

Even the most brilliant creative thinkers fit into the fundamental framework of science. Einstein's general theory of relativity, the treatment of gravitation that many critics regard as an extreme example of scientific individualism, fits the mold according to Morrison. After all, he explains, other physicists recognized that traditional gravitational theory was insufficient to explain the observed world entirely, and some were even working—albeit ineffectively—on new approaches to the subject in the half decade up to 1916, when Einstein published his general theory. "So, although Einstein was brilliantly original, his problems were not original," summarizes Morrison. "He was working in the broad framework."

A supreme example of the deep roots of a major find, says Morrison, is the buildup to the discovery of quasars, the remarkable starlike objects that still amaze astronomers. The story really started in 1932, when Karl Jansky first observed that the sky emitted radio waves. "He was doing it for practical purposes," recalls Morrison. "His company was studying the operation of short-wave telephone links across the ocean and wanted to know the cause of all the noise. Jansky discovered that most of it came from distant thunderstorms, but some did not—it emanated instead from outside the Earth. That was the start of radio astronomy—a direct answer to a question put on a technical basis."

Jansky didn't stun the world of science and technology. Only one other paper on radio astronomy was published in the 1930s. But then came World War II during which many physicists became familiar with and interested in radio technology. They noticed, for example, that radio noise from the sun interfered with military radar signals. After the war, as radio astronomy burgeoned into a full-scale science, astronomers realized that the sun and other stars were notably weak radio emitters. Other objects in the heavens—nebulas, remnants of supernovas, and distant galaxies—put out radio signals with billions of times the strength of the sun.

Identifying the strong emitters precisely in the optical spectrum became a major challenge. Radio astronomers had to give their optical counterparts such exact locations for their sources that the optical astronomers could spot them unerringly in fields filled with stars and other objects. That exercise produced a major puzzle: a group of strong radio emitters appeared to be stars in the optical spectrum, despite the growing dogma that stars don't emit strong radio signals.

That wasn't the only mystery. "The spectral lines of the 'stars' were of a kind never seen before," relates Morrison. "It was very exciting, but in a way frustrating."

Then in 1963 Maarten Schmidt, a young Dutch astronomer working at the Mount Palomar Observatory, had an insight. The ratios of the spectral lines from the strange stars were quite normal, he postulated, but the lines were in unusual positions. The lines came from hydrogen, the basic element in all stars, but they were shifted toward the red end of the spectrum by a factor of a few thousand. "If Schmidt had waited six months, someone else would have done it," says Morrison. "Jesse Greenstein, a very respectable spectroscopist who sat in the office next to Schmidt's, said it was dumb of him not to have seen it."

Nevertheless, the glory went to Schmidt for identifying the red shift, which implies that the objects now known as quasars lie billions of light years from Earth. "We now know 1,500 of them and know from sampling that there must be a few million," says Morrison. "Ever since, we've been trying to understand what quasars are and how they work. That's the typical way that a mixture of new instruments, obvious continuity, and brilliant understanding of some anomaly produces a discovery."

Science doesn't stand still in its methods, any more than in its discoveries. As Morrison sees it, the pattern of discovery is probably

already altering. "I think that great contributions will come less and less from single, brilliant figures and more and more from interactions of very able people working together in teams," he predicts.

Philip Morrison is himself an embodiment of the consummate scientist—a man of letters *and* data who is as comfortable writing book reviews as he is reviewing astrophysical calculations. His philosophical nature allows him to see science in its proper perspective in the world of information, without forgetting that science is at its essence a human endeavor.

This book sets out to show modern science in all its manifestations—great individuals, impressive teamwork, notable discoveries. Based on the Public Broadcasting System's extraordinarily successful "NOVA" series, the book examines the world of science and scientists from the submicroscopic regions of the atom to the enormity of the universe, stopping en route to view wonder drugs, lasers, language, space exploration, and other topics from science, technology, and medicine. "NOVA" says Morrison, "succeeds in showing the milieu and the people at work in science." This book has precisely the same objective.

Part I

O R I G I N S

Scientists are constantly exploring our heritage, searching for clues that will unravel the fundamental mysteries of life: Where did we come from? Why are we here? Unlike philosophers, who embrace the enigma whole, scientists attack this massive puzzle of existence piece by piece. They analyze rock and animal fossils, study genetics, observe our fellow inhabitants of Earth, and dig deep into the planet to understand how it works. In the process their work becomes a channel for our own curiosity, our own questions of life.

To exist means to be able to offer stories that begin, "Once upon a time. . . ." Modern science is concerned, in large part, with finding and recording these narratives. Part of its success is due to the ingenuity and persistence of the researchers themselves, but they have had an ally. We live in a universe with a remarkable memory and a clear concern for the integrity of the record. Everywhere–in the cambium layer of trees, in the cosmic fluff of comets, in mineralized sediments and skeletons, in the patterns of radiation broadcast by the stars–the flow of events has been written down, in one code or another, and secreted away like so many time capsules. Over the last several centuries we have begun to crack these codes and listen to the stories that the patterns, forms, relations, forces, and objects of nature have to tell about their origins. What follows are a few of these.

Neutrino tracks recorded in the bubble chamber at Fermilab in Batavia, Illinois, a step in the search for the elementary particles of matter.

A Whisper from Space

When we look into the sky on a clear, moonless night, the stars appear as bright white specks against a vast black canvas. And if we look through powerful telescopes, the surrealistic spirals and gaseous clouds of distant galaxies leap into focus against a backdrop of black. This broad expanse, often stretching tens and hundreds of light years between two celestial bodies, looks to be the ultimate void, a desert of silence and darkness.

But if we could turn a knob and adjust our eyes to see far beyond the rainbow of colors in the visible part of the spectrum—the radiation we know as light—we would see that space is not a black void at all. It is aglow with radiation. This radiation is so pervasive and of such significance that its discovery was the most important event in the study of cosmology in the past fifty years. It has yielded crucial information to scientists who ponder the creation of our universe, providing able support for the theory that it all started about 20 billion years ago with a Big Bang.

An edge-on view of a spiral galaxy, M104, also called the Sombrero Galaxy, in the Virgo cluster.

Its discovery is also a fascinating scientific detective story, in which astronomers and physicists used space-age equipment and techniques available only in the past two decades or so to make their observations and then applied well-established principles of physics to interpret them. Like many break-

throughs in science, this one happened through a combination of careful planning and sheer luck, and because investigators, after initial reluctance, followed the trail of evidence even though it shattered some cherished assumptions along the way.

The story opens in 1965, when Arno Penzias and Robert Wilson, two radio astronomers with the Bell Telephone Laboratories, began some experiments with a microwave horn perched on a hill, in Holmdel, New Jersey. The horn, an antenna shaped like a giant ear trumpet, was originally constructed for satellite communications, but Penzias and Wilson wanted to use it to measure the intensity of radio waves emitted in portions of the Milky Way galaxy. Initially, they set the antenna for a seven-centimeter wavelength, a point on the spectrum that they and other radio astronomers had assumed was relatively free of galactic radiation and would provide a quiet base line for comparing other results.

"We had purposely picked a portion of the spectrum where we expected nothing or almost nothing, no radiation at all from the sky," Penzias explains.

"Instead," adds Wilson, "what happened is that we found radiation coming into our antenna from all directions, just flooding in at us, and it clearly was orders of magnitude more than we expected."

In fact, neither scientist believed the results. Their first thought was that a coating of dung deposited by a pair of pigeons who had made a home inside the horn had distorted the antenna's reception. They dismantled the instrument, cleaned it, and reassembled it. Their reading was virtually the same. Then they considered that the sheet metal structure of the horn was defective and admitting unwanted noise. They carefully resealed every joint with metal tape. Still, the signal remained, its source a mystery.

"We were stuck with the sky beyond, which was not easy for us to accept," Penzias said. "But there was no alternative." They checked dozens of possible sources, includ-

Robert Wilson *(left)* and Arno Penzias of the Bell Laboratories, with the horn-reflector antenna in Holmdel, New Jersey, that first detected unexpected background radiation from deep in space, signals emanating from far beyond the Orion Nebula, pictured behind them.

Wilson and Penzias check the interior of the Holmdel antenna. At first they thought the signal they had detected was caused by defects in the antenna. But the antenna checked out, and the scientists concluded that the signal came from an unknown source in deep space.

ing the sun and planets in the solar system at various times of day and seasons of the year. They focused the antenna at different points in our galaxy. They even checked other galaxies. They found no variation in the signal. Nothing offered an explanation for their mysterious findings.

"We were left with the astonishing result," Penzias explains, "that this radiation was coming from somewhere in really deep cosmic space, beyond any radio sources that any of us knew about or even dreamed existed."

Meanwhile, however, another research group at Princeton University, just thirty miles away from Holmdel, had independently launched a similar project and was coming at the problem from the other direction. These researchers, who included professors Robert Dicke and James Peebles, indeed dreamed that such radiation existed; they wondered if they could find it.

At the time, two dominant theories of the universe were the subject of broad scientific speculation and debate. One held that the universe was in a steady state, infinite in its dimensions and unchanging in its nature. The other, the Big Bang theory, held that it was expanding and in constant change. This

second notion suggested that if you had a film of the development of the universe and ran it backward through a projector, everything would move closer together. Thus, something must have exploded at the beginning to send everything flying apart. Dicke theorized that the explosion was the result of such extraordinarily high temperatures that remnants of the radiation emitted then would still be present in the universe.

Detecting these remnants would pose a major problem, Peebles reasoned, since the sky is full of sources of radiation. Many are within our own atmosphere, solar system, and galaxy, cluttering the front of the stage with so many characters that the background radiation they were searching for could be lost in the confusion. Even if they could find this radiation, they wondered, how would they prove that the signal wasn't really something else?

Nevertheless, Peebles performed calculations that indicated they had a chance to pick up the signal. While the group designed its experiment, Peebles was invited to a scientific meeting to talk about the implications of the group's work. He attended, certain that no other scientists were as advanced in their thinking and research as the Princeton group. What he didn't expect was that other researchers might have already done the experiment and not understood the results.

Neither Penzias nor Wilson was at that meeting, but through the academic grapevine, another scientist familiar with their work soon learned of Peebles's presentation. He called Penzias and suggested that the two groups compare notes. They did, and the result was a remarkable match of hypothesis and experimental observation. Scientists strive for such a union as a fundamental part of their inquiry, but this case was an unusual and delightful piece of luck because one group of researchers came up with the observations entirely independently of the group that devised the theory. Their joint conclusion: the Holmdel horn was picking up a signal whose source was far beyond everything else in space.

But just how could scientists reach the conclusion that this pervasive signal's message was so important, that they had found something from the background of the universe? How could they be certain this wasn't just so much galactic static, akin to the annoying interference that obscures the signal from a fading radio station? Indeed, physicists and astronomers had plenty of new questions to ask, and the initial conclusion about the source of this signal became yet another hypothesis, one to be tested in part against well-established principles of physics.

Radiation is a general term that includes all forms of electromagnetic radiation commonly characterized according to their wavelengths on a spectrum. The longer radio waves and microwaves are on the low end of the spectrum, and, as waves get progressively shorter, we have infrared light, visible light, ultraviolet light, X-rays, and, finally, the very short-wavelength gamma rays. (Visible light composes only a small fraction of the spectrum. If the spectrum were represented by an octave on the piano keyboard around middle C, the radio and microwave region, which is the domain of the radio astronomer, would fall well beyond the end of the keyboard.)

While Penzias and Wilson puzzled over the signal, Robert Dicke and James Peebles, working independently at Princeton University, posed a theory that such a signal existed and that its source was the radiation emitted by the Big Bang that created the universe. They searched for such a signal, using the antennas pictured here, and confirmed the Bell scientists' findings.

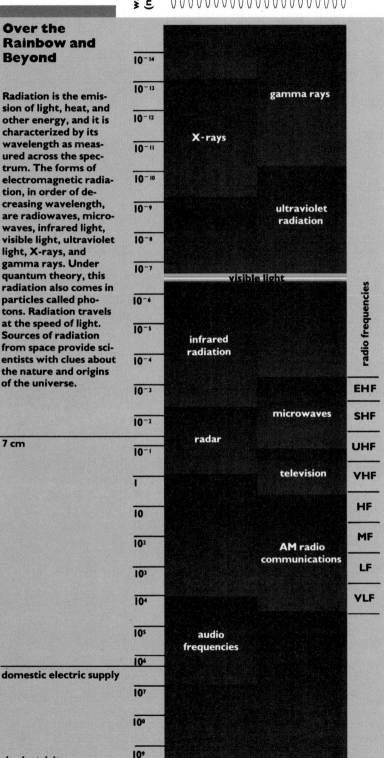

Over the Rainbow and Beyond

Radiation is the emission of light, heat, and other energy, and it is characterized by its wavelength as measured across the spectrum. The forms of electromagnetic radiation, in order of decreasing wavelength, are radiowaves, microwaves, infrared light, visible light, ultraviolet light, X-rays, and gamma rays. Under quantum theory, this radiation also comes in particles called photons. Radiation travels at the speed of light. Sources of radiation from space provide scientists with clues about the nature and origins of the universe.

Radio astronomers also measure radiation by the intensity of its heat and describe it in terms of an "equivalent temperature." Everything that has a temperature above absolute zero emits heat radiation. Thus, the radiation the Holmdel horn detects is much like that emitted by, say, a piece of iron fired to a red-hot glow. We can see the glow of that radiation, but the same piece of iron, cooled to its natural black and not even warm to the touch, emits radiation as well. We can't see it, but the right piece of equipment can detect it with ease and can record the curve of its equivalent temperature in the spectrum. Scientists have learned through calculation and experiment that each temperature traces its own individual curve, as unique as a fingerprint, through the spectrum. The peak or hump of the curve corresponds to the wavelength where the radiation is most intense.

Penzias and Wilson determined through calculations that the radiation picked up at Holmdel had an equivalent temperature of about three degrees Kelvin, or three degrees above absolute zero. This was consistent with Peebles's estimates and, as it happens, with the work of several other physicists who mathematically predicted the existence of this radiation in the 1940s and 1950s, but who for various reasons never performed the experiments to verify their calculations. (Their pioneering work was lost to science and rediscovered only after Penzias and Wilson's discovery.)

But the Holmdel antenna was set to record radiation at a wavelength of seven centimeters only and, as a result, provided only one point along a three-degree temperature curve. Despite the apparent reliability of mathematical projections, these scientists were working on the edge of our knowledge of the universe. One point hardly constituted proof. A few months later, Princeton added a second point to the curve, and then a group of radio astronomers from England came up with a third. All three seemed to fit the three-degree curve, but all were well below its peak. Even three points, if too far

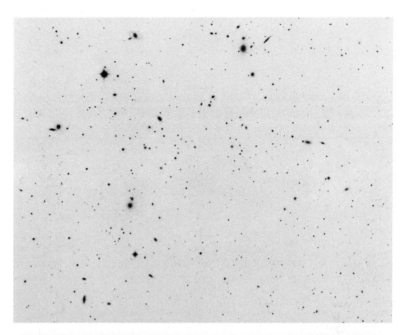

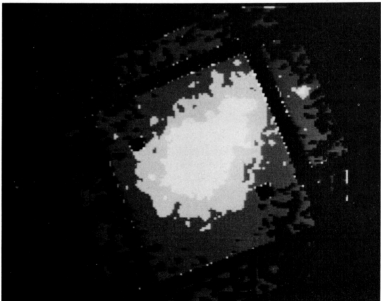

from the hump, could actually represent some other curve. Scientists wanted some points at the peak.

Mathematically, the scientists knew exactly where they wanted their results to fall. But they had a problem: radiation at the wavelengths they expected cannot penetrate the Earth's atmosphere. They had to take their experiments above it.

The first results came back from a rocket flight, and they were far above the predicted hump of the curve. Experimenters were confounded until they determined that the equipment on board had monitored radiation not only from the far regions of space, but from the heated parts of the rocket itself. So scientists switched to balloons launched into the stratosphere, high enough to avoid most of the interference from the Earth's atmosphere. Several labs became part of the effort to find other points on the curve. Each experiment meant a couple of years of work and, usually, the development of more sophisticated detecting equipment. Finally, one night in 1977, a group from the University of California at Berkeley sent a balloon twenty-five miles into the air above Texas. It collected the data the researchers had long awaited: it traced a three-degree temperature curve up and over the hump and down the other side.

In addition, the results of these experiments confirmed another piece of the Holmdel evidence—that the radiation was uniform in all directions. It is like the radiant glow when the inside of a kiln and its batch of pots reach a steady, even temperature: the glow fills the whole space, washing out the detail of the individual pots. The three-degree background radiation is 99 percent of all the radiation in the universe, washing out by comparison that emitted by bright stars, galaxies, supernovas, and even quasars, the most powerful individual emitters of radiation in the sky.

What we have, then, is a tremendous amount of energy emanating from a source that we can't see, but that is everywhere,

A cluster of galaxies (top) known as Abell 1367, and the same view (bottom) as seen through an X-ray telescope that detects the presence of gas in the space between the galaxies. The irregular appearance of the gas indicates that the cluster is in an early stage of its evolution. According to the Big Bang theory, galaxies were formed after the universe had cooled to the point where electrons and protons could join to form atoms of hydrogen. That is also the point that radiation and matter were separated, and it is marked by the background radiation that Penzias and Wilson detected.

Austrian physicist Christian Doppler (1803–1853).

at the very low temperature of three degrees above absolute zero. It seems to have a fundamental quality to it. We can verify that supposition by looking back 150 years, to the time of an Austrian physicist named Christian Doppler, whose work with sound waves led to the principle known as the Doppler effect. We know this principle by the apparent change in the pitch of a whistle as a train approaches and passes by: as it nears, the pitch is higher, indicating shorter sound waves; when it passes, the waves are longer and the pitch drops.

Doppler worked only with sound, but he suspected and it was later established that the effect is the same for light and radio waves. The shift is noticeable only at speeds much faster than that of a passing train since this radiation travels at the speed of light. But at high speeds, the Doppler effect has proven crucial to our understanding of the sky. It is testimony to the power of physics that a principle that explains something as commonplace as the changing pitch of a train's

whistle also helps us understand the nature of the universe.

Several astronomers detected the Doppler effect in the sky during the nineteenth century, but it was about the time of World War I and into the 1920s that first V. M. Slipher and then Edwin Hubble made the most dramatic observations. They looked not just at stars, but at whole galaxies, and they found enormous Doppler shifts indicating that celestial bodies were rapidly moving away from us. Since these shifts are indicated by longer waves, or a variation toward the red part of the spectrum, they are called "red shifts." Changes to shorter waves, indicating approaching objects, are called "blue shifts." Hubble determined that the faster cosmic objects recede—that is, the larger the red shift—the more remote those objects are. That principle, known as Hubble's Law of Red Shift, is the foundation of modern cosmology; it is, in effect, the tool we use to examine the origins of the universe.

Astronomers continuing Hubble's work,

Doppler's Dilemma

The Doppler effect is a change in the length of sound, light, and other waves when the source and observer are moving relative to one another—as in the changing pitch of a moving train's bell. As the train nears, the pitch is high (short waves); as it passes, the pitch drops (long waves). Radiation from celestial objects shows similar effects—toward the blue (short wave) end of the spectrum for approaching objects and toward the red (long wave) end for receding ones. Astronomers have detected important red shifts throughout the sky, evidence that the universe is expanding.

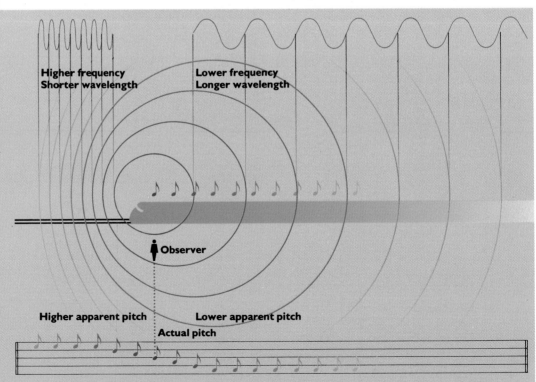

Higher frequency
Shorter wavelength

Lower frequency
Longer wavelength

Observer

Higher apparent pitch

Lower apparent pitch

Actual pitch

for example, have found that the farthest galaxies in space have a red shift equivalent (to return to the piano analogy) to an octave on the piano scale. Quasars, which are receding even faster, have a red shift of perhaps a couple of octaves. But the three-degree temperature radiation detected by Penzias and Wilson is red-shifted so far that its proportional equivalent runs far off the end of the piano.

This background radiation, then, is the glow of a most ancient heat, deep in the universe and farther away than anything else we can see.

If, by magic, we were able to follow the radiation from our planet back to its source, we would travel back in time as well as out in distance, back toward the moment roughly 20 billion years ago when the Big Bang happened. We would fly quickly past the moon, past Jupiter, to the bright stars in the Milky Way, then to the edge of our own galaxy and into the distant realm of the other galaxies. Finally, we would pass the outermost galaxy into a region beyond the reach of telescopes, to the time when galaxies themselves were being formed. And then we would reach the source of the Holmdel whispers, a misty wall of white-hot hydrogen gas.

We would now be so deep in space, we would have traveled back more than 19 billion years. Within this white-hot wall, known as a bland plasma, the temperature is so hot that atoms of hydrogen, the universe's most basic element, are broken into their constituent electrons and protons. It is an opaque mist of matter and radiation caught in a state of equilibrium.

As the universe expanded, however, this plasma cooled slightly, to the point where on just this side of the wall, its temperature was not high enough to keep the electrons and protons apart. So each electron joined with a proton to form an atom of hydrogen. Once that happened, the mist disappeared. Hydrogen is transparent, and so is the universe. Radiation, set free from matter, has traveled through the universe unimpeded ever since. Back on Earth, we see this radia-

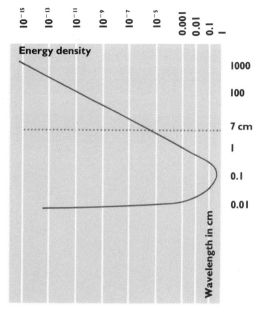

Energy density

Wavelength in cm

The 3° Kelvin temperature curve. The curve shows the expected relation between wavelength (vertical axis), and energy density (horizontal axis) for radiation at a uniform temperature of 3° above absolute zero. The dotted line is the original measurement at 7.35 cm made by Penzias and Wilson in 1965. Measurements at other wavelengths (particularly below 0.3 cm, which required sending rockets and balloons above the Earth's atmosphere) confirmed that the entire universe was radiating at this temperature.

tion profoundly red-shifted and cooled to three degrees Kelvin.

Thus the three-degree temperature marks the time in the formation of the universe when matter and radiation were separated; the radiation now brings us news of this event. The time is the same everywhere in the universe, adjusted in proportion to the red shift of its wavelength. Thus it serves as a kind of cosmic Greenwich mean time by which we here on Earth could synchronize our watches with those of other observers, if there are any, anywhere else in the universe.

This white-hot wall is probably the limit of what we can see through electromagnetic radiation. To learn about events even further back in time, approaching the instant of the Big Bang itself, astronomers, physicists, and cosmologists will have to rely on other means—gravity, neutrinos, and other elements of particle physics—and other paths perhaps yet to be discovered. Meanwhile, these explorers can hope to learn more about the visible universe through further and detailed study of this radiation, looking perhaps for minute differences or lumps in the otherwise uniform radiation. Such lumps might yield new information about the formation of the "lumpy" part of the universe, the stars and the galaxies.

Astronomer Edwin P. Hubble (1889–1953), whose law of red shift is the foundation of modern cosmology.

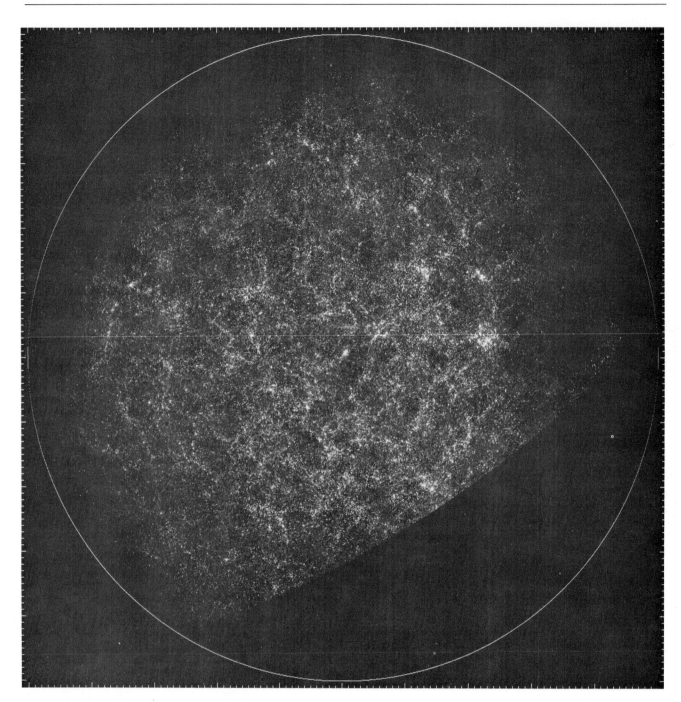

A map of a million galaxies as visible by telescope from the Earth's northern hemisphere and within one billion light years of Earth. The survey was compiled over twelve years by scientists from the Lick Observatory. The picture shows that galaxies are not distributed smoothly or evenly through space, but rather appear in clusters and clouds across the sky. The crescent on the lower right of the map shows no galaxies because the view from the observatory was limited by the Earth's horizon.

Indeed, more experiments are already planned. The space shuttle, for example, may soon launch a special satellite to monitor the cosmic background radiation in fine detail from high above the Earth, completely free of its interference, and for years without interruption.

One experiment, carried out thirteen miles above the Earth in a U-2 spy plane converted for study of the cosmic radiation, has already found one intriguing glitch in the uniformity of the radiation. After recording the results of several flights, Richard Miller of a research group at the University of California at Berkeley noted a signal with increased intensity that seemed to be coming from the vicinity of the constellation Leo. This was countered by a slightly weakened signal that appeared to emanate from the area of Aquarius, precisely 180 degrees away.

What this group had charted was the Earth's own Doppler effect as it travels through space. The radiation ahead of us is slightly hotter, or blue-shifted, while that in our wake is red-shifted. These measurements have allowed physicists to compute a remarkable number—the speed of the Earth (and the sun and the solar system) through space: 250 miles a second. All motion is relative, but this speed is relative to the greatest benchmark imaginable—the average speed of all the matter in the universe.

Penzias and Wilson's discovery of cosmic background radiation has thus provided much new information in our quest to understand the workings of the universe. In fact, it was significant enough to earn them the Nobel Prize in physics. But as it answers some questions, it raises others and gives new meaning for those long unanswered. What is the future of the universe? Will its expansion continue unimpeded so that it becomes more and more dilute and finally ends at a temperature of absolute zero? Or will the forces of gravity catch up with the scattered matter in the universe, slowly collapsing it back together again until the temperature increases to the point of yet another Big Bang? Science doesn't yet know the answers to

those questions, nor does it know fully about the deep past. But the cosmic background radiation will enable cosmologists of the future to go further in both directions and give new detail to our sketchy understanding of creation.

Long ago, the sages of ancient India told how the god Vishnu, with a churn made of a mountain, churned an ocean of milk. Out of this nourishing material came the divine cow, an eight-headed steed, the moon, and other wonders. We have made great strides in our knowledge since that mythology of creation was written, but we are still churning a great sea, seeking more and more material with which to write and rewrite our own mythology of the universe.

A U-2 plane is fitted with a sensitive antenna to detect cosmic radiation while in flight thirteen miles above Earth. With this plane, scientists charted the Earth's Doppler effect as it moves through space and calculated the Earth's speed through space as 250 miles per second.

Vishnu, assuming the form of a tortoise, bears the weight of Mount Mandara as the gods use it as a churn to obtain life from the milky ocean.

The Hunting of the Quark

It is the physicist's unique and challenging quest not only to understand the beginnings and behavior of the largest objects in the universe—the planets, stars, and galaxies—but also to search for the smallest, to try to discover the fundamental unit from which all matter is made. Nearly 200 years ago, that search led scientists to the discovery of the atom. Then, 140 years later, they looked inside the atom to find the nucleus, inside which they soon found the proton.

But a unit is not *fundamental* unless it is constructed of nothing but itself, unless it has no smaller parts. In the past three decades even the proton, which is one ten-millionth of a centimeter in diameter, has fallen before this definition. Inside the proton, scientists believe, is a bizarre, pointlike particle called the quark.

The first cyclotron *(above)*, measuring 27 inches, was invented in 1931 at the University of California at Berkeley by M. S. Livingston *(left)* and Ernest O. Lawrence, who stand next to the second version of their invention, which measured 37 inches.

The search for this particle is a fascinating scientific journey, conducted by some of the world's most brilliant theoretical physicists, including Murray Gell-Mann and Richard Feynman, and the world's most able experimental physicists, including Burton Richter—all of whom have won the Nobel Prize for their work in this area. The most important vehicle on this journey is the largest and most expensive piece of laboratory equipment in the world: the accelerator.

Accelerators, or atom smashers as they are popularly called, are gigantic, powerful machines that fire protons and other subatomic particles at nearly the speed of light through vacuum tunnels more than two miles long, where they ultimately smash into one another, shattering into other particles. These particles are too small to see, of course, but by means of sensitive equipment physicists can record their tracks on film. The patterns, lengths, and shapes of the tracks yield clues about the lives and properties of the particles, some of which live only a few billionths of a second.

Accelerators re-create as closely as is possible at present the high-energy collisions that occurred during the first moments after the Big Bang, when, scientists believe, the building blocks of matter were created. Thus the search for the quark is a chapter in the story of the origins of the universe.

The search began at least as early as 1947, when physicists sent photographic film aloft in balloons to record the high-energy cosmic rays that bombard the Earth from space. When the protons in the cosmic rays struck atoms in the sensitive film, they created unusual patterns that to physicists were evidence of unexpected behavior in the realm of the atom. The cosmic rays were, in effect, nature's own atom smashers, and what they had done on the film was split atoms into particles the physicists had never seen before.

Wanting to know more, scientists rallied government support for the construction of their own atom smashers, and over the next two decades, miles of precisely designed circular tunnels, lined with huge magnets to speed the particles through them, were built in such places as Stanford, California; Geneva; Long Island, New York; and Illinois. As these machines were put into operation and scientists began firing protons and other particles at each other, they began to learn much more than they wanted to know. Two protons can create as many as twenty or more protonlike particles when they collide, particles that are not fragments of protons, but

(Left) By 1945–1946, Lawrence and his staff had increased the dimensions of the cyclotron to 184 inches. *(Below)* Today, accelerators like the one at the Fermi National Accelerator Laboratory near Chicago *(inset)* are as long as two miles and usually built in circular tunnels. Particles are accelerated through the tunnels by giant magnets. This tunnel, the "Main Ring" at the Fermilab, houses two rings of accelerator magnets. The upper ring operates at an energy of 400 billion electron volts.

Периодическая система элементов

Периоды	Ряды	I	II	III	IV	V	VI	VII	VIII	O
I	1	1. H Водород 1.008								2. He Гелий 4.00
II	2	3. Li Литий 6.,40	4. Be Берилий 9.02	5. B Бор 10.82	6. C Углерод 12.010	7. N Азот 14.008	8. O Кислород 16.000	9. F Фтор 19.00		10. Ne Неон 20.183
III	3	11. Na Натрий 22.997	12. Mg Магний 24.32	13. Al Алюминий 26.97	14. Si Кремний 28.06	15. P Фосфор 30.98	16. S Сера 32.06	17. Cl Хлор 35.457		18. Ar Аргон 39.944
IV	4	19. K Калий 39.096	20. Ca Кальций 40.08	21. Sc Скандий 45.10	22. Ti Титан 47.90	23. V Ванадий 50.95	24. Cr Хром 52.01	25. Mn Марганец 54.93	26. Fe Железо 55.85 27. Co Кобальт 58.94 28. Ni Никель 58.69	
	5	29. Cu Медь 63.57	30. Zn Цинк 65.38	31. Ga Галлий 69.72	32. Ge Германий 72.60	33. As Мышьяк 74.91	34. Se Селен 78.96	35. Br Бром 79.916		36. Kr Криптон 83.7
V	6	37. Rb Рубидий 85.48	38. Sr Стронций 87.63	39. Y Иттрий 88.92	40. Zr Цирконий 91.22	41. Nb Ниобий 92.91	42. Mo Молибден 95.95	43. Ma Мазурий	44. Ru Рутений 101.7 45. Rh Родий 102.91 46. Pd Палладий 106.7	
	7	47. Ag Серебро 107.88	48. Cd Кадмий 112.41	49. In Индий 114.76	50. Sn Олово 118.70	51. Sb Сурьма 121.76	52. Te Теллур 127.61	53. J Иод 126.92		54. Xe Ксенон 131.3
VI	8	55. Cs Цезий 132.91	56. Ba Барий 137.36	57. La* Лантан 138.92	72. Hf Гафний 178.6	73. Ta Тантал 180.88	74. W Вольфрам 183.92	75. Re Рений 186.31	76. Os Осмий 190.2 77. Ir Иридий 193.1 78. Pt Платина 195.23	
	9	79. Au Золото 197.2	80. Hg Ртуть 200.61	81. Tl Таллий 204.39	82. Pb Свинец 207.21	83. Bi Висмут 209.00	84. Po Полоний 210	85. —		86. Rn Радон 222
VII	10	87.	88. Ra Радий 226.05	89. Ac Актиний 227	90. Th Торий 232.12	91. Pa Протактиний 231	92. U Уран 238.12			

R₂O	RO	R₂O₃	RO₂	R₂O₅	RO₃	R₂O₇	RO₄	
Высшие солеобразующие окислы				Высшие водородные соединения				
			RH₄	RH₃	RH₂	RH		

*Лантаниды

| 58. Ce Церий 140.13 | 59. Pr Празеодим 140.92 | 60. Nd Неодим 144.27 | 61. — | 62. Sm Самарий 150.43 | 63. Eu Европий 152.0 | 64. Gd Гадолиний 156.9 |
| 65. Tb Тербий 159.2 | 66. Dy Диспрозий 162.46 | 67. Ho Гольмий 164.94 | 68. Er Эрбий 167.2 | 69. Tu Тулий 169.4 | 70. Yb Иттербий 173.04 | 71. Cp Кассиопей 174.99 |

Russian chemist Dmitri Mendeleev (1834–1907), who devised the periodic table of the elements (top) when only sixty of the ninety-two elements were known. He left blank spaces on his chart where he expected the yet-to-be-discovered to fall, and he predicted the physical properties of three elements with a high degree of accuracy. Physicist Murray Gell-Mann similarly made order out of subatomic particles in a theory called "the eightfold way" that predicted the existence of particles later discovered by experimental physicists.

brothers and sisters of protons and as complex as the proton itself. These experiments produced such a profusion of new particles that physicists had to carry a booklet around just to keep track, and the effort to find a fundamental particle was thrown into crisis. (Jeremy Bernstein, writing in *American Scholar,* provided a glimpse of the frustration in the scientific community when he reported that he once overheard Robert Oppenheimer suggest awarding a Nobel Prize to the first physicist who *didn't* discover a new particle in a given year.)

"No one believes that all of these are in any sense fundamental particles," says Murray Gell-Mann, who is a professor at California Institute of Technology. "We theoreticians have during the same period of time been hard at work trying to discover a simple picture. If there are fundamental entities, they must be few in number and arranged in some simple and beautiful pattern. The search for that pattern led us to the notion of the quark."

The first task was to find some pattern among the particles themselves. Nature has a way of grouping other things like butterflies or fish into families with similar characteristics, the theorists noted, so why should particles be any different? A century ago, the Russian chemist Dmitri Mendeleev, searching for a way to teach his students the properties of the elements, organized the elements into family groups according to their related characteristics. The chart he devised is what we

now know as the periodic table of the elements. At the time, only sixty of the ninety-two stable elements were known, but Mendeleev expected there would be gaps in the pattern, and he actually predicted to a remarkable degree of accuracy the physical properties of three of the missing elements.

About 1960, Gell-Mann sought to make similar order out of all the particles, and he worked out a theory called "the eightfold way," which predicted that some particles would come in families of eight. It was a fascinating prediction because, like Mendeleev's chart, it was based in part on particles that had not yet been discovered. Eventually, they were discovered, and they fit nicely into Gell-Mann's scheme.

But he still had not found the fundamental unit of the particles that composed this family, that explained the basic nature of its construction. In 1963, he and, independently, George Zweig (also of Cal Tech, though both men were on leave in different parts of the world at the time) came upon the quark. Actually, Gell-Mann first nicknamed it "the strange"—which it was indeed—but he officially called it a *quark,* a reference from James Joyce's novel *Finnegans Wake.*

"Quarks behave as if they have fractional electric charges, in units of the proton's charge," he explains. "Namely, two-thirds, minus a third, and minus a third—a very queer notion."

Up to that point, no one had considered that an electric charge could be split up. But in these fractions, three quarks can be put together to form all the proton family. A configuration of minus one-third, plus two-thirds, and plus two-thirds leaves a charge of one—a proton. A different arrangement, minus one-third, minus one-third, and plus two-thirds, makes a charge of zero—the neutron. And so on. The whole pattern is formed by different arrangements of these three simple entities.

At least that is how the theory went then. Its publication sent experimental physicists scurrying back to their accelerators and other equipment in search of these strange

particles with only a fraction of a charge. But several years of experiments—including careful examination of thousands of photographs of particle collisions and a search for free-floating quarks, possible remnants from the Big Bang—turned up nothing. That did not necessarily disprove the theory, however.

"It's increasingly unlikely that these fractionally charged quarks exist as separate entities," Gell-Mann explained, "but that was never a necessary part of the scheme, and that was clear from the beginning."

Scientists continued their search inside the proton, looking not for one single quark, but for a cluster of three. In one experiment at Stanford, scientists fired 20-billion-volt electrons into protons contained in a tube of liquid hydrogen and measured the energy the electrons lost as they deflected off the protons. They figured if the electrons lost energy, it meant they had collided with moving parts of the proton, possibly quarks. Measurements indicated the electrons did indeed lose significant amounts of energy.

Other experiments showed that the parts inside a proton have a property called spin, in precisely the amount theoreticians had predicted quarks should carry.

But that still didn't establish the actual number of moving parts within the proton. At CERN, the gigantic laboratory of the European Organization for Nuclear Research near Geneva, scientists worked out a plan to use yet another particle, the neutrino, to look for quarks. The neutrino is another bizarre particle; it has no weight and no elec-

trical charge, and it can pass through millions of miles of matter without interacting with the atoms. In one CERN experiment, scientists made eight million photographs of neutrinos colliding with protons and shattering them into other particles. By measuring the tracks of the particles, the experimenters were able

Viewing a projected, enlarged photograph made of neutrino tracks in a bubble chamber. Only one in a trillion neutrinos will strike another particle and leave a record like this one.

Key quark hunters include *(left to right):* **Murray Gell-Mann receiving his**

Nobel Prize; George Zweig of California Institute of Technology; Burton Richter of

Stanford University; and Richard Feynman of California Institute of Technology.

The search for the building blocks of nature has led from the atom, considered the elementary particle of matter at the turn of the century, to ever smaller particles as physicists peeled away the layers of the atomic onion. It now appears that a half-dozen or so varieties of "quarks," together with their "antiquarks," account for most of the matter in the universe.

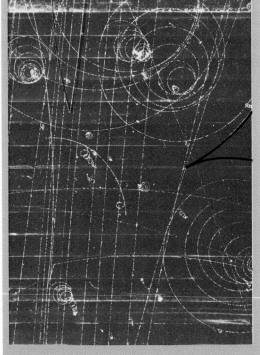

Atom

Nucleus

Protons and Neutrons

Quarks

Up   Down
Strange   Charm
Bottom   Top

?

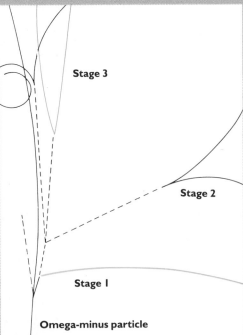

Stage 3

Stage 2

Stage I

Omega-minus particle

The discovery of the so-called omega-minus particle, one of the most celebrated photographs in the history of physics. It confirmed Gell-Mann's theory of the patterns of particles. The photograph and the accompanying drawing show the omega-minus particle at the lower left breaking up in three stages, shedding one dose of "strangeness" at each stage and finishing as a proton.

to calculate the net number of parts inside the proton. Within an acceptable range of experimental error, the number was three, exactly as predicted by the quark model. And further measurements seemed to verify that each part came in a fractional charge – again, just as Gell-Mann had figured.

Thus, one by one, the experimental physicists seemed to confirm the expectations of the theoretical physicists, which is just how scientists like things to happen. They could say with increasing confidence that quarks did indeed exist and that they appeared to be a fundamental unit of matter.

By early 1973 new experimental data were constantly being adding to the picture. Sometimes they provided important confirming evidence; sometimes they yielded insignificant information; and sometimes they raised important contradictions. One such additional piece of information resulted from an important experiment conducted by Burton Richter at Stanford in which electrons and positrons (the electrons' antiparticles) were fired at each other. In such collisions, the particles, whose properties are exactly opposite, annihilate each other in a tremendous burst of energy that creates new particles. This burst of energy is, in effect, a very small fireball, similar in its properties to the primordial fireball that exploded in the Big Bang. But these tiny fireballs at Stanford raised hard questions about Gell-Mann's theory; they produced many more particles than the quark model predicted.

"The results are a delight to an experimenter and a plague to a theorist," Richter said, "because they simply don't agree with anybody's model."

But was this enough to send the quark model to the scrap heap? Hardly.

"This is really a very exciting experiment," Richter continued. "It's a very exciting time for all of us who are doing it and also for the theorists who are trying to understand it. You have a theory which agrees with a whole bunch of data, and along comes another experiment and this beautiful theory doesn't agree at all. There sits a contradiction. Now you must resolve it, and in the resolution of that contradiction, you hope to make some major advance in your understanding of nature."

Indeed, since then scientists have worked with these and other results and determined that they were not actually contradictions. But they did force some modifications to Gell-Mann's original model—changes that accommodated the new evidence without undermining the original premise of the theory.

In fact, scientists now believe there are not three but six quarks, to which they have assigned such whimsical names or "flavors" as *up, down, strange, charm, bottom,* and *top.* In addition, they now believe that each quark comes in three "colors," slight variations that determine how different quarks combine.

In 1980, scientists working at the Cornell Electron Storage Ring in Ithaca, New York, announced the discovery of the fifth, or "bottom," quark. The sixth, so-called "top," quark still eludes experimental physicists.

"We may, of course, never have the final picture, the final understanding of what a neutron or proton is built from," Donald Perkins of Oxford University said in 1973. It is not yet possible to change that assessment. But so far, the theory of quarks works to make order out of the chaos in the minuscule world of particles and to lead physicists to the conclusion that they have discovered a fundamental unit of matter.

Theorists hope that further research will also bring them closer to another long-elusive goal. These particles, like all matter in the universe, are governed by four basic forces: gravity; electromagnetism; the so-called strong force, which holds the nucleus of an atom together; and the weak force, which governs radioactive decay. These forces seem to behave differently from each other, in effect establishing different rules for the behavior of matter within the atom than in other areas of the universe. It means that Einstein's theory of general relativity and Planck's theory of quantum mechanics are at odds with each other in

some areas of the universe. For the physicist, this is the ultimate contradiction. Thus the search is on for one grand unifying theory that explains, without such a paradox, the forces that govern the behavior of all matter in the universe.

For that reason and others, the hunt for the quark is much more than some kind of subatomic boxing match in which tiny bits of matter are tossed into the ring and hurled at each other to see what comes out. "It's the universe we're looking at," Richard Feynman explained. "We're not just exploring a little thing. It's only small in dimension. As far as the universe is concerned, it's all-encompassing. So it's a tremendous adventure. It's apparently important. And it's impossible to stop."

In 1900, German physicist Max Planck proposed that energy is not infinitely divisible but must exist in discrete units, or quanta. This idea so revolutionized physics that, together with Einstein's relativity theory, it began a flood of new ideas and discoveries that continues to the quark theory of today.

Four Forces

Physicists have determined that there are four basic forces at work in the universe, each governing a different aspect of the vast cosmic drama.

1
The strongest of these forces, appropriately known as the strong or nuclear force, binds the particles that constitute the nucleus of an atom. This force is transmitted by particles referred to as mesons, which race back and forth between particles such as neutrons and protons while holding them together in the nuclear unit. Recent experimentation has led scientists to conclude that the strong force is actually a secondary manifestation of a more fundamental force within the nucleus. The discovery of quarks, the most basic of atomic particles (from which both

mesons and other nuclear particles are constructed), has set physicists off in search of the ultimate force linking the quarks themselves. The as-of-yet-undiscovered particles that transmit such a force have already been named gluons, in honor of their adhesive nature.

2
The electromagnetic force is $1/137$ as powerful as the strong force, but its strength can be transmitted over a much greater distance. This force binds electrons to the nucleus of an atom and consequently governs chemical reactions. Particles known as photons transmit this force through the process of electromagnetic radiation.

3
The third strongest force, called the weak force, governs the decay of certain types of elementary particles and is a mere 10^{-13} as

powerful as the strong force. Physicists postulate that this force is also transmitted by a particle such as the meson or photon, but such a courier remains undiscovered. Like the gluon, though, it has already been named; it will be known either as an intermediate vector boson or a W particle.

4
The weakest force in the universe is gravity— a mere 10^{-39} the strength of the strong force. But gravity maintains its force over enormous distances and is therefore the dominant influence on the vast scale of the cosmos. Scientists are not sure how gravity is transmitted; they are looking for particles (to be called gravitons) as well as for gravitational waves. Like so much else in the mind-boggling realm of physics, however, gravity is still quite a mystery.

A Conversation with Steven Weinberg

In 1979, Steven Weinberg received the Nobel Prize for his work in theoretical particle physics. With chairs at both Harvard University and the University of Texas at Austin, he is recognized as one of the leading scientists currently working on the unification of the forces of nature with one grand theory.

Even as a young man, he was captivated by the fact that scientists are able to draw conclusions about the real world through logical patterns of thought with the aid of mathematics. "I thought that it was most remarkable to be able to sit at your desk and use your knowledge and experience to describe the nature of the universe." He found even more amazing the ability of a scientist to describe things that depart from normal areas of intuition. This, he thinks, is what led him into the realm of elementary particle physics. Not only are the sizes of elementary particles far smaller than anything in common experience, but one must consider extremely short time intervals as well.

Weinberg has also held a fascination for the other end of the spectrum of reality—the size and age of the universe. He cites cosmology as another example of the power of the human mind to consider and make relevant contributions despite unaccustomed distances and time units.

Weinberg thinks that some of the most interesting work that has been done in physics during the lifetime of "NOVA" includes the mathematical elaboration and experimental verification of the unification of the electromagnetic force with what is called the weak interaction. The weak interaction is one of the nuclear forces (the other is the strong interaction, which holds the nucleus together) and plays an important role in nuclear reactions. In fact, the nuclear reaction that provides the sun with most of its energy is a result of the weak interaction.

The unification of the weak and electromagnetic forces had been previously proposed as a theory, but Weinberg's work and that of his colleagues completed the picture. The theory bears his name—the Weinberg-Salam theory.

A very important development in physics, in his opinion, has been the progress made in the field called quantum chromodynamics. As far as scientists can tell, this field puts the strong interaction on the same theoretically sound basis as the weak and electromagnetic forces. Experimentally, Weinberg recognizes the importance of the verification of the whole quark situation that underlies quantum chromodynamics.

Up to this point in his career, Steven Weinberg has found most satisfying a realization of something that had been earlier appreciated by other people: namely, that nature has symmetry, which implies a great deal of simplicity in the physical laws at the fundamental level, but which is often hidden from view by a process called symmetry breaking. As an example of symmetry breaking, Weinberg describes some characteristics of water. Water is the same in all directions, but when it crystallizes, the directional symmetry breaks down and the crystal lattice forms in certain directions only. Symmetry breaking occurs throughout the universe.

This notion of symmetry breaking fascinated Weinberg for almost two decades, and eventually he applied it to the weak interaction. His work led him to a rational understanding of that nuclear force and its relationship with the electromagnetic force. He states, "To me, the most exhilarating thing was realizing the existence of symmetry beneath the surface appearances." Now, he says, "we must come to an understanding of how and at what energy scales the symmetries with which we are currently familiar in theory on the fundamental level become broken."

In the scientist's search for the fundamentals of nature, symmetry often makes the job more manageable. The assumption that symmetry exists leads people to look for certain things, as patterns are sought and equivalents developed. It is the extent of the symmetry that Weinberg wants to understand. He feels that it is largely an experimental problem. The construction of a larger, more energetic generation of particle accelerators will enable scientists to study proper-

ties of particles that have heretofore been unexplored. Theoretical work will shed light on the question of why there seem to be vastly different energy ranges at which the symmetry breaks down. In addition, Weinberg hopes that in the near future theorists will be able to develop a quantum theory that will make more sense mathematically than the existing one. Some very exciting times are ahead.

With the operation of new particle accelerators, he looks forward to the experimental determination of just how many generations of quarks and leptons there are. "At present, up to the highest energies available in the laboratory, experimentalists find new elementary particles. Is this going to continue as our technology extends our observational range?" he wonders. As a theorist, his challenge is to explain the existence of all the different elementary particles.

Steven Weinberg considers the Nobel Prize he was awarded for his work a confirmation of the contribution he has made to science, but warns that recognition may or may not follow achievement. In his case, as with others, the awards come much later, but the reward is personal and immediate.

Scientific work is frustrating, in his opinion, and a good scientist must have the ability to keep on working. Since most of what a scientist does turns out to lead nowhere, he must be able to recognize the false directions and forgive himself. The problem that each scientist faces is that usually he does not know what direction to take in choosing research topics and approaches. Weinberg explains it as follows: "If you have a very difficult calculation to do, such as predicting the future position of a comet, at least you have in sight what you want to know despite the complications of the mathematics. The difficulty, however, often arises when you don't know the problem you should work on. You have to realize that most of the time, you'll be thinking about the wrong problem."

Steven Weinberg thinks that a good scientist has to have a sense of what kind of physical theory really represents progress.

Unfortunately, this cannot be taught, but it is an absolutely essential attribute that warns a scientist when he is working in the wrong direction.

An area of concern for Weinberg, in addition to that of symmetry, has been the mathematical problem of nonsensical results surrounding the meaning of infinity in equations. "Infinity," he states, "has been a plague to the theoretical physicist since its exposition by Oppenheimer in the 1930s." Weinberg has labored with this problem throughout his work on unifying the weak and electromagnetic forces, and in fact he succeeded in eliminating the mathematical terms bearing infinity.

Thus, Dr. Weinberg has known what is important to study. He has been able to sort out the numerous projects in physics and focus on what will make progress in the field of elementary particle physics. He has emerged honored and respected as one of the finest theoretical physicists of our time.

The Ring of Fire

The borders of the Pacific Ocean form an enormous "ring of fire," a zone of unusually great volcanic activity. Of more than 450 volcanoes active on Earth today, almost two-thirds are situated along the Pacific rim. This area is further characterized by an inordinately high level of seismic activity. Each year, more than 80 percent of all energy released by earthquakes throughout the world is released here. Why should volcanoes and earthquakes be correlated? And why should they be so concentrated along the margins of the Pacific? For an answer, scientists turn to the theory of plate tectonics.

Until late in the nineteenth century, most geologists agreed that disruptions of the Earth's surface resulted from the contraction of the planet. Proponents of this hypothesis argued that the aging Earth was cooling and actually shrinking in size, causing the fixed continents to buckle and break under the constantly increasing tension. Around the turn of the century, though, geologists such as Eduard Suess of Austria and Frank Taylor of the United States began to consider evidence that the continents have not always occupied such fixed positions, but have drifted about the face of the Earth over the course of millions of years.

The greatest advocate of this hypothesis, a German scientist named Alfred Wegener, worked from 1912 to 1929 to collect data supporting his contention that the Earth's continents originated from a single land mass that broke apart into drifting segments. Evidence to substantiate Wegener's notion came from studies of land forms, past climates, and the distribution of plants and animals, but most scientists rejected continental drift because no mechanism had been proposed that could adequately account for the motion.

By the 1930s the theory of continental drift (or "Wegener's Hypothesis," as it came to be called) had fallen into complete disfavor, and the scientific community once again turned to variations of the contraction model to explain the instability of the Earth's surface. Interest in the drift theory did not revive until the early 1950s, when surprising evidence, collected during the mapping of the ocean floors, generated renewed debate about the origin of the continents. The discovery of a mammoth mid-Atlantic mountain range running the length of the ocean led to further research, which revealed a continuous system of undersea ranges stretching through every ocean on Earth. Other studies also in-

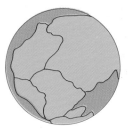   

In the eighteenth century world maps became accurate enough to show the unusually striking fit between the continents, particularly Africa and South America. Scientists of the day concluded that the continents had once been a single land mass and were somehow torn apart. In a classic

case of an idea that refused to die, the theory reappeared periodically, receiving its most famous treatment in Alfred Wegener's *The Origin of Continents and Oceans,* published in Germany in 1915.

Wegener named the ancient supercontinent Pangaea, meaning "all lands," and he surmised from fossil evidence that it began to break apart some 200 million years ago *(far left).* With the various pieces moving only inches a year, the continents have gradually assumed their present

positions *(center left and right).* Scientific hypotheses often carry hidden bonuses, and recently it has become clear that continental drift, now known as the theory of plate tectonics, very nicely accounts for the observed distribution of volcanoes and earthquakes.

This is most apparent along the margins of the Pacific Ocean, where colliding continental and oceanic plates have produced a tectonic "ring of fire" *(above).*

dicated that most oceanic earthquakes occur along these ridges and that water temperatures above the undersea mountains are significantly higher than those of the surrounding sea.

These findings, coupled with complex paleomagnetic data, led in 1967 to the theory of plate tectonics, which finally provided the explanation of continental drift that had eluded Wegener thirty years before. W. Jason Morgan and Dan McKenzie, the geologists who advanced the theory, demonstrated that the surface of the Earth is divided into a number of huge plates, separated by a network of midocean ridges that represent fractures in the Earth's crust. Molten material continuously rises through these great cracks and solidifies. The constant addition of new material pushes away the sea floor to either side of a ridge, and the plates so pushed carry with them their continental burden.

When the leading edge of a moving plate encounters the edge of another, one is subducted, or pushed under and down, into the Earth's mantle. This collision line between plates is usually marked by a deep ocean trench into which the edge of the subducted plate dives, and by a chain of young mountains where the overriding plate's edge is uplifted. As the edge of the subducted plate melts, the molten rock may well up again as lava, vented through the volcanic mountains. When the edges of two plates meet they may also slide past each other. This motion, called transform faulting, doesn't always proceed smoothly. Plate edges often catch against each other, even

The lands bordering the Pacific Ocean are the scene of much of the world's volcanic and seismic activity. In western North America, the Cascade chain of volcanoes (including Mount St. Helens, shown in the inset in its May 1980 eruption) and the San Andreas Fault to the south are spectacular examples of the consequences of the movement of the Earth's plates by the mechanism of sea-floor spreading.

while the plates themselves continue to move. Enormous amounts of potential energy are stored when plate edges become locked. In time, the stresses become too great, the straining edges unlock and snap forward, and the energy suddenly released causes the Earth to quake. The faulting and subduction of the boundaries of the Pacific plates, where they abut adjacent plates, result in earthquakes in California, Japan, and New Zealand, and volcanic eruptions in the Pacific Northwest, the Andes, and the Philippines—the ring of fire.

California is prime earthquake country, with a level of seismic activity ten times the world average. Earthquakes of magnitude 3 or greater on the Richter Scale occur at the rate of 200 a year. The extensive and complicated San Andreas–Garlock fault system, which stretches from Eureka in the north to the Mexican boundary, guarantees future earthquakes of a magnitude equal to or greater than the famous San Francisco quake of 1906, in which the energy released was equal to one million bombs the size of the one dropped on Hiroshima. These future earthquakes will occur as the edge of the Pacific plate, which carries on it most of coastal California and Baja California, periodically rebounds after locking against the edge of the North American plate. Many Californians remain cheerfully indifferent or doggedly fatalistic in the face of inevitable catastrophe, but geologists and geophysicists throughout the state actively monitor hundreds of sites in their efforts to predict the next earthquake.

Shallow-focus earthquakes, like the ones in California, originate as ruptures in the rocks that lie within forty miles of the surface. A rupture sets up waves that travel through

According to the theory of plate tectonics, volcanoes are often formed when oceanic rock, flowing out of an undersea rift, is over-ridden by the edge of a continental plate.

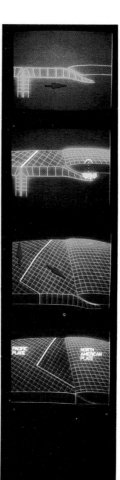

The friction energy in the subduction zone melts millions of tons of rock and strains the whole coastal region, allowing some of the melted magma to escape to the surface.

Just such a situation once existed in the northeast Pacific Ocean. The Pacific plate moved north as the plate in the middle continued to spread southeast. And the rift line had a huge right-angle turn, where it stopped spreading and slipped sideways instead.

Fifty million years ago, the Pacific plate itself made contact with the North American plate. Since the relative movement between these plates was purely north-south, progressively more of the coast became a slipping zone instead of a subduction zone.

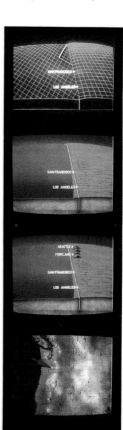

This is the situation today, with Los Angeles and San Francisco sitting astride a slip-zone 630 miles long, the famous San Andreas Fault, still slipping at a rate of 2½ inches each year. Eventually, the strip of coast west of the fault will shear off and drift northward, to founder in the Aleutian Trench off Alaska in about 50 million years.

North of the fault, near Portland and Seattle, are the last remaining bit of a once-huge oceanic plate and the remaining active young volcanoes—the Cascade Range—less than a million years old.

The Cascades stretch from Mount Lassen in northern California to Mount Baker near the Canadian line. Mount St. Helens is about two-thirds of the way up and somewhat to the west of the others.

the rock masses and that can be picked up as sound by sensitive instruments called seismometers. Small earthquakes occur daily, and the waves that they generate are recorded. (One theory, promising for a while but of little practical use so far, was that a drop in the speed at which the waves are traveling could indicate an increase in tension in the rocks below, a tension that could build until released in a large earthquake.) Other instruments, called creep meters, monitor the speed at which points to either side of a fault line are moving relative to each other, on the lookout for signs of a speed-up. Tilt meters measure the deformation of surface rocks. And magnetometers record changes in the rocks' magnetic fields. All of these phenomena—the speed-up of fault movement, the bulging of surface rocks, the shift in magnetic fields—are looked to as signals that

In 1935, Charles Richter devised an accurate system for measuring the magnitude of earthquakes. Dissatisfied with the lack of precision offered by intensity scales, which depended on subjective assessments by local witnesses, Richter came up with the scale, named for him, in which each earthquake is classified according to a function of its largest seismic wave as recorded by instruments.

Measuring the Quake

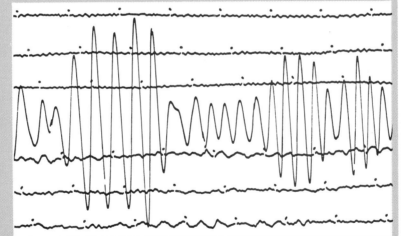

An earthquake can be measured qualitatively or quantitatively. The quality of an earthquake is determined by its perceived intensity, so a single earthquake can give many different measures of intensity, depending on the observer's distance from the quake's center and the nature of the local ground. The scale commonly employed is the Modified Mercalli Intensity Scale, which, on the basis of the reactions of both living things and inanimate objects, distributes earthquakes among a dozen subjective categories. In category I are small tremors, hardly detectable by people, which may cause animals to become uneasy, trees to sway, and doors and chandeliers to swing slowly. At the other end of the range, category XII consists of fearsome catastrophes that produce general panic and nearly total destruction. A more precise method is the Richter Scale, a quantitative measure of earthquake magnitude based on the logarithm of the height of the seismic wave *(inset),* in which each unit designates a quake ten times greater than the unit before. A one-unit increase in magnitude corresponds to an approximately thirtyfold increase in the amount of released energy at the quake's focus. Earthquakes of magnitude 2.5 or less can scarcely be felt by local witnesses. Structural damage starts to occur at Richter 3.5, and quakes of magnitude 6 or more can prove very destructive. The San Francisco earthquake of 1906 and the even more devastating Alaska quake of 1964 (with respective magnitudes of 8.3 and 8.4) were two of the most severe seismic events ever recorded.

With an estimated magnitude of Richter 8.3, the San Francisco earthquake of April 18, 1906, was one of the strongest tremors of this century. Buildings toppled and great fires swept the city. Shifts of more than 15 feet occurred along the fault line, severe ground shaking was experienced as far away as Eureka and Monterey Bay, and more than 450 people lost their lives.

Designing for Chaos

The earthquake-induced effects that cause building damage are ground rupture, ground failure, tsunamis (tidal waves), and ground shaking. Since there is little the architect or engineer can do to proof structures against the first three, most efforts to produce quake-resistant buildings concentrate on coping with the crushing, twisting, and shearing forces generated by the erratic, multidirectional, often violent motion of the ground. Because forces tend to become greater as they are transmitted upward, concentrating a building's mass close to the ground will keep damage to a minimum. This was the guiding principle in the design of San Francisco's Transamerica tower *(right)*, whose stable, pyramidal configuration also resists torsion. To protect their buildings from wrenching ground motion, designers of certain edifices in Vancouver and other western cities have actually had them suspended by giant straps from external supports. In most new buildings erected in zones of high earthquake risk, extensive use is made of hinge joints, slip joints, and resilient materials, which absorb stresses and retard their transmission between structural and nonstructural components like windows, doors, facade panels, and nonbearing walls. A building's occupants, as well as the people in the street below, are safer when architects avoid parapets, cornices, and other elements likely to be shaken loose during a quake. Barring the discovery of some miraculous construction material that is both infinitely flexible and endlessly rigid, and given the inevitability of a quake so cataclysmic that even the most pessimistic design assumptions are exceeded, the likelihood of developing a completely earthquake-

proof building is nil. But with the conscientious application of rational design principles, architects can come up with buildings able to withstand many or most of the life- and property-threatening forces unleashed when the Earth is thrown into spasm.

stress is building up and that the conditions that produce earthquakes are ripening.

Data from the network of sensors are fed into computers and analyzed. Significant departures from the normal are noted. They provide the basis for predicting an imminent earthquake. Confidence in the applied science of earthquake forecasting, still in its infancy, will grow as knowledge of fault behavior grows and instrumentation and techniques of data analysis become more refined.

But even if given warning, how would the citizens of San Francisco, Los Angeles, and other cities fare in the event of a major earthquake? Evacuation of such large urban centers would be extremely difficult. Most older buildings weren't designed to withstand sizable quakes, and the collapse of many is a virtual certainty. Secondary reinforcement of existing structures is possible but very expensive (Los Angeles recently embarked on such a program), and even new buildings, erected to conform to modern standards of earthquake resistance, have failed in severe quakes. Massive firestorms, which often sweep cities shaken by large earthquakes, pose another hazard. And the liquefaction of the ground itself, a particular danger wherever structures stand on unconsolidated sand, silt, or fill, is an extremely difficult engineering problem. Forewarned may be forearmed, but there seems little reason at present to be optimistic about the abil-

ity of urban Californians to cope with an earthquake of major dimensions.

As we have seen, however, earthquakes are but one of the products of plate tectonics and the ring of fire. North of California, in the Cascades of the Pacific Northwest, looms Mount St. Helens, once celebrated as a rival to Japan's Fujiyama in the perfect symmetry of its snowcapped cone. On the morning of May 18, 1980, a violent eruption removed the top 1,300 feet from the mountain's peak, devastated completely the blast zone to the north, felled mature trees as far away as fourteen miles, sent torrents of mud, debris, and meltwater hurtling down the valleys that radiate from the mountain, and shot scorching blasts of gas and steam, laden with rock fragments and volcanic ash, more than 65,000 feet in the air. A series of further eruptions spewed additional gas, ash, pumice, and liquid rock, in all more than a cubic mile of material. An area larger than 160

square miles was utterly transformed by the blasts, and forests, farms, and towns throughout the region were buried in ash.

Mount St. Helens had been quiet since its last eruption in 1857. Still, its eruption was not unexpected. Two months before, nearby sensors had begun to register signs of renewed seismic activity. The frequency of earth tremors mounted, and several minor eruptions—steam explosions—occurred that opened up small craters on the summit. During this time, an ominous bulge grew rapidly along the north slope of the mountain. There was no question that a larger eruption was on its way. State authorities restricted access to the immediate environs of the volcano, and the Forest Service closed its access roads. Most area residents were evacuated.

Yet when an earthquake of magnitude 5 finally triggered the eruption, no one was truly prepared for the great force of the lateral blast. The quake caused the bulging

On May 18, 1980, a violent eruption rocked Mount St. Helens in southwestern Washington. The beautiful symmetry of the volcano was forever destroyed by a terrific blast that removed more than 1,300 feet from the mountain's peak and shot torrents of gas, steam, rock, and ash more than 12 miles into the air.

(Above) In the path of the blast, almost all the trees that cloaked Mount St. Helens's slopes were felled. Those not toppled were scorched to death, but a few hardy individuals, like this subalpine fir, battered but still green, somehow managed to endure.

(Right) Chaos reigned in the valleys below. The astonishing force of the debris flow overturned trucks, scattered enormous logs like matchsticks, and buried timbering camps under tons of mud, rock, and uprooted trees.

north slope to fail and slide down the valleys. With the north face gone, the mountain "uncorked" and vomited forth its mammoth cloud of gas, ash, and rock.

Most living things, including sixty-three people and thousands of deer, elk, and bear in the immediate vicinity, perished instantly. On the sides of the mountain that sustained the least damage, life has returned almost to normal. The fight for life is a little tougher elsewhere, however, as the soil has been robbed of its nutrients. The soil also now either has an impenetrable crust that shuts out light and air or is so porous that water simply runs through it. A University of Washington scientist has tried, with minimal success, to grow lettuce in soil samples taken from the blast site. And other University of Washington scientists say that for some sections severely affected by the eruption, it may take 500 or 1,000 years for the terrain to return to normal.

Still, biology is destiny in some cases. Not two months after the initial blast, biologists found blue-green algae and metal-eating bacteria already swarming in the debris-laden pools and a few hardy plant survivors poking through the ash. In the time since, scientists have made much of the unique op-

Braving dangerous and unsettled conditions, geologists risked their lives by returning to study Mount St. Helens soon after the first great eruption. They probed the layer of ash and made detailed measurements of the new-formed crater in their attempt to ascertain the volume of material removed by the blast. Their finding: two-thirds of a cubic mile of the peak had been blown away!

The eruptions sent up huge clouds of ash that blotted out the sun. In the aftermath, distressed citizens were left to cope with the thick layer of fallen ash that blanketed their towns and fields. In Yakima, eighty miles northwest of Mount St. Helens, more than 600,000 tons of ash were deposited.

Perhaps science's chief task—certainly its most immediate—is to monitor the belching behemoth and give ample notice to the populace at large should there be an impromptu repeat performance. The mountain is ringed with geophysicists' instruments to record and warn of even the slightest tremble. Its temperature is monitored, and teams of scientists regularly enter the crater to examine the dome. Planes occasionally fly overhead carrying photographers to record the still-smoking cone. There isn't a scientist clairvoyantly confident to hazard a guess as to when Mount St. Helens may erupt again, although all think another major blast is unlikely in the next few decades. The instruments should provide a few days' warning, at least.

There are some people, inculcated in the prideful notion that nature is there to be subdued, who respond to phenomena like earthquakes and volcanic eruptions with peevish anger. Most witnesses to such events, however, react with appropriate humility and awe when faced with the majesty and power of the forces of nature.

portunity to study the literal revival of the desolate posteruption landscape, observing the immigration of a host of pioneer creatures and the succession of biotic communities becoming reestablished on an ecological *tabula rasa*.

Five months and several major eruptions after the initial blast, Mount St. Helens continued its activity, building a lava dome and venting steam. To its desolate slopes, life slowly returned. Pioneers, like this lupine poking through six inches of ash, initiated the restoration of a community of life to the devastated neighborhood of the volcano.

Cosmic Catastrophe

One of the most famous of all fossil finds was *Archaeopteryx*, discovered in a Bavarian slate quarry in the middle of the last century. The 140-million-year-old remains of this oldest known bird revealed a mix of avian and reptilian features.

The extinction of a species, like the death of an individual, is a fact of life. Of all the species that have ever existed on Earth, the vast majority are today extinct, and no one would deny that every species now alive (including our own) will, sooner or later, join the ranks of the departed.

Almost all our knowledge of the life of the past comes from the discovery and study of the remains of once-living organisms preserved in sedimentary rocks. Examining these fossils is science's chief means of giving voice to our long-silent ancient past. Now the likelihood is rather small that an individual plant or animal will be buried in silt or clay or limy mud, in sand or wind-blown dust. Also small is the likelihood that the sediment, with its included organic remains, will harden into stone and will endure through countless years of further burial, uplift, and erosion. Smaller still is the chance that the sediment will eventually be exposed and that its embedded fossils will come to the attention of people who study such things. Despite these great odds, fossil remains of ancient life are sufficiently abundant and accessible that geologists and paleontologists, students of the Earth and of ancient life forms, have been able to reconstruct the evolution of life with a good deal of accuracy.

If that evolution had been a gradual and even affair, we might expect to find that species have become extinct at a pretty

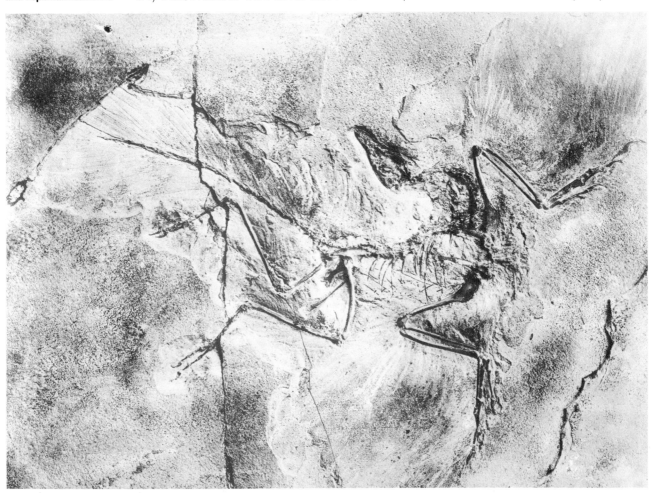

constant rate. The fossil record would reveal, at every level, about the same number of species making their final appearance, just as the obituary pages of a newspaper in a city with a constant population would always report roughly the same number of deaths, from day to day and year to year. In fact, the fossil record frequently reads like the obits during a major war or epidemic, when more deaths than usual are reported. Several times in Earth history, a significant fraction of all the species then living seem "suddenly" to have gone extinct. (Remember that events that might have taken tens or hundreds of thousands of years, or even a few million, seem sudden in the context of the 600-million-year history of complex life, or the three billion years since the first living things appeared on this planet.)

One of the most puzzling and dramatic of the several great extinctions happened about 65 million years ago, at the end of the Cretaceous period. Puzzling, because until recently no altogether satisfying explanation has been offered to account for the widespread extinction of a variety of organisms in all habitats, on land and in the seas. Dramatic, because numbered among the casualties were the dinosaurs.

Did a large asteroid collide with the Earth 65 million years ago? Did the dust resulting from the impact enshroud our planet in a dark cloud that suppressed photosynthesis and caused the starvation of dinosaurs and many other forms of life? New findings suggest that dinosaurs and company may indeed have been dealt the _coup de grace_ by a (perhaps) identified flying object.

Because large animals are rarely found fossilized intact, paleontologists must often reconstruct ancient creatures from the incomplete remains of several individuals. When the great fossil hunter Othniel Marsh excavated the headless skeleton of a brontosaur in 1879, he blithely added to it another animal's skull, found 400 miles away. About 50 years later, a brontosaur skeleton, head and all, was unearthed. Marsh's mosaic specimen, which had served as the model for many subsequent restorations, was discovered to be an unnatural hybrid, and mounted museum brontosaurs, like the one above at Chicago's Field Museum, had to be decapitated and fitted with new, proper heads.

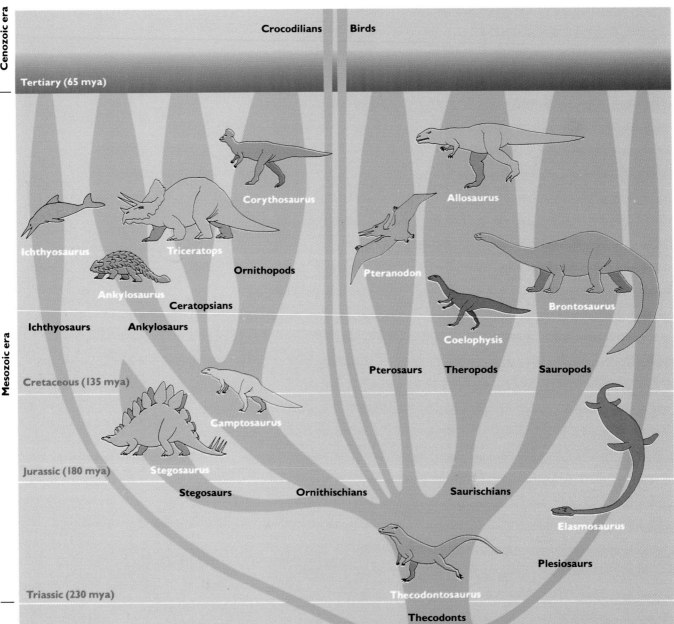

Cenozoic era

Crocodilians   Birds

Tertiary (65 mya)

Corythosaurus

Allosaurus

Ichthyosaurus

Triceratops

Pteranodon

Brontosaurus

Ornithopods

Ankylosaurus

Coelophysis

Ceratopsians

Ichthyosaurs   Ankylosaurs

Mesozoic era

Cretaceous (135 mya)

Pterosaurs   Theropods   Sauropods

Camptosaurus

Stegosaurus

Jurassic (180 mya)

Stegosaurs   Ornithischians   Saurischians

Elasmosaurus

Plesiosaurs

Thecodontosaurus

Triassic (230 mya)

Thecodonts

Paleozoic era

Permian (280 mya)

Cotylosaurs
(stem reptiles)

The term *dinosaur* is rather misleading, since it implies the existence of a single, natural group of related animals. The name is actually applied indifferently to members of two quite distinct orders. The Saurischia included the gigantic, plant-eating sauropods such as *Brontosaurus* and the bipedal, carnivorous theropods, which ranged from small, ostrichlike forms to huge beasts such as *Tyrannosaurus.* The Ornithischia comprised several major lines of plant eaters, many of them variously adorned with body armor, plates along the back, or bizarre head crests. Related to these two groups were the flying pterosaurs ("Rodan" was one). Several groups of even more remotely related aquatic reptiles, such as the plesiosaurs and dolphinlike ichthyosaurs, are also frequently, if inaccurately, termed dinosaurs.

The dates given in the chart refer to the beginning of each period of geological time. *Mya* means millions of years ago. This whole group of animals abruptly disappeared about 65 million years ago at the end of the Cretaceous period–possibly as the result of the impact of an asteroid.

Thanks in no small measure to their representation in cartoons like "Alley Oop" and "The Flintstones," and in films like *Godzilla, Rodan,* and *One Million Years B.C.,* dinosaurs have become as familiar as any animal still living. Many people would recognize a tyrannosaur more easily than a wombat or an anglerfish. Museum exhibits of dinosaur remains are well attended. Dinosaur toys, coloring books, and model kits sell like hotcakes, and names like "brontosaurus" fall as trippingly from the lips of children as "hippopotamus." Except for new finds of human and prehuman fossils, no paleontological discoveries are as certain to be picked up by the popular press as those of dinosaurs.

The fascination with dinosaurs infects scientists as well. Active investigation of the appearance, physiology, and even habits of these long-dead creatures is ongoing, and a number of exciting new ideas, among them the possible warm-bloodedness of at least some of the dinosaurs, have received considerable attention and enlivened scientific debate in recent years. Yet the latest (and, to date, most interesting) explanation of dinosaur extinction does not come directly from dinosaur specialists. As so often happens in science, the explanation comes obliquely, this time from a father and son, one a physicist and the other a geologist, who started by seeking little more than a method to figure the rate at which sediments were deposited.

Paleontologists had for years studied the limestones exposed in a deep gorge in Gubbio, a small town in central Italy. Some of the limestone was formed during the Upper Cretaceous, when the land where Gubbio now stands lay under water. In that water lived tiny protistans (one-celled organisms) of the order Foraminifera; geologists fondly dub them "forams." Forams secrete shells of calcium carbonate (lime), and the accumulation and compression of millions of these little shells fallen to the sea bottom is one of the ways in which limestone forms. The Cretaceous limestone at Gubbio is composed of the shells of many different species of forams. But capping the limestone, right at the

Gubbio, Italy.

The Message Is in the Rocks

Certain chemical elements "decay," or transform into others, over time. Uranium, for example, decays at a known rate to produce a form (isotope) of lead, and rubidium changes to strontium. By determining the ratio of a parent element to its descendant, scientists are often able to estimate with fair precision the age of a sample of native Earth rock, and of moon rocks and meteorites as well. Some "creation scientists" hew to the biblical line, which dates the Earth's origin at 6,000 years ago. Most scientists, however, have accepted the compelling evidence of radioactive dating, which suggests an age of 4.6 billion years for our planet.

(Right) **A close-up of the rock strata exposed on a hillside in Gubbio, Italy. A narrow vein of clay, red in its upper half and gray in its lower, lies at the boundary between the white Cretaceous limestone below and the red Tertiary limestone above. The Cretaceous limestone contains a variety of forams, including some that are relatively large *(below)* and disappear abruptly at the boundary. The earliest Tertiary foram fauna, preserved in the clay layer, are much less diverse, consisting only of the tiny species *Globigerina eugubina (above).***

boundary between the Cretaceous and the Tertiary period, is a peculiar layer of reddish and gray clay. The clay appears on gross inspection to contain no forams at all, but examination under a microscope reveals a great many forams, all much smaller than the ones in the underlying limestone, and all belonging to a single species. Above the clay layer, limestone resumes, but now it is composed of forams obviously descended from the tiny, lone species found in the clay.

Of course, it is no accident that the layer of clay occurs at the transition from Cretaceous to Tertiary times. It is just these kinds of abrupt changes in the sedimentary record, and the changes in the species embedded as fossils, that permit geologists to recognize discrete geological eras, periods, and epochs in what is, after all, unbroken time. Geologist Walter Alvarez was curious about the clay layer because he knew that it was deposited at a time when animals all over the world, and not just forams in the shallow seas over prehistoric Gubbio, had become extinct.

He showed a sample of the limestone-clay-limestone sandwich to his father, Nobel Prize–winning physicist Luis Alvarez. They were interested in discovering how long it took for the half-inch clay layer to be deposited. They knew only that during that time, the diversity of forams had become severely depressed.

Since several factors conspire to determine how quickly sediment accumulates, the thickness of a sedimentary layer is only a rough indicator of deposition time. Walter estimated that a half-inch layer of clay would form in about 5,000 years, but he and his father sought greater precision. They devised an ingenious method of investigation, based on the known deposition rates of rare elements. Luis knew that the sun and planets of our solar system had formed from an enormous primordial cloud, remnants of which filled the system with a fine cosmic dust. About 100,000 tons of that dust still fall to Earth each year as a sort of "cosmic rain."

When our planet congealed, the constituent elements of the formative cloud be-

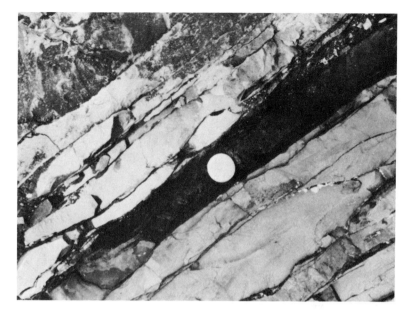

haved differently. Some remained in the crust, while others sank to deeper layers. Such elements as platinum, osmium, and iridium were among those which sank and are therefore rare in the Earth's crust. Since the cosmic dust is still unconsolidated, it contains a homogenized mix of all the original elements. For this reason, when rare elements are found in sediments on Earth, they are assumed to have arrived in the constant rain of cosmic dust. From several candidate elements, the Alvarezes chose iridium to study because it is relatively easy to detect. Assuming a constant rate of cosmic dust fallout,

At the end of the Cretaceous period, 65 million years ago, all land vertebrates heavier than 50 pounds, and a host of marine animals, both large and small, suddenly became extinct. Clues to the reason for their disappearance were found in this thin layer of iridium-rich Italian clay, only one centimeter (less than half an inch) in thickness. The coin shown for scale is smaller than a dime.

Successive layers of fossiliferous sedimentary rocks provide a record of the Earth's former life. In its Grand Canyon, the Colorado River has cut down through rocks whose ages span a billion years. The Precambrian rocks exposed at the bottom are so ancient that whatever fossils they may once have contained have been erased by eons of heat and pressure. The sediments at the top, along the South Rim, were deposited during the Permian period, when the ancestors of the dinosaurs were flourishing, a mere 250 million years ago.

Quantity of iridium found in sediment at levels above and below the presumed asteroid impact.

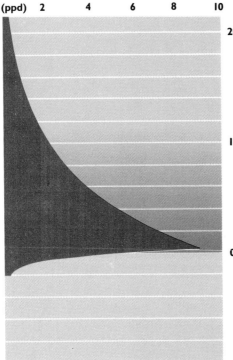

(ppd)   2   4   6   8   10

20 cm

10 cm

0

Tertiary period
••••••••••••••••••••
Cretaceous period

they figured that a high incidence of iridium in the clay layer would indicate a long period of deposition, while a low incidence would mean the clay accumulated rapidly.

The Alvarezes enlisted the aid of Frank Asaro and Helen Michel, nuclear chemists and colleagues of theirs at the University of California at Berkeley. Samples of the clay were submitted to a procedure called neutron activation analysis. In a nuclear reactor, samples were bombarded with neutrons, which excited the atoms in the clay and caused them to decay and give off gamma rays. Each element in the clay produced a unique gamma-ray "fingerprint." Iridium was indeed found to be present. More than present, it was abundant! The concentration of iridium was about 30 times higher than expected, even though all other elements were present at values close to what is considered ordinary. Tests on other samples were also puzzling; a sample from Denmark tested 160 times higher than normal for iridium.

The University of California team that unraveled the mystery of the great dying at the end of the Cretaceous period: *(left to right)* **Helen V. Michel and Frank Asaro, nuclear chemists at the Energy and Environment Division of the Lawrence Berkeley Laboratory; Walter Alvarez of the Department of Geology and Geophysics; and Luis W. Alvarez, professor emeritus of physics and recipient of the 1968 Nobel Prize for physics.**

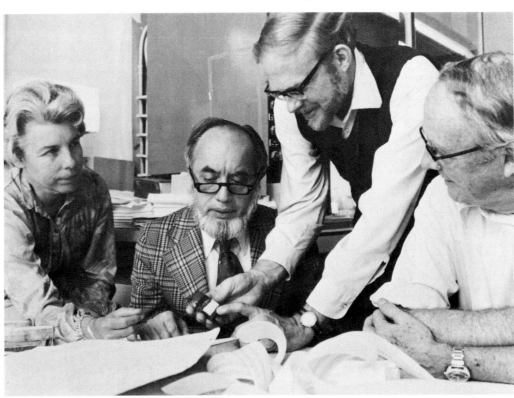

What could have caused such high concentrations? For the thin layers of clay to have accumulated so much iridium relative to other sedimentary materials, either a long period must have occurred during which the normal settling of material on the sea bottom had unaccountably stopped, or else the iridium must have been "dumped" by something other than the cosmic rain. The team considered and rejected many possible but unlikely explanations. At last, Luis offered a hypothesis that might explain not only the high iridium levels but also the widespread, simultaneous extinction of forams, dinosaurs, and many other forms of Cretaceous life: the iridium had been carried to Earth in an asteroid.

Most asteroids, small planetary bodies about one to ten miles in diameter, orbit the sun in a belt between Mars and Jupiter. But an estimated one to two thousand, called Apollo asteroids, have orbits that at times bring them relatively close to Earth. One of these may have collided with Earth 65 million years ago and spelled doom for a host of species. Given the number of Apollo asteroids and the great age of the Earth, collisions must be inevitable.

The Alvarezes suggest that an asteroid six miles in diameter crashed to Earth. The force of impact caused the asteroid to disintegrate. This same enormous force tore from the Earth's surface an enormous chunk of material equivalent to perhaps fifty asteroids. A huge cloud of pulverized asteroid and Earth material was thrown up; most of it entered the stratosphere as a thick layer of dust that enveloped the Earth and blocked the sun's rays. The phytoplankton, tiny photosynthesizing plants in the oceans, died immediately. Land plants died soon thereafter, leaving only their seeds to revegetate the Earth once the black cloud thinned enough to permit the sun to shine again. With the plants gone, plant-eating animals perished, and their predators died in turn.

This decidedly catastrophist scenario could not so easily have been proposed just a few decades ago. Uniformitarian geologists,

Uniformitarianism versus Catastrophism

Throughout the nineteenth century and well into the twentieth, Earth scientists subscribed to one of two contending schools of thought. Uniformitarianism—whose dean was the great geologist Charles Lyell (*above*), author of *Principles of Geology,* a book that influenced Darwin profoundly—held that geologic change proceeds slowly and evenly, that processes of change known to be active at present can explain all changes in the past, and that the Earth has remained largely the same since its formation. Opposing the uniformitarians were Georges Cuvier (*below right*), Louis Agassiz, and others who proclaimed the doctrine of catastrophism. Like their adversaries, the catastrophists also favored currently operating processes as agents of past change. Unlike the uniformitarians, they maintained that other forces, now quiescent or acting in a different fashion, might also be invoked to explain certain ancient events. They believed particularly in the possibility of rapid, cataclysmic change, which to the uniformitarians was unthinkable. Although the catastrophists came to be stigmatized and much reviled as dupes of the unscientific theological establishment, their theory in fact sprang from careful observation of the hard evidence of events in Earth's history. The earliest advocates of catastrophism argued that the Biblical Flood was the major event in the shaping of the Earth's surface. Cuvier refined this idea into a theory of periodic catastrophic change. Certainly, features like fractured, grossly twisted rock strata, huge plateaus formed by the outfall of volcanic eruptions of apparently enormous power, and a fossil record testifying to the sudden disappearance of entire communities of life all gave credence to the belief in the violence and rapidity of past change. The catastrophists have been at least partly vindicated by the findings of contemporary geologists, who no longer reject out of hand hypotheses as radical as collisions with death-dealing asteroids to account for the extinction of the dinosaurs.

What Done It?

© Walt Disney Productions

Most paleontologists would agree with the rather circular statement that a species dies out when it fails to adapt to changing conditions. Such eminent students of the past as George Gaylord Simpson and Edwin Colbert have candidly admitted to ignorance of the specific causes of dinosaur extinctions. Some of the theories that have been advanced at different times are listed below. A few are downright silly, but most are plausible. Some seek to explain broad-spectrum extinctions, while others apply mainly to the dinosaurs. None is entirely satisfactory.

1

The "Too Much of a Good Thing" theory: evolutionary trends, once in motion, assume a life of their own and, like a car out of control, cannot be halted; the unhappy possessors of features resulting from these trends are driven extinct by the sheer bad design of those awkward features. This peculiar notion (once dignified as "orthogenetic" or straight-line evolution) would lead one to recognize "unstoppable" trends toward gigantism or elaborate head ornamentation in different dinosaur lineages, trends that resulted in creatures too ponderous or encumbered to survive.

2

The "Lennie Small" theory: dinosaurs, like the hapless character in Steinbeck's *Of Mice and Men*, were big and stupid and therefore incapable of making it in the modern world. Well, some dinosaurs were rather small, but they became extinct at the same time as their larger cousins. As for the putative stupidity of dinosaurs, measurements of dinosaur cranial capacities and extrapolation from the brain sizes of living reptiles demonstrate that the brains of dinosaurs were exactly as big as one would expect in animals of their size. And mounting evidence of a complex social life in some dinosaurs, and of the active, agile style of some (warm-blooded?) others, chips away at the traditional view of dinosaurs as doltish and dim-witted.

3

The "All about Eve" theory: while dinosaurs, the superannuated Great Stars, were paying no attention, the ambitious and ruthless mammalian understudies slunk in from the wings and snatched away the plum roles. Actually, mammals had been on the scene throughout more than three-quarters of dinosaur history without making much of an impression. The explosive radiation of mammals took place only after the dinosaurs had made their exit.

4

The "Grand Mal" theory: dinosaurs were eliminated by a terrible plague. Disease organisms, however, are usually highly host-specific and attack members of only one or several closely related species. To wipe out the diverse array of Cretaceous reptiles, the simultaneous descent of an army of assorted germs would have to be invoked.

5

The "It's Too Darn Hot" theory: dinosaurs perished because of a global warming trend that turned their favored habitats into barren deserts. This idea got a big push from Walt Disney; in his cartoon feature *Fantasia,* parched dinosaurs vainly seek sips from evaporating pools, gasp for air, and finally keel over and die under a blazing sun. Disney's vision notwithstanding, evidence for late Cretaceous climatic change favors a world growing cooler, not hotter. Go to #6.

6

The "Baby, It's Cold Outside" theory: dinosaurs perished when the climate changed from constant subtropical warmth to a seasonal regime with cold winters. But if some dinosaurs were warm-blooded, why didn't they endure? One suggestion is that the warm-blooded ones lacked an insulating coat of hair or feathers; they therefore lost vital body heat during the increasingly rigorous cold seasons. Seeming to contradict this is the persistence of cold-intolerant large crocodilians at high latitudes (Canada and Argentina) in late Cretaceous and early Tertiary times.

7

The "No Home on the Range" theory: the late Cretaceous was a time of progressive habitat loss, on land and in the sea; rich and varied habitats were replaced with monotonous and uniform ones; the reduction in resources led to increased competition; since large animals are usually early casualties when complex ecosystems are disrupted and collapse, the dinosaurs never had a chance. This very reasonable theory almost certainly accounts for some of the Cretaceous extinctions. One problem is the extent of disagreement among experts about the causes of habitat loss. Some advocate active mountain building as a disruptive factor, while others claim a cessation in mountain building; some invoke marine incursions onto low-lying land, but others say the epicontinental seas retreated.

8

The "It Came from Outer Space" theory: a supernova exploded close to the sun, flooding the Earth with cosmic rays that killed the inhabitants outright or caused them to spawn hideous and unviable mutants; those not done in by the cosmic rays froze to death when X-rays in the upper atmosphere brought about global climatic deterioration. This theory is rejected because of the absence in sediments of concentrations of certain elements that would have followed from a supernova source, and because the probability of a supernova's exploding so near the sun as to drop the amount of iridium found in the Gubbio and Danish clays is practically nil.

So back we come to the asteroid theory, an "outer space" hypothesis, to be sure, but a far more plausible one than the supernova theory. Of all the explanations outlined, the ones that name climatic deterioration and habitat loss seem sensible. Some of the groups destined for extinction at the end of the Cretaceous were already in decline, probably because of these factors. Certainly the sauropod dinosaurs, the ichthyosaurs, and the ammonite mollusks had already seen their heyday and were by the late Cretaceous much less diverse than in earlier times.

Our planet is by no means isolated from the rest of the universe in which it floats. We are bombarded constantly with radiation from space. Tens of thousands of tons of fine cosmic dust rain down on us every year. And the meteorites and asteroids (like the one whose trail is seen in this time-lapse exposure) that occasionally collide with our world provide scientists with (literally) solid evidence for their studies of the history and composition of the universe.

with their cherished belief in the gradualism of past change wrought by agents still in operation at present, would not have received very kindly such a frankly violent and cataclysmic interpretation of past events. But classical uniformitarianism has been shaken by new discoveries about plate tectonics and continental drift, the Pleistocene glaciations, enormous landscape-transforming floods, and volcanic eruptions of an intensity sufficient to alter the world's weather. Catastrophism, after more than a century of academic resistance and disdain, is again respectable.

If we grant that the contemporary intellectual climate allows the suggestion of catastrophist explanations, we then must ask: does the asteroid theory explain the Cretaceous extinctions better than other hypotheses? And is there evidence that can be marshaled in support of asteroid overkill?

We find in the fossil record evidence of more than just a gradual, uniform dwindling of species diversity in the late Cretaceous. We also find evidence of a sudden catastrophic event, one that extinguished life forms throughout the world and in all major communities, both terrestrial and aquatic. The oceanic phytoplankton disappeared so pre-

cipitately that the event is marked in the record as a "plankton line" well known to geologists. Also gone from the seas were the majority of forams and coccoliths, another group of one-celled organisms. Among mollusks, all the ammonites, all the belemnites (ancient cuttlefish relatives), and all the strange, reef-forming rudistid clams disappeared forever. And, of course, of all the dinosaurs that roamed the land, all the pterosaurs that soared in the air, and all the mosasaurs, ichthyosaurs, and plesiosaurs that swam in the seas, none lived to see the dawn of a Tertiary day. Of all the theories proposed, only the Alvarezes' asteroid theory explains the sudden and simultaneous disappearance of all these disparate groups.

One wonders, though, how anything managed to survive the cataclysm. A number of important groups, among them the land plants, the insects, the teleost fishes, the mammals, and the birds, pulled through. If the asteroid theory is correct, then an adequate explanation of this differential survival is surely needed.

Another problem with the theory concerns the point of impact of the asteroid. A body six miles in diameter hurtling into the

The most recent aster- oid impact on Earth's moon occurred almost 4 billion years ago. It left a crater many hundreds of miles across, which can be seen above in a photo- graph made by the Rus- sian spacecraft *Zond 8.* Weather has erased what must have been similar craters on Earth, and today the best preserved is Me- teor Crater in Arizona *(center),* a mere 25,000 years old and less than a mile across. *(Bottom)* Fourteen-mile-wide Phobos, one of the two moons of Mars, is thought to be a cap- tured asteroid that will one day leave a gi- gantic crater on the red planet.

Earth would punch out a crater one hundred miles across. Yet no crater of the requisite age and size is known, on either land or ocean floor. But what if the asteroid landed on a midocean ridge, one of the fractures in the Earth's crust where molten material wells up and pushes apart the great plates that carry on them the drifting continents? If the asteroid had indeed landed on such a ridge, it would have punctured the crust (which is relatively thin along these zones of fracture) and tapped into the vast reservoir of lava below. A volcano on a grand scale, with the potential for millions of years of activity, would have formed, and from all this volcanic activity a new land mass would have been born.

But is there such a land mass sitting astride a midocean ridge anywhere in the world? Yes is the answer, and Iceland is the island. What is more, the proximity of Denmark to Iceland (and therefore to the site of the asteroid crashdown) might explain why the iridium content of the Danish clays that the Alvarezes tested was so much higher than even the anomalously high levels in the Italian clays. Finally, the age of Iceland is estimated at 50 to 60 million years, which jibes very nicely with the arrival, 65 million years ago, of the asteroid that capped the Cretaceous period.

Although there exists controversy over the Alvarez work and some puzzling questions remain, such as the reasons for the selective survival of certain groups, their theory has been greeted with excitement in scientific circles. The elegant experimental method, and the serendipitous connection between an investigation of sedimentary deposition rates and the formulation of a novel theory of mass extinction, together represent scientific activity at its best. Whether in years to come the theory is strengthened and improved or falsified and discarded, it will stand as an audacious and imaginative effort to explain a phenomenon that has long stumped some of the greatest minds in the field.

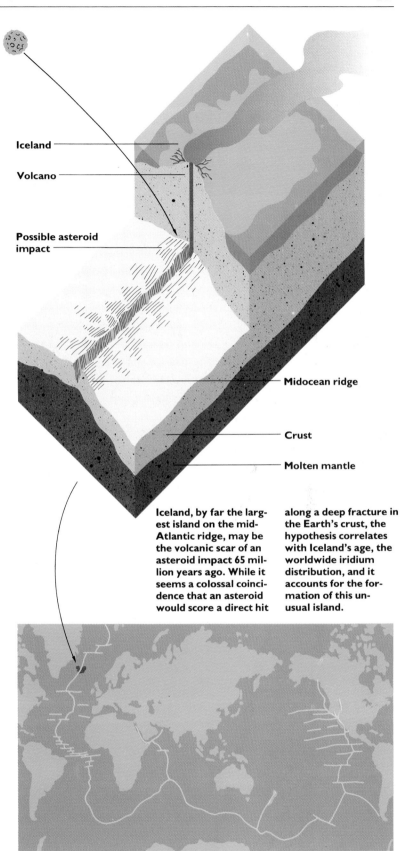

Iceland — Volcano — Possible asteroid impact — Midocean ridge — Crust — Molten mantle

Iceland, by far the largest island on the mid-Atlantic ridge, may be the volcanic scar of an asteroid impact 65 million years ago. While it seems a colossal coincidence that an asteroid would score a direct hit along a deep fracture in the Earth's crust, the hypothesis correlates with Iceland's age, the worldwide iridium distribution, and it accounts for the formation of this unusual island.

The Green Machine

You began three billion years ago, a single-celled oceanic alga. Evolution persuaded you to take on the hazards of mobility, respiration, and sexual reproduction. Drying your limbs considerably, you stepped onto land. Mosses came first, then lacy ferns. Last to reach the new world were seeders and flowerers. Their tardiness, though, was no sign of weakness; since their arrival, they have tyrannized the rest of the plant kingdom with their numbers.

Your seedlings are brash and tough, shoveling through dirt with a force of up to 450 pounds per square inch. Roots grow up—or down—at a rate of four inches a day, using gravity-sensing devices for direction; tip and leaves are directed by light detection. Five hormones, each containing hundreds of compounds, set the pace for growth, stimulating or inhibiting as necessary. You are pushed down and pulled up as a by-product of their battles and imbalances. In the heat of their combat, you can reach heights—as trees do—of up to 300 feet. Water nourishes your

crow's-nest branches from anchoring roots by means of a massive suction system: you pull liquid up your woody veins at terrific pressures.

The sun is the patron, the provider. Through a process called photosynthesis, you convert its energy into glucose, a sugar, and give thanks by releasing oxygen. Minuscule pores on your leaves draw in carbon dioxide and free oxygen—a reversal of the respiratory sequence used by the stolid humans who prune, garden, and trample over you. They don't realize their existence is by your leave—literally.

Inside each leaf cell circulate streams of green particles called chloroplasts. Inside each chloroplast are tiers of tiny light collectors, each collector itself made of thousands of molecules of chlorophyll. Chlorophyll is a magnesium-based compound for which you alone hold the patent. It traps light that—over many complicated steps—energizes the reaction between carbon dioxide and hydrogen to produce glucose. At the same time, water is being split into its component parts, hydrogen and oxygen. From light and liquid and a dead gas, you supply the Earth with energy and food.

Another molecule, also yours alone, acts as a botanical light meter to "read" the different wavelengths in the spectrum. Phytochrome responds to colors and is especially sensitive to light from the far-red region, well beyond the limits of human vision. Given a flash of far-red light, your growth speeds up ten times. You have, it is clear, a favorite color.

You have other humanlike characteristics as well: musclelike properties and nervelike impulses that can be measured electrically in the snapping of a Venus's-flytrap or the stretching of a morning glory. Some people even believe your powers exceed theirs. They say you possess an intuitive sense that can be charted empirically on a lie-detector machine: you respond not only to damage done to your own tissue, but also to damage done to any living tissue near you.

You even respond to voiced or silent in-

Photosynthesis is a complex process that converts carbon dioxide (CO_2) and water (H_2O) into glucose ($C_6H_{12}O_6$). A by-product of this reaction is oxygen, essential for all animal life; animals in turn produce carbon dioxide as a by-product, and that is one of the ingredients plants use to make glucose. This exchange begins when chlorophyll molecules within leaf cells are activated by sunlight to trigger a series of reactions that break down water into its components, hydrogen and oxygen. Oxygen escapes from the leaf through thousands of tiny openings called stomates. As the CO_2 permeates the cell's walls, it binds with the hydrogen atoms to form glucose.

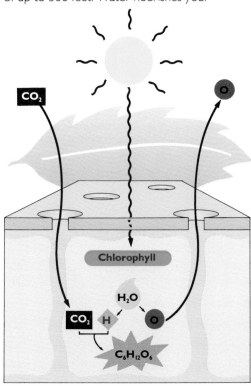

Moss cells with chloroplasts, magnified 400 times. Inside each chloroplast are tiers of light collectors, each made up of thousands of molecules of chlorophyll, a magnesium-based compound produced by the plant. Chlorophyll traps light that energizes the reaction between carbon dioxide and hydrogen to produce glucose.

tentions. Some people are convinced that positive reinforcement—prayer, especially—encourages you to lavish growth, while discouragement or verbal dressing-down sends you shrinking to the ground. Energy fields, they explain, are associated with all living organisms. Why shouldn't yours and theirs find a level of communion? They are setting up trials and machinery for experiments; they need to be assured of these things in these ways.

Much to their benefit, though, it is enough for you to continue existing. Your life is taken up in caring for their needs and supplying their requirements. You have no time left for scientific studies.

You have been at the business of survival—yours and theirs—for three billion years. You know perfectly well who you are.

A Tangled Web

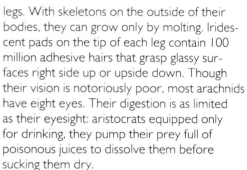

O nce in Ancient Greece, Arachne the weaver challenged Athena to a match. When the proud goddess ruined her tapestry, Arachne tried to hang herself. Athena turned the rope into a web and the artist into a spider. Arachnology is the study of her unblessed descendants—the spiders. Its focus is more than mythological: in the lives and livelihoods of spiders, scientists hope to disentangle several threads from the mysterious web of interactions that characterizes all living organisms.

Arachne's progeny have fused heads and chests, separate abdomens, and eight legs. With skeletons on the outside of their bodies, they can grow only by molting. Iridescent pads on the tip of each leg contain 100 million adhesive hairs that grasp glassy surfaces right side up or upside down. Though their vision is notoriously poor, most arachnids have eight eyes. Their digestion is as limited as their eyesight: aristocrats equipped only for drinking, they pump their prey full of poisonous juices to dissolve them before sucking them dry.

Webs are the warp and woof of spider life. Frenetic construction is followed by long lounging periods until company arrives. Some webs take an hour to construct and are rebuilt each day; others, more manorly, need up to four nights to complete but last up to six weeks. Some are built in air, some underground; some are orblike, some tentish; the silks are woolly or sticky, thick or fine.

Tension on the threads controls the direction in which a spider moves; each web creates a growing force field, and its master builder responds automatically. Most spiders have three pairs of spinnerets containing hundreds of spigots, each connected to an internal silk gland. Liquid silk solidifies as it leaves the body, and a spider grasps its own threads by means of claws on the tips of its legs.

Buzzing guests trip off a doorbell in the form of vibrations that run down the web threads. Motionless in a corner, the spider responds to its bell with a reflexive jerk, entangling diner and dinner still further. One politic bite paralyzes the insect, who is then wrapped—finest silks only—and eaten, or else reserved for another menu. Spiders were adept at mummification long before two-legged creatures claimed the art.

As strategic as the events involved in feeding are those that lead to reproduction. A male deposits sperm from his abdominal opening onto a silk pad, then suctions it into his pedipalps and strikes out, a mobile, hopeful sperm bank. Plucking and drumming on the threads of a female's web, he enacts an ancient courting rite, less out of chivalry than self-preservation. His rhythm is a terse plea: Don't shoot! Family here.

A black and yellow argiope (Argiope aurantia) in its web.

The female falls into a trance. When she is safely immobile, the male approaches and inserts his pedipalp into an opening on her abdomen. The length of trance is unreliable, though, and a lingerer may pay for his mistiming with several limbs. The male black widow pays with his life. After mating, he reluctantly assumes another paternal responsibility; his offspring become, quite literally, the flesh of his flesh. Males die soon after depositing sperm, anyway. For most, though, it is not by the fangs, or in the arms, of an intimate.

The female stores sperm until her eggs are ripe. Then she prepares a protective sac and drops the eggs into it. Variations in climate, hatching times, and species account for the differences in sacs. Some spiders suspend them; others carry them between their jaws and will not eat until they hatch.

When the babies, opalescent bits of life, break through the egg sac, an exhausted female performs one final act on their behalf. Directing her digestive juices inward, she liquefies herself. The spiderlings begin sucking at her joints, where the skin is soft. Her heart and lungs keep working until the end.

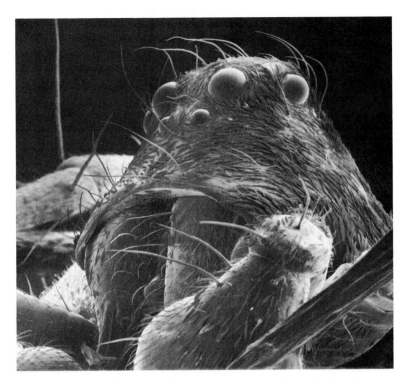

Thirty-five thousand spider species have been identified since the time of luckless Arachne. They form only a quarter of the total number. Are the rest too busy weaving to look up for a head count? Or have they learned from the lesson of their origins and refused to be caught in their own webs?

A small lynx spider photographed through a scanning electron microscope (magnification: 90X).

Animal Imposters

In this dog-eat-dog world, there are clear advantages in fooling one's potential enemy or prey. Nature is rife with examples of creatures that benefit from looking like something else. The phenomenon called camouflage—whereby an animal is colored to blend with its surroundings or patterned in a way that breaks up its outline and makes it "disappear"—is well known. Many animal species, especially among the insects, are disguised by a protective resemblance to twigs, leaves, thorns, or even bird droppings. Certain harmless species mimic the appearance of poisonous or noxious species (like monarch butterflies, wasps, coral snakes, and skunks) that advertise themselves with striking warning colors, often a combination of black, red, white, or yellow. Predators can't tell the difference and leave the harmless imposters alone. Even plants get into the act. The large and successful orchid family includes many species whose flowers resemble insects, luring "mates" that attempt copulation and, in doing so, ensure pollination.

First Signs

Washoe uses a weeding tool while her adoptive parents, Allen and Trixie Gardner, look on.

Meet Washoe. With a flick of her wrists, she is overturning the belief that humans alone are capable of language. This chimp speaks, but the words are not in her voice; Washoe was raised in the noiseless world of Ameslan, American sign language. Equipped with a vocabulary of more than 130 hand signs, she chains words together in their proper sequence, asks questions of her human companions, and even—like any profound thinker—talks to herself.

The consequences may shake some foundations in the carefully constructed worlds of anthropology, psychology, and linguistics. These fields agree: our uniqueness

trace their acquisition of language and meaning and to offer them, ultimately, liberation.

Born wild, Washoe exchanged her African citizenship for American papers when she was ten months old. For four years she lived in the home of Allen and Trixie Gardner, psychology professors at the University of Nevada. They treated her as parents would treat any normal child—not an unreasonable approach, since the early development of chimpanzees closely parallels that of humans, biochemically and chronologically. Washoe was provided with all the requisite suburban pleasures: playroom, backyard, natty wardrobe, and constant hu-

as human beings is predicated on our ability to acquire language. But Washoe breaks ground in another field, too: she may enable researchers to map the process by which thought and communication are linked, perhaps to understand the way in which deaf children learn to sign. Through her example, scientists may learn to enter into the silent world of the unhearing and, once inside, to

man companionship. The Gardners reasoned that if they didn't become Washoe's good friends, she wouldn't talk to them—and if she didn't lead an interesting life, she would have nothing to talk about.

Chimps can learn to do almost anything with their hands. Since their imitative ability is nearly flawless and they are unequipped for speech, gestures are more suitable a form

When asked by Roger Fouts, a former assistant to the Gardners, to name the object in his hand, Washoe correctly replies "book."

of communication for them than speech. To ensure no mixing of linguistic mediums, Ameslan was the only language used in Washoe's presence. She was exposed to it continuously, just as deaf children are. It is a complex language, in which many signs translate conceptually; at the same time that one gesture may represent several words, several gestures together may represent one word. Symbolic interpretation goes far beyond aping.

Washoe tackled the challenge like a chimp off the old block. After eighteen months, she had mastered a range of simple concepts: from "sweet" to "toothbrush"—a logical hygienic response—and from "open" to the frequent, gleeful command "tickle." Soon she was able to distinguish between herself and her teachers, using the idea of self and nonself; and eventually she was able to identify different images on slides.

Animal IQ

Even within our own species, the measurement of intelligence—a quality whose satisfactory definition has so far eluded scientists—is a task fraught with difficulty. Since the assessment of intelligence usually rests on a number of culture-specific assumptions, comparisons among people from different societies are highly suspect. How much harder, then, it is to investigate braininess in other animal species. The results of experiments with chimpanzees and gorillas suggest that these

creatures may possess an intelligence akin to our own. Researchers who tested their ability to acquire and use symbolic language have declared the simians to be able linguists. Their widely publicized findings have set off an acrimonious exchange among ethologists (students of animal behavior), psychologists, an-

thropologists, and linguists. Critics dismiss the researchers' conclusions, insisting that the apes are just that— clever mimics with a talent for aping their human instructors. In response, supporters cite the animals' spontaneous coining of new words, like "candy-fruit" for watermelon and "water-bird" for duck, and the unprompted transmission of signs from "educated" apes to naive ones, as dramatic evidence of the apes' genuine grasp of at least some features of language, heretofore regarded as a unique human phenomenon.

Animal Athletes

The feats of animals aren't limited to mimicry, of course. A cheetah can cover short distances at speeds that approach sixty-five miles per hour, and some horses have traveled twenty miles in an hour's time. A flea can jump to heights 130 times greater than its own, equivalent to a person's leaping tall buildings in a single bound. An archerfish, taking aim from below

the water surface, can spit at (and hit!) a target in the air fifteen feet away. A sperm whale can dive to 7,000 feet, holding its breath for more than eighty minutes. A salmon can leap more than eleven feet out of the water. A stooping peregrine falcon can exceed eighty miles per hour in its dive. A monarch butterfly can make a nonstop Atlantic crossing in three days. A South American reptile aptly, if irreverently, named the Jesus Christ lizard can run on water. A human being seems rather feeble when compared to these natural athletes.

Next, Washoe learned to combine words in proper order, rearranging them into requests, commands, or confirmations. Her most animated discussions centered on meals: those that were imminent, and also those she felt ought to be imminent. Finally, her attention turned to the art of conversation. "Who's stupid?" an assistant would sign, and Washoe responded forthrightly with an opinion.

She was gaining a sense not only of the power of language, but also of its acquisitional difficulties. New assistants complained regularly of the humiliation of being signed to at a subnormal pace. Their pupil, turning the tables solicitously, was slowing down for their benefit.

Project Washoe reached an official end

The Silence Barrier

Few hearing people have any notion of what it means to live in a world of perpetual silence. Most able-bodied Americans, if forced to choose between loss of sight and loss of hearing, would probably elect to be deaf. Yet severe hearing impairment must count as one of the most profound afflictions among members of our social species, who place a premium on the ability to communicate well. The isolating effects of deafness, most particularly for those born deaf, go beyond the difficulty of learning words. Our very thoughts and experiences, our manners, our knowledge of our place in society, our ability to assert ourselves in situations where our rights our challenged—all depend directly on the possession of a consciousness shaped by language, the acquisition of which is one of the most vexing problems that face people unable to hear.

when its object, at age five, retired to the pastures of a primate station outside of Oklahoma City. Retirement, though, has dampened none of her zest for American living. Days are filled with ministering to the younger chimps in the station, and Washoe has continued to learn and practice Ameslan on the side. Does she realize that, on the basis of her achievements, research was initiated with other chimps in several primate centers across the country? Does she know that the Gardners themselves are training a new generation of babies in sign language? Does she ever miss the suburbs?

It's difficult to tell. Along with a new language, Washoe may have acquired an even more delicate societal skill: she keeps her thoughts to herself.

At age five, Washoe was using Ameslan (American Sign Language) to name many objects. When Susan Nichols, a research assistant showed her a lollipop, Washoe called it "sweet" *(opposite page)*. Having been shown a picture of a cat, Washoe correctly identified it *(top left)*. A woolen cap is referred to as "hat" *(bottom left)*. And in answer to Roger Fouts, Washoe naturally called the van "car" *(above)*.

A Conversation with Noam Chomsky

It was 1953, and Noam Chomsky, a twenty-four-year-old Harvard junior fellow in linguistics, was on a boat to Europe and feeling seasick. Still, his mind was on his work. Since his undergraduate days at the University of Pennsylvania, he had been strenuously applying the principles of traditional linguistics to the study of modern Hebrew. At the same time, however, he had been consumed with the pursuit of a "hobby"- a new theory of language that sought to answer questions that went beyond the mere description of grammar.

During that boat trip, Chomsky recalls today, "it suddenly struck me that the work I was doing myself was in the right direction." The aim of traditional (or "structuralist") linguists, he explains, was to describe a language, to apply mechanical methods and "come up with the grammar of that language."

"I assumed that was the right approach, and I was trying very hard to make it work," he says in a voice that is quiet and friendly, without a hint of arrogance. "But I kept running into a wall. Finally I realized that the whole question should be studied from another point of view, one that I had been working on quite a while on my own."

That new approach, which Chomsky further developed over the next few years, culminated in a book, *Syntactic Structures,* published when he was just twenty-eight years old. It also created a revolution. Chomsky's impact on the cognitive sciences—linguistics, psychology, and philosophy—has been profound not because of any single discovery but because of his approach. Not since Freud has there been such a radical shift in these fields. Chomsky's effect on the cognitive sciences has been compared to Einstein's on the study of physics.

Chomsky, a theoretician, sought to answer three questions: How is it possible for human beings to understand intricate and specific sentences that they have never heard before? Why is it that children all over the world show such similarity in the manner in which they learn languages? And why is it that the 4,000 languages spoken today (many of which are not historically related) share such similar structures? Chomsky reasoned that language, a uniquely human ability, must be based on an "innate mechanism" in the brain, a "biological endowment" that is genetically determined.

He further speculated that there is encoded in our brains a finite set of rules, a "universal grammar" that applies to all languages, from Armenian to Zulu. He set as his task the discovery of what these linguistic universals are. (Many of them are abstract, even mathematical.)

Chomsky has been on the faculty of the Massachusetts Institute of Technology since 1955, the year he received his doctorate from the University of Pennsylvania (as a result of his work at Harvard). But although his work revolutionized linguistics, he first entered the field because of what he calls "a series of accidents."

Avram Noam Chomsky was born and reared in Philadelphia, in a family of politically radical Jewish intellectuals. His own interest in politics began in childhood. He recalls reading articles about the Spanish Civil War at the age of ten and browsing by himself in the secondhand bookstores of New York City's Fourth Avenue at age twelve.

When he was a sophomore at the University of Pennsylvania, he met Zellig Harris, a professor of linguistics there who shared his radical political views. "My main interests were political at that time," says Chomsky, "and I was pretty well disgusted with college. I was seriously going to drop out. I could have ended up living on a kibbutz in Israel. I probably would have gone back to college eventually, but I have no idea what I would have studied."

Because of his affinity with Harris, however, and because of a "family familiarity" with language—his father was a Hebrew scholar—Chomsky persevered in school and took up linguistics. But, he stresses, "I had a very unconventional academic training. I never really had a systematic training in linguistics." Harris encouraged his students to work independently, and thus the young genius was able to develop the "hobby"

that was to become the foundation of modern linguistics.

Today Chomsky and his followers are immersed in theoretical work toward a refinement of his concept of universal grammar. But for Chomsky, the larger purpose of linguistics is the study of mind. "Language is apparently unique to humans and connected in an intimate way to human thought and action," he says. "Linguistics gives a unique insight into the nature of certain aspects of the human mind."

Human beings are the only creatures with a true ability to use language, he says, although scientists have taught sign language to chimpanzees, including Chomsky's well-known namesake, Nim Chimpsky. "A principle that was put forth centuries ago," says Chomsky, "is that language involves the infinite use of finite means. The sentences we produce and that we read are unbounded in complexity and scope. That property is completely missing in all the systems that are laboriously imposed upon chimpanzees and other animals.

"It would be an amazing biological miracle if some other species had the capacity for language and had never used it," he continues. "It would be as if *we* had the capacity for flight but had just never used it."

Scientists may someday discover the "neurophysiological structure" in the brain that Chomsky is certain accounts for our language ability. Ultimately, he says, the linguistic and physiological approaches to the study of cognition will converge, and linguistics will be assimilated into the discipline of biology.

In the 1960s Chomsky became an intellectual hero of the New Left, well known for his anti–Vietnam War activism. He remains a vocal spokesman for human rights and a critic of United States foreign policy, of what he sees as American imperialism in the Third World.

In his numerous books and articles on politics, he speaks not as a scientist but as a private individual. Scientists, he says, have the same obligation that anyone has to be socially concerned. "Everyone must be responsible for the predictable consequences of their

actions," he says, "and sometimes these consequences arise from your scientific work. If you are acting in such a way that human life and welfare are affected, you should not fail to pay attention to those consequences."

Chomsky's political activism has perhaps brought him more media attention than his work in linguistics. "I would prefer that the world were such that I could be left alone," he says of that attention. "It certainly would be more peaceful."

Being famous is not important to him. "I would prefer that the import of my political work be well known," he says. "My linguistic work is important intellectually, but in terms of human life and welfare, my political work is more important. It deals directly with problems of central human concern."

Balancing his personal and professional lives has been difficult. He and his wife, Carol Chomsky—also a linguist—have three children, and in his limited spare time he likes to relax with his family, read, tend his garden on Cape Cod, and listen to classical music. "The real problem," he explains wistfully, "is that I have two professional lives, both of them more or less full time."

In spite of the intellectual revolution that Chomsky has fomented, he remains thoroughly unassuming. When asked what he would most like to be remembered for, he laughs quietly, as if surprised at the suggestion that he *will* be remembered. After a long pause he replies, but not in terms of innate mental structures or linguistic universals. He says simply that he would like future generations to think "that I contributed a bit toward making a life a little less full of sorrow and travail than it usually is."

Why Do Birds Sing?

Listen.
Those sounds are neither tuneless nor aimless. They're not from heaven, but their origins aren't human, either. One feather-breasted listener hears an assertion in the notes; another, a disputation; and a third, an attraction. Here is the mating call of a wistful single; there, the machismo of a territorial claim and the volatile backtalk it generates. Simple people hear simply birdsong, but the artists themselves are listening to a music that orchestrates their lives.

Birdsong is practiced almost exclusively by males. Each species has a repertoire of from one to one thousand pieces, and each piece is a proclamation of both general heritage and specific identity: I'm a yellow warbler, and what's more, pardner, I'm *this* yellow warbler. Because different species respond only to their own songs, up to thirty

Peter Marler, professor of biology at Rockefeller University in New York and pioneer investigator of bird songs, with one of his subjects, a male swamp sparrow.

can coexist without controversy in the same field. The songs themselves have distinguishing dialects, in the same way that American accents audibly identify American subpopulations.

Multigifted artists, birds produce and perform their works. Birdsong is complex, like any other composition with multiple meanings. Dependent on both technical structure and vocal interpretation, it expresses seduction one moment and verbal swordsmanship the next. Rearranging the order of the notes, while retaining their shadings, still draws the same response from an avian audience.

Communication between different species is limited to warnings. The aggressive mob call is familiar to all birds: a sharp, machine-gun-like tatting, it blasts an outraged general call to wings in the face of invasion. Conversely, a hawk overhead elicits the thin flee call: this ventriloquial wisp of a whistle advises retreat—with all discreet haste.

In general, the relative importance of visual aids in avian communication depends on the habitat. In wide-open fields, birds rely on sight as much as sound to identify brethren. Their songs are illustrated: some tail flicking, a bit of neck craning, a good show of colors. In dense forests, the songs become increasingly elaborate; there, sound alone brings females to the roost and keeps wolves from the door.

Songs are learned as well as inherited, as much mimicry as genetics. Songbirds are born with a crude template of their song already in their brains. During a critical period shortly after birth but before the age of seven weeks, the bird must be exposed to a live performance of its song, which sharpens and refines its template. One spring later, when the fledgling begins to sing, it gropingly compares its sound with the internal model, until the two agree. Birds deafened before seven weeks sing a crippled version of their song; birds raised in isolation also produce a limping singsong. Birds deafened after they have begun to sing, though, are unaffected musically.

A male indigo bunting, whose song is shown below in sonogram (pitch is on the vertical axis, time on the horizontal). In an effort to find the essential components that make the song recognizable to the species, ethologist Steven T. Emlen manipulated various aspects of the song by splicing the audio tapes in new patterns to see how the birds responded. He found that it was the *duration* of sounds, rather than the *sequence* of sounds, that made the song recognizable to the bird.

Pitch

Seconds   1   2   3

As with human language, birdsong is dependent for control more upon one side of the brain than the other, a phenomenon called neural lateralization. In adults the left side dominates; cutting the nerve leading to this part of the brain causes song to disintegrate into gibberish. However, in young birds still apprenticed to their templates and the examples of their elders, severing the nerve has no effect; fledglings are able, eventually, to burst out in full and accurate voice.

The warble of a nightingale, the rasp of a crow; our music, their messages. Some humans quantify; they measure, tape, and experiment. Others whistle and drum, or play a mean bass or a wailing saxophone. But none can throw back their heads, open their throats, and let tumble forth a vocal history recorded in their genes and refined in their environment. On the subject of song, at least, birds still have the last word.

Calls of the Wild

Animals communicate with members of their own species through a combination of sounds, movements (including body posture, facial expression, and physical contact), odors, and color changes. In these ways animals signal information related to food, defense, territorial boundaries, courtship and mating, and care of the young. A particular animal's means of communication depends on its sense organs and apparatus for producing signals. Bees, for example, produce wing noises and hear with specialized organs on their legs. Bees have also been found to communicate the direction, distance, and quality of food sources by means of distinctive dances. Marine mammals communicate chiefly by sound, perhaps because of the poor visibility and good sound-transmitting characteristics of water. Some whales, such as the humpback, sing melodious songs lasting up to thirty minutes. These songs are believed to be repeated messages associated with breeding.

(Below) An excerpt from a humpback whale song as recorded off Bermuda by Roger Payne. This is the theme used as the melody of the "Lullaby from the Great Mother Whale for the Baby Seal Pups" composed by Whale, with Paul Winter and Jim Scott. Paul Winter records whales, wolves, and other mammals. He then orchestrates around this "living music," as he calls it, to produce songs that are truly calls of the wild.

Darwin Revisited

The publication in 1859 of Charles Darwin's *On the Origin of Species* was one of the most significant events in the history of science. Like the assertions of Copernicus and Galileo more than three centuries earlier that the Earth revolved around the sun and not vice versa, Darwin's theory of evolution has meant a profound change in the way we have looked at ourselves: we humans are not a breed apart; we are intimately and inextricably interconnected with the rest of life on Earth.

Simply stated, Darwin maintained that all species of plant and animal life, including humans, have descended gradually over the eons from ancestors with significantly different physical traits. His theory suggested that feathered birds arose from reptile ancestors, that humans evolved from apes, that pine trees developed from some other form of plant, and so on. This view conflicted sharply with the prevailing religious belief of the day, that our human ancestry can be traced back only to Adam and Eve, whom God put on Earth in fixed and unchanging form along with the rest of the planet's myriad varieties of life. Not unexpectedly, Darwin's views were condemned as heresy in many quarters, a view steadfastly held today

by religious fundamentalists who insist that scientific evidence supports the theory of divine creation (see "Darwin Defended," pages 76–77). Nevertheless, all 1,250 copies of the first edition of his book were sold on the day of publication; the time was ripe for Darwin's theory, and it quickly gained acceptance.

Today it is considered a watershed in the history of biological science, as fundamental to biology as the theories of Newton and Einstein are to physics. Its publication spurred a flurry of new research that continues in earnest today in such fields as paleontology, molecular biology, and genetics. Much of this research is aimed at proving or disproving Darwin—a theory, after all, is meant to be tested—and while his basic premise, scientists universally agree, remains untarnished, some scientists have recently raised questions about some of its important elements.

The heart of Darwin's theory—the mechanism by which evolution works—is known as natural selection. From the evidence he compiled during his five-year around-the-world expedition on the *Beagle,* and from his reading of Thomas Malthus's views on population, Darwin made two basic observations: one, that individual organisms within the same species have varying characteristics, and two, that most organisms produce far more offspring than will actually survive. His conclusion: organisms that vary in ways most suitable to their environment will survive and reproduce; those that don't will die. Thus nature selects organisms with the most favorable characteristics, which the survivors pass on to future generations. Through these gradual changes, a species adapts to its environment.

Darwin believed that natural selection was a creative force, that it built a fit species while eliminating unfavorable traits. But the variations themselves were random, not possessed of intent. That is, a giraffe did not grow a long neck in order to reach the fruit in a tree. Rather, among variations of longer and shorter necks in giraffes, natural selection "picked" the ones with long necks, ena-

Charles Darwin, English naturalist (1809–1882), photographed the year before his death.

Darwin *(lower left)* was in his early twenties when he sailed around the world on the HMS *Beagle (upper left)* for a five-year scientific expedition, during which he made most of the observations that led to his book *On the Origin of Species* (1859). While on the Galápagos Islands, Darwin observed thirteen different species of finches (plus a fourteenth on Cocos Island). Now known as Darwin's finches, they have similar appearances, calls, nests, and eggs, but decidedly different beaks that seemed adapted to their feeding habits and habitat—some live in trees, others on the ground. The presence of so many species in an isolated area prompted Darwin to speculate about their origins and led him to the theory of evolution by natural selection.

bling them to reach the fruit in the trees and thus survive to produce more giraffes with long necks. Thus, as Stephen Jay Gould writes in *Ever Since Darwin,* "Evolution is a mixture of chance and necessity—chance at the level of variation, necessity in the working of selection."

Darwin's theory works well in explaining individual differences among organisms within closely related species—geographic variations, for example, in which animals that live in colder climates may be larger or develop thicker fur than their counterparts in warmer areas; or in which a flower developed a particular scent to attract a local insect for pollination.

The Case of the Midwife Toad

Most scientists abandoned their belief in Lamarckism (which invokes the inheritance by offspring of characteristics acquired by their parents during their lifetime) after Darwin proposed his theory of evolution by mutation and natural selection. One who didn't was Paul Kammerer, a Viennese whose experiments with amphibians early in this century seemed to support Lamarckian notions of evolution. Kammerer's provocative experiments were received with violent scorn by orthodox Darwinians, and a specimen of midwife toad offered as crucial evidence was found to have been a phony. Whether the toad was doctored by Kammerer himself or by a saboteur who sought to discredit him is unclear. But the unhappy Kammerer died a suicide, and the theory he championed is no longer seriously considered. Still, it flourished for much of this century in the Soviet Union, whose ideologues found its tenets compatible with Marxist-Leninist social and political thought, if not with the facts of biology.

But there are broader questions. How does a feathered bird evolve from a reptile? Darwin's theory suggests there should be transitional life forms to make the bridge between species, to explain the development of new species.

Niles Eldredge, paleontologist at the American Museum of Natural History, whose study of trilobites turned up little change over millions of years of fossil records. What evolutionary change there was came in spurts, followed by long periods of stability. This observation conflicts with Darwin's belief that evolution was a gradual process.

Niles Eldredge, a paleontologist at the American Museum of Natural History in New York City, looked for patterns of evolutionary change among a group of trilobites, now-extinct marine animals, for his Ph.D. dissertation. To his surprise he found little change over millions of years of fossil record.

"What little change I did find within this long period of time," he explained, "seemed to be concentrated in rapid bursts of evolutionary change and then tremendous stability which set in again until there would be another episode." Even some subtle changes in trilobite eyes came not gradually, but in fits and starts, his research showed.

Indeed, for paleontologists who have dedicated careers to confirming Darwin's theory, the fossil record is discouragingly void of the transitional life forms that might explain the progression from one species to the next. Darwin himself acknowledged in his day that the record did not yet support his belief in gradual change, but he was convinced that further digs and more fossils would eventually prove him right. However, no explicit examples of the kind of evolution Darwin predicted between species have turned up; the record is filled with gaps or "missing links" that seem to undermine his theory. What the fossils do indicate is that the pattern Eldredge found in the trilobites is the rule rather than the exception. Thus, much evolutionary change seems to occur in only a fraction of an organism's life span and is followed by millions of years of stability or stasis

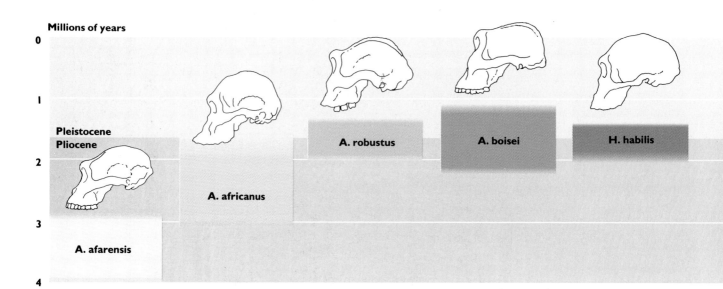

Millions of years

0

1

Pleistocene
Pliocene

2

A. robustus

A. boisei

H. habilis

A. africanus

3

A. afarensis

4

before it is wiped out and replaced by another organism.

In 1972, Eldredge and Gould collaborated on a paper in which they called this process "punctuated equilibrium" and maintained it was prevalent enough that paleontologists should accept it and quit the search for the transitional species. Such a proposition represents a significant revision of the Darwinian view of natural selection, and it has sparked heated and lengthy debate among students of evolution.

Indeed, their view could hardly be described as universal. Some scientists argue that the fossil record is far from reliable, that it includes mainly the hard skeletons or shells of animals and not the softer parts, which might change dramatically (and gradually) over time while the animal's basic structure remains the same. Others maintain that much transition may take place in isolated areas—far from the paleontologist's beaten path—where new traits that might be swallowed up in a species's mainstream could compete more successfully for dominance. (Theodosius Dobzhansky, a geneticist known for his pioneering work with fruit flies, argued that if

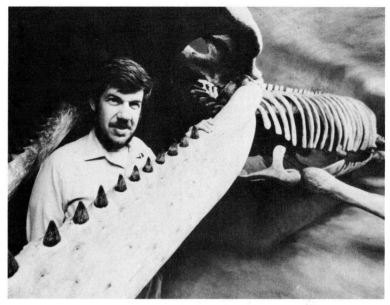

part of a species is to split off to form a new species, it *must* be geographically isolated to prevent breeding with its parent stock.) Still other disagreements arise from differences in disciplinary perspective: a paleontologist considers the evolution of a new species in 100,000 years to be an instantaneous event—accurate in terms of geologic time, but unbelievably slow and gradual to a fruit-fly geneticist, who can produce several generations and many new varieties in the laboratory within a year.

Stephen Jay Gould, paleontologist and Darwin expert, who collaborated with Niles Eldredge on a controversial paper that contradicted Darwin's notion of gradual progression from one species to the next.

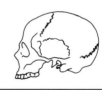

"Archaic H. sapiens"

Neanderthals

H. sapiens

H. erectus

This chart shows the time ranges in which various species of fossil hominids were prevalent. Notice that each species persists for some period of time without a major evolutionary change.

A genetics researcher works with DNA. Long threads of DNA are spooled onto a glass rod for treatment with enzymes that will clip away all but one gene, then insert the gene into the genetic material of a bacterium.

The stops and starts in the fossil record could also suggest that major catastrophes have played an important role in evolution. "At the big mass extinction at the end of the Paleozoic era about 250 million years ago, conceivably as many as 96 percent of the species then living died out," David Raup of the Field Museum of Natural History in Chicago says. "There are perhaps half a dozen mass extinctions of this magnitude in our geologic record. . . . There is chemical evidence in the rocks for the fall of a very large meteorite at just the time the dinosaurs went extinct. And so the difference between survivors and nonsurvivors may be more a matter of good or bad luck than good or bad genes." (See "Cosmic Catastrophe," pages 46–57.)

Controversy, debate, speculation, and revision are nothing new to Darwin's theory. Parts of it have fallen in and out of favor several times in the century since his death. Punctuated equilibrium is only the latest volley in a long-running disagreement between the gradualists and those who were called saltationists and believed that evolution jumped from one species to the next. And

the idea of natural selection gained little following at all until the 1930s and 1940s.

The most significant development in this history, however, has been the impact of genetics, a field that was virtually unheard of in Darwin's day. For a while, key genetic discoveries seemed to conflict with Darwin's theory, but gradually, a synthesis developed between two divergent schools of thought, ultimately providing Darwin with impressive support. For example, the random variations that Darwin observed within species during the *Beagle* expedition are the result of mutations or of different combinations of existing genes within an organism. The changes they cause in the genetic structure usually have only a small effect, thus their impact is gradual, in accord with Darwin's belief. Recently, molecular biologists have discovered that all species use the same genetic code, providing striking support for Darwin's then-daring proposal that all species descended from one or very few ancestors.

Today, genetic researchers are making startling advances in our knowledge of DNA— the genetic code—and how it works. They can read the code much as we read a book, and by comparing the DNA of one organism to that of another, they can pinpoint the spots on a strand of DNA where mutations have occurred. They have discovered that bits of DNA from higher organisms can be removed and stored in the DNA of bacteria in culture, and that pieces of DNA can be removed, flipped around, and replaced backward within the same genome (the set of chromosomes). They have found that DNA seems to jump or shift within the genome and even multiply en route to a different position. And they have discovered that new genetic material can be passed from one cell to another by viruslike carriers, raising the possibility that the barrier designed to protect sperm and egg cells from change can actually be breached, making the next generation vulnerable to some rapid evolutionary change. What effect these and other developments might have on evolution and our understanding of it is the subject of current research. And while genes seem to be the agent for evolution, scientists still aren't sure just how they work.

"We don't know how to connect what's going on down there at the level of genes to what's happening to the organism itself," Richard Lewontin, a geneticist at the Museum of Comparative Zoology at Harvard, says. "There's a huge gap between genes on the one hand and the organism on the other." It's not unlike, perhaps, the unexplained gaps between forms of plant and animal life that hold the answer to our questions about just how we and the vast number of other living things on this planet managed to get here.

What scientists do know about Darwin's theory is this: despite important unanswered questions, despite important disagreements among experts on the subject, evolution remains a strong and powerful theory. Every major advance in the field of biology fits within the broad parameters that Darwin defined. When new evidence has conflicted with it, the theory not only has accommodated the discovery without retreating from its essential position, but has come out stronger. As Ernst Mayr of Harvard described

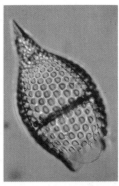

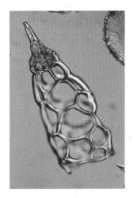
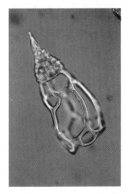
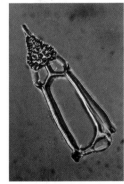
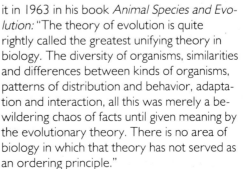

This sequence shows the evolution of a complete genus of radiolaria over a span of 10 million years. *Pedocyrtis papalis (upper left)* is about 50 million years old; *P. sinuosa (upper right)* is 48 million; next is *P. mitra*, 45 million; *P. chalera* (early form) and *P. chalera* (late form), 42 million; and *P. goetheana* (early) and *P. goetheana* (late), about 40 million. The samples were obtained by Dr. W.R. Riedel of the Scripps Institution of Oceanography.

it in 1963 in his book *Animal Species and Evolution:* "The theory of evolution is quite rightly called the greatest unifying theory in biology. The diversity of organisms, similarities and differences between kinds of organisms, patterns of distribution and behavior, adaptation and interaction, all this was merely a bewildering chaos of facts until given meaning by the evolutionary theory. There is no area of biology in which that theory has not served as an ordering principle."

Darwin Defended

When Duane T. Gish, assistant director of the Institute for Creation Research in San Diego, considers the disagreements among scientists who grapple with Darwin's 125-year-old theory of evolution, he virtually rubs his hands with delight. In his view, the inability of scientists to reach a consensus on evolution is prima-facie evidence of its unreliability as a theory.

"We object philosophically to the idea that we are the random products of chance," Gish says, "that, as evolutionists maintain, there is no God, that . . . we are nothing more than a mechanistic product of a mindless universe. I believe that this is wrong. I believe the scientific evidence is clearly in contradiction to that."

An artist's view of the biblical story of the Creation: *The Ancient of Days Striking the First Circle of the Earth,* **a watercolor by William Blake.**

Gish and his fellow creationists believe in divine creation, based on a literal interpretation of the Book of Genesis. Essentially, this view, which Gish describes as "creation science," is that God created the world and the rest of the universe between 6,000 and 10,000 years ago and populated it with all of its present and presumably extinct forms of life in one week. Such a belief, Gish and his followers insist, is well grounded in scientific evidence and deserves equal time alongside the Darwinian version in our nation's textbooks and classrooms. Gish and his supporters, some of them university-trained scientists with Ph.D.s after their names, have argued this view before school boards and state legislatures around the country, demanding laws that grant them equal time. In 1981, legislatures in Arkansas and Louisiana complied with their wishes, making Darwinism without creationism illegal in the classrooms of public schools.

But creationism has no basis in science; moreover, it is in absolute conflict not only with Darwin's theory, but also with major theories that are the foundation of astronomy, modern physics, and the Earth sciences as well.

Nevertheless, the creationists push on. One of their key pieces of "evidence" is also one focus of controversy among evolution scientists: the apparent gaps in the fossil record that suggest some organisms did not evolve gradually, as Darwin proposed, but in rapid jumps, followed by long periods of little change.

"The fossil record," Gish maintains, "shows a sudden explosive appearance of very highly complex creatures for which we've never been able to find any ancestors. And the fossil record produces systematic gaps which would be predicted on the basis of creation . . . but not evolution." Gish sees the efforts of scientists to accommodate these gaps into Darwin's overall theory as "born out of desperation."

In 1982, the Arkansas "equal time" law was overturned in a court challenge, raising hopes among scientists and educators that the creationists will fail in their efforts to

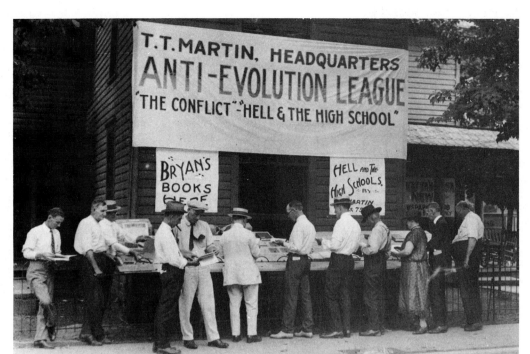

(Above) **John T. Scopes.**
(Left) **Antievolution
headquarters in 1925,
at the height of
the furor over evolu-
tion caused by the
Scopes trial.**

force their religious beliefs into public school
curricula. But even in their failure, the creationists have had a chilling effect on the quality and content of science education,
reminiscent of the days of the Scopes trial in
1925, in which the court upheld the firing of a
Tennessee teacher who had taught evolution in the classroom. (That law was not re-
pealed until 1967 when another Tennessee
high school teacher was fired for the same
reason. In 1968, the U.S. Supreme Court finally struck down an Arkansas antievolution
law as unconstitutional.) In the face of the re-
surgence of the creationist movement,
countless teachers skim over or ignore this
fundamental tenet of science. Several pub-
lishers have already shortened or omitted
sections on the subject in their textbooks to
guarantee school adoptions in the uncertain
political climate fostered by the creationists.

Scientists themselves are hardly protected against the power and determination
of those in the creationist movement. Niles
Eldredge, a paleontologist at the American
Museum of Natural History, and others who
have questioned aspects of Darwin's theory
find their views used by creationists to raise
doubts about the validity of evolution.
"They're trying to eliminate or at least modify the teaching of evolution in schools,"
Eldredge said. "I construe this as an attack on

science education in general in this country.
And I've been used."

The debate continues, each side feeling confident in its evidence, secure in its
mission. Most Darwin defenders point to the
larger issues at stake, however, that lend
weight to their argument: if one hundred
years of scientific progress can be dismissed
so readily, what measures of scientific censorship are to follow?

Clarence Darrow *(left)*
**and William Jennings
Bryan, opposing attor-
neys at the Scopes trial.
Darrow defended
Scopes, and his court-
room ability to shatter
Bryan's arguments
against evolution won
him wide acclaim in
some circles, but he
lost the case.**

A Boy for You, a Girl for Me

It is this generation's fondest ambition: social progress. Over the last century, we have routed racial and ethnic constrictions from most front parlors in the same way we once removed buggies from most back roads. They serve us no purpose anymore, those prejudices; they are obsolete, and we paste their foolish images into old scrapbooks, labeling them part of the unenlightened past.

But in the midst of our progressive times, one American group continues to bump down back roads in antiquated carriages. American parents still prefer boys. Twice as many mothers prefer sons as prefer daughters; the bias among fathers is even greater; and couples with girls are statistically more likely to continue having children than those with at least one boy. From a swaddling age, children are shown differential treatment whose impact on them continues throughout their lives.

Despite the remarkable sex role changes that have occurred in American life over the last two decades, our attitudes about males and females have not changed that much. Like other stereotypes, fixed ideas about what is masculine and what is feminine have not been radically changed by experience. Scientists have found that what we think of as normal, appropriate behavior for girls as opposed to boys has become what we think of as entirely natural for them.

In an ideal scientific laboratory situation, investigators could place a boy and a girl infant in identical settings, treat them exactly the same, and discover whether the differences between boys and girls are the result of biological or sociological inheritance. But evidence collected by psychologists across the country shows that even if this kind of experiment were feasible, it would not bear ideal fruit. Regardless of an infant's genetic

The baby's born, the swaddling appears: blue for a boy, pink for a girl. Whence pink, why blue? Their origins are Western, yet their significance remains unclear. Blue is a primary color, associated with elemental power: the shades of sky and water; the risqué joke; the unexpected source; the strictest moral laws; the strongest personal unhappiness. Pink is not a primary color, but a delicate mixture of hues. It is associated with softness, frivolity, health, and perfection: the blush of the rose; the color a teased cheek takes; the condition of top health.

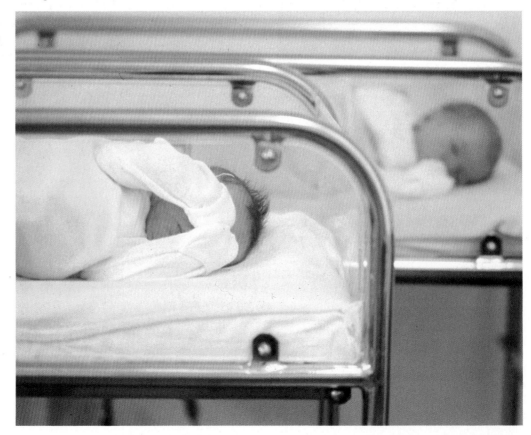

inheritance, the way it is received into the world is determined solely on the basis of its sex. The implications of this discovery are enormous.

Within twenty-four hours after birth—enough time, perhaps, for a single feeding or several glass-tapping sessions by the nursery window—both parents already hold expectations based only on the gender of their baby. When tested and scored, they both agree that the baby is, at the advanced age of a single day, better coordinated, more alert, and stronger if it is a boy, while they regard a girl as softer, more finely featured, and also—somehow—more inattentive.

Dr. Zella Luria, co-director with Dr. Jeffry Rubin of the studies that provided this analysis, is fascinated that such stereotypes can persist in absolute defiance of both women's progress and biological research. The truth of the matter, says Dr. Luria, is that "there is a greater vulnerability in the male sex than there is in the female sex. There are more accidents for young boys than for girls. There are more diseases that kill off boys than girls. Nonetheless, the stereotyping related to the male role is that boys are less vulnerable and girls are more vulnerable. This flies in the face of biological reality; nonetheless, the stereotype holds."

The tenacity of stereotypes such as this continues relentlessly from birth on, regardless of how well any individual child may or may not conform to the mold. It is testimony to the power of these ideas that twenty years of converging sex roles, emerging consciousness, and "female firsts" in every field have barely made a dent in how most parents raise their young. For the most part, researchers contend, parents themselves are unaware that they are differentiating. Scientist observers, attuned to the smallest details of behavior, have identified significant differences in the ways parents act with their newborn children.

Until the child is two, fathers touch and talk more to boys, mothers to girls. Fathers play more bouncing games, more rocking games; mothers prefer watching games, ver-

Little girls' play stresses structure and closeness, while little boys are encouraged, in competitive sports, to tussle with life and learn self-reliance.

bal games. When the child begins to walk and investigate its familial turf, father pulls out a baby buggy for his daughter, but doubles his fists to the delight of his son. Mother makes less of a distinction: she seems equally comfortable offering her son toy cars or toy curlers.

But by the age of three, boys and girls have begun to make consistent choices on their own: boys race for outdoor toys, while girls settle in with a good set of plastic china or a miniature broom.

Children take their next set of subtle clues from preschool teachers. Albert, knocking over Ann's blocks, receives a loud, clear reprimand. Action stops. Everyone stares. The floor is his. Reprimand is a petty price for attention. Albert develops a flair for demolition and a passion, in particular, for Ann's construction work. When Ann reciprocates the favor on Albert's tower, she is ignored or quietly cautioned: she should know better. And, eventually, she does: the closer she remains to the shepherd, the more he will stroke her fleece. Proximity and goodness, not aggression, bring her rewards. She attaches herself literally and emotionally to the teacher. She learns not to stray from the flock—while Albert is simultaneously painting his fleece black and rampaging mightily. Indeed, studies have verified that the physical distance between a girl and her teacher directly affects the amount of attention she receives; boys receive equal amounts whether they remain near or not.

At the University of California in Berkeley, Drs. Jeanne and Jack Block spent eleven years on a study of personality development in 130 boys and girls from a spectrum of family flocks. From the age of three onward, each child was interviewed, tested, and videotaped regularly. The children were raised in an age of social realignment, when, it seemed, traditional male and female roles were converging and when parents were the first to declare themselves nondiscriminatory. It soon became apparent to the Blocks that, along with children, they were studying differences in parents as well: mothers and fathers cherished vastly different dreams for their sons and daughters and directed their expectations, consciously or not, toward those ends. For sons, an emphasis was placed on achievement: they were expected to be independent, to hone intelligence in the interest of ambition, to control their feelings and, with time, their world. Self-reliance was stressed, particularly by fathers. The emphasis for daughters was on behavior—preferably ladylike. Girls were loved for the emotional qualities they possessed instead of the cerebral potential they might develop. Parents described these relationships as physically closer; they placed more faith in the trustworthiness and truthfulness of their girls. They stressed dependency, lovingly but firmly.

"Everything tends to focus on the little girl in the house," Jeanne Block explains. "Parents of girls encourage girls to stay closer to them. It's important to recognize that the embeddedness of the girls in the family network from a very early age insulates them from experience."

In contrast, parents of boys "encourage them to play competitive games. They value their independence and want their boys to be self-reliant. These are not patterns stressed in girls." Because girls develop in a more restrictive, supervised, structured environment, they have less opportunity to engage in active experimentation, the kind of trial-and-error learning that benefits problem solving—and that boys typically experience.

Duplicate DNA

Nine-banded armadillos routinely give birth to identical quadruplets, and an Irish setter that whelped fewer than eight or nine puppies might be considered something of a flop as a mother. But among people (and most other primates), multiple births are rare: only one birth in eighty results in twins, and only three or four births in a thousand are identical twins—a pair of individuals who develop from the same fertilized egg. Twins are much studied by psychologists, physiologists, and many other

scientists. Investigations of identical twins who were reared separately are especially fruitful in efforts to elu- cidate the relative contributions of nature and nurture—heredity and environment—to human development.

Serious business: play is taken seriously, with goals, strategies, and rewards for this little boy.

Fascinated by these observations, Jeanne Block devised a test to measure the separate roles parents assume in teaching children to solve problems. Working first with a son, then a daughter, parents were asked to guide the placement of differently shaped and colored blocks within a complex diagram. Fathers emphasized the cognitive aspects of the task to their sons; they clarified the goals, the game plan, the strategies, and the rewards. Serious business. With daughters, though, fathers subordinated the role of coach to that of companion; they had fun. The task itself paled beside an opportunity it offered for parent and child to share a pleasant puzzle. Mothers gave gratuitous help to both children, but particularly to girls.

Any life is shaped by the forces around it. Under the influence of external signals, boys and girls arrive at different understandings of the worlds within themselves. Deliberately sheltered from the wolves of experience, little girls develop few resources with which to conquer the new and unexpected. The games and play of their adolescence take place in small groups. Rules abound. Deviations are not tolerated. Given a game that causes consistent disagreement, girls will often abandon it in favor of a new, less controversial version.

Little boys are encouraged to tussle with life, to grab it by the throat and shake. They develop improvisational gifts; they enjoy the unfamiliar. They gravitate toward team sports and learn to seize the moment, the flag, the ball. They look ahead, and never over one shoulder. Peeking is unmanly.

Out of her study, Jeanne Block identified eight major areas of sex difference in childhood. These liberated times notwithstanding, boys still show greater aggression, activity, curiosity, and impulsiveness. Girls still show a higher degree of anxiety—and its corollary, obedience—and cherish more intimate relations with fewer friends. Boys continue to assume they will do better than girls; and, though girls blame poor performance on themselves, boys in the identical situation are quick to point out the impossibility of the problem, not to mention the ambiguity of those darn instructions.

Chicken and egg, egg and chicken. Which came first: expectations or their manifestations? Will boys be boys because they are, or because their world assumes they ought to be? Is the sugar and spice of girls physiological, or is it sprinkled on by the well-meaning hand of society? Anatomy may indeed dictate its own destiny; but the dictation may be only psychological. The deepest genetic truths foretell, quietly, another set of fates.

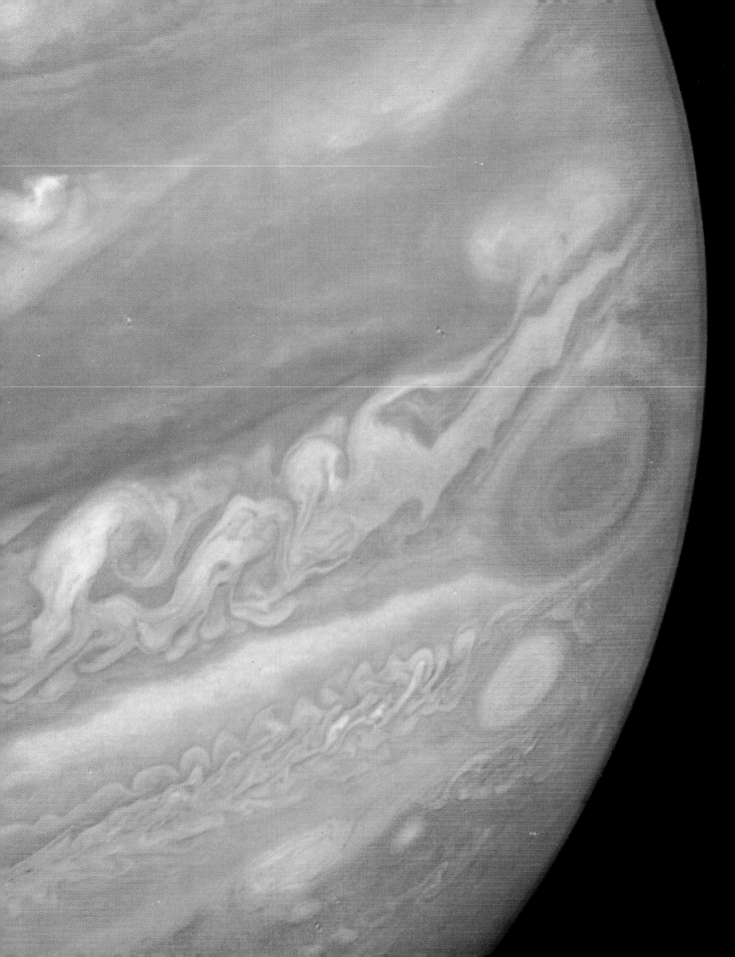

As a rule, animals stay well within the borders of their own habitat. One does not encounter penguins in the Sargasso Sea or tigers in the Arctic. The reason for staying at home is simple: environmental adaptation is complicated and specialized, and it requires time to develop. What animal would be so reckless as to venture into a totally different habitat? Taking such a risk would seem to defy the whole point of evolution, which is the progressive refinement of strategies for dealing with specific environments.

Yet humans—and scientist humans in particular—persist in ignoring the logic of this reasoning and set off in search of new habitats. In doing so they have pushed the frontiers of our world beyond the borders of our own planet. Not content to stay at home any longer, we have sent humans and metal into the air—not once, but repeatedly in the last twenty years. Our aims? To see what (or who), if anything, exists in the hinterlands of our universe, to learn more about how our own home was created by studying the other planets, and to search for new ways to help life on Earth. All very sound—and true—reasons. But one can't help wondering whether a simpler motivation applies: we feel compelled to reach for the stars . . . because they are there.

The Great Red Spot of Jupiter, a hurricane that could swallow several Earth-sized planets, is seen in this photograph made by the *Voyager* spacecraft.

One Small Step

Projects Mercury, Gemini, Apollo, and Skylab. Astronauts Shepard, Glenn, White, Stafford, Lovell, Borman, Armstrong, Scott, Aldrin, and Grissom. We recall the names of these world-famous space missions and international heroes from the early days of the United States space program. But what is remembered best are not the missions or the men, or even the scientific ingenuity that made space shots possible or the political motives that made them seem necessary. The most enduring legacy of the early space flights is the change in perspective that was inspired by the opening of the last frontier.

Space walks, moon walks, and *Skylab* experiments, events that only thirty years ago would have been science fiction, demonstrated that for the first time in history humans were not confined to planet Earth. And now, as space shuttle flights are scheduled regularly, as the public witnesses gliderlike landings by space ships, and as plans for solar satellites and space colonies are considered, Americans are even taking for granted the idea of living in space. What began in the 1950s as a space race with the Russians for the moon was, two decades later, a race

against time to exploit extraterrestrial resources to benefit an overpopulated Earth.

The Russians had said in 1955 that they would launch a satellite, but the American public did not believe them until the sounds of *Sputnik* beeping in orbit were picked up by American ham radio operators in October 1957. *Sputnik's* mocking beeps shattered the cozy myth of American superiority over Soviet Russia and marked the beginning of the space race. According to Richard Lewis, a journalist who has covered the space program from its start, "*Sputnik* had an enormous impact on our whole society and launched us to the moon. Without *Sputnik,* we might not have reached the moon yet."

America's answer to *Sputnik* was the *Explorer* satellite, rushed to a launching only two months after *Sputnik.* It was to be hurled into space by the Vanguard, a Navy research missile. Whereas the Soviets had used a military rocket, a missile, for their space shot, President Eisenhower wanted America's entry into space to have the appearance, at least, of a nonmilitary enterprise. But the Vanguard failed at liftoff, and as humorous European newspaper articles called the rocket "puffnik" and "flopnik," American embarrassment mounted.

To the rescue of American pride came the man who had built the V-2 war rocket for Hitler, Wernher von Braun. Von Braun's new job was to upgrade the V-2 rocket to create intercontinental nuclear missiles. After the Vanguard disaster, Eisenhower turned from research rockets to one of von Braun's military rockets, and on January 31, 1958, a spare Jupiter-C missile smoothly boosted a U.S. satellite to orbit. America, too, had entered the space race, but in second place.

Two years later, President Kennedy was elected in a campaign that had stressed the "missile gap" between Russia and the United States. The time had come to demonstrate to the world America's determination to open a new frontier. The National Aeronautics and Space Administration (NASA) was created, and the goal was now, clearly, a man in space.

President Kennedy and rocket scientist Wernher von Braun meet during the early days of planning for the U.S. manned space program.

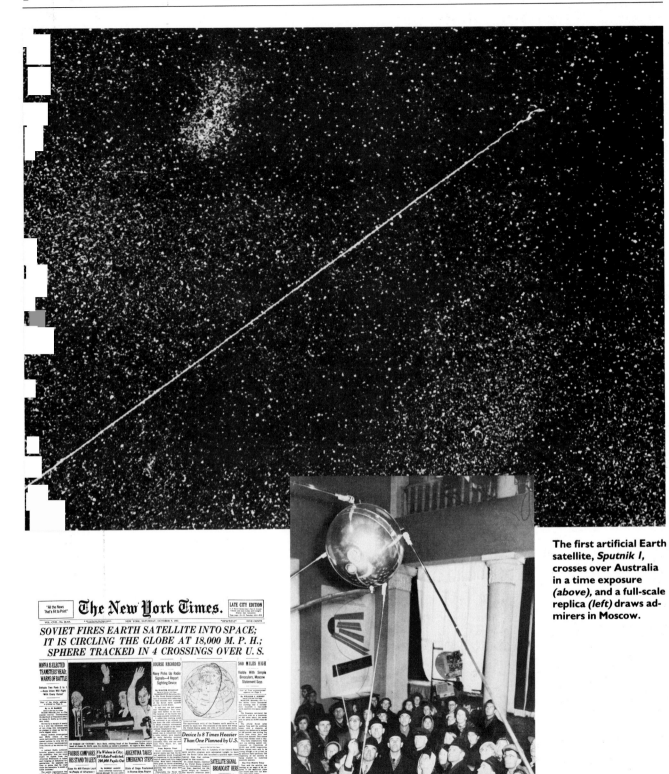

The first artificial Earth satellite, *Sputnik 1,* crosses over Australia in a time exposure *(above),* and a full-scale replica *(left)* draws admirers in Moscow.

The beginnings of Project Mercury were inauspicious. The launch escape tower took off instead of the new Redstone rocket, another von Braun missile. The payload was tugged from the unmanned *Mercury* space capsule and bumped unceremoniously to the ground. Due to be the payload on the next test flight was Ham, a chimpanzee, and fortunately the Redstone rocket lobbed him safely into space.

Ham seemed healthy on his return, but NASA doctors remained dubious about man's ability to survive in space. Doctors were concerned that in the absence of gravity, such body functions as breathing, swallowing, and food absorption would be impaired. These worries, however, were not shared by the Russians. Two months after Ham's flight, the Soviets sent into orbit a twenty-seven-year-old fighter pilot named Yuri Gagarin.

The Russians had once again seized the initiative in space and celebrated jubilantly this further proof of their superiority. Needless to say, Gagarin's flight only increased U.S. determination. That the Russians put the first man in space, says Lewis, had "not only political and military effects, but economic effects. We were trying to sell machinery in Europe. We were patting ourselves on the back as the technological leaders of the world, particularly at the end of World War II. The Rus-

sians were coming up fast and moving ahead." Gagarin's obvious good health after his orbit around the Earth also galvanized NASA's doctors. Space was not the dangerous environment imagined.

Three weeks after Gagarin's flight, a Redstone rocket awaited the arrival of America's first astronaut at Cape Canaveral, Florida. Alan Shepard's arrival at the launch pad on the morning of May 5, 1961, was conspicuously different from Gagarin's. He was met by an eager press corps and television cameras. America, unlike Russia, would conduct its space program in public. NASA officials were taken aback by this policy because of the dangers associated with missile launches and the possibility that any failure would be witnessed worldwide.

Shepard's five-minute entry into space showed that an American could survive there as well as a Russian. But even before the first American went into orbit around the Earth, the United States was committed to the moon. Kennedy's new administration, only four months old, was in trouble over the economy, the abortive Bay of Pigs invasion, and the Gagarin flight. Kennedy urgently needed proof of his New Frontier. His vice-president, Lyndon Johnson, was an ardent supporter of the space program. In a memo, Kennedy asked Johnson to identify a goal in space that America had the best chance of attaining before the Russians. Johnson's answer was prompt: only going for the most difficult goal of all would allow America time to overtake the Russian lead. In a speech before Congress, Kennedy announced that goal: to land a man on the moon and return him safely to Earth before the decade was out.

Congress applauded and agreed to the moon commitment without even taking a vote, initiating the largest single project other than a war ever undertaken by a nation. NASA now had a clear objective and, for the moment, a blank check. NASA had pushed for the moon program, but when Kennedy's challenge came, many in the agency were aghast. Christopher Kraft, who directed most

Alan Shepard *(right)* strides to the Redstone rocket *(above)* that carried him on a brief ride into space, making him America's first astronaut. This cautious step into space followed three weeks after the more daring Earth orbital flight of Russia's Yuri Gagarin.

of America's manned space flights, recalls that "when he said that we were going to land men on the moon and then bring them back safely by the end of the decade, some of us thought that was biting off a bit more than we could tolerate. Here we were in the throes of still trying to fly our first orbital flight, and somebody said we were going to go land men on the moon."

In fact, it was almost a year before NASA was ready for its first orbital flight. The astronaut chosen was John Glenn. A new, more powerful rocket, the Atlas, an inter-continental missile, had been converted into a manned space booster. Glenn was blasted into space on February 20, 1962, and he cir-cled the Earth three times. He returned to find himself an instant national hero. The Mercury program included three more manned flights, and while the astronauts rode in ticker-tape parades and toured nationally, the first doubts about the moon program be-gan to surface. Did a space race really exist?

It was true that the Soviets had grabbed another first in March 1965 when cosmonaut Alexei Leonov left his spacecraft for the first space walk. But were the Russians competing for the moon? Indeed they were, reports Lewis, who spoke with Leonov several years later. "It was his opinion that they were going to the moon. He was quite definite about this, and he hoped that we would all go to the moon together. But there was no doubt in what he said, that the Russians were considering a manned lunar landing."

Meanwhile, uncertain of Russia's inten-tions or capabilities, NASA persevered, and with Project Gemini, the space race en-tered a new phase. *Gemini* was a two-man spacecraft, launched by another modified intercontinental missile, the Titan. Its first manned flight came five days after Leonov's space walk, but within ten manned *Gemini* mis-sions, NASA learned how to get to the moon.

On the second manned *Gemini* flight, astronaut Ed White stepped out into space—for no very good reason except to show than an astronaut could walk in space as well as a cosmonaut. Back on the ground, less

spectacular but more important progress was being made. Spacecraft checkout proce-dures were refined to eliminate delays, and the launch control and mission control opera-tions were developed. Going to the moon would entail the continuous monitoring of tens of thousands of components and actions. And it was during the *Gemini* mission that the *way* to get to the moon was chosen.

Initially it was thought that the astronauts would go to the moon and back in the same spacecraft, a very complex engineering problem. A simpler way would be to take a main vehicle up, put it into orbit around the moon, and then descend to the lunar surface in a second vehicle. This made a lunar landing a much more reasonable engineering task because only a small portion of the main spacecraft need descend to the moon and land. Soft-landing a lunar module would be much easier than landing the whole space-craft, saving fuel and weight. The same would be true on takeoff. But the method involved the lunar module's finding and docking with the command ship, if its crew were to return safely home, and orbital rendezvous was an unknown, untried science.

Astronaut Ed White floats 100 miles above the Earth, using a hand-held propulsion unit to move about. White's space walk occurred during the *Gemini* se-ries, designed to teach astronauts how to live and work in space.

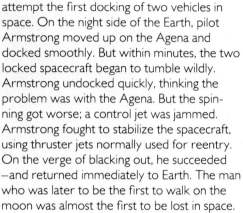

(Above) American astronauts train for an emergency landing in the desert, 1963. *Front row, left to right:* **Frank Borman, James Lovell, John Young, Charles Conrad, James McDivitt, Edward White.** *Back row:* **Training Officer, Thomas Stafford, Donald Slayton, Neil Armstrong, Elliot See.** **(Below)** Soviet cosmonauts in 1965. Yuri Gagarin is at top row, far left.

The first attempt at an orbital rendezvous, between *Gemini 6* and a small Agena rocket, was postponed when the rocket failed to reach orbit. *Gemini 7* was launched on schedule, and a new plan emerged. Eleven days later, *Gemini 7* was circling the Earth for a rendezvous with *Gemini 6,* which was still in orbit. The ability to rendezvous and maneuver, crucial for a lunar landing, was proved. The *Gemini 6* pulled away, leaving Jim Lovell and Frank Borman alone again on *Gemini 7.* The crew of *Gemini 7* returned to Earth after fourteen days in space, twice as long as previous U.S. flights and as long as a lunar expedition was expected to take. It was another vital link in proving a moon landing was possible: weightlessness clearly was not the killer it was feared to be.

Three months later, another Agena target vehicle was launched. This time, it achieved a perfect orbit. Later the same day, the crew of *Gemini 8,* Neil Armstrong and David Scott, caught up with the Agena to attempt the first docking of two vehicles in space. On the night side of the Earth, pilot Armstrong moved up on the Agena and docked smoothly. But within minutes, the two locked spacecraft began to tumble wildly. Armstrong undocked quickly, thinking the problem was with the Agena. But the spinning got worse; a control jet was jammed. Armstrong fought to stabilize the spacecraft, using thruster jets normally used for reentry. On the verge of blacking out, he succeeded —and returned immediately to Earth. The man who was later to be the first to walk on the moon was almost the first to be lost in space.

Gemini 10 finally accomplished a successful docking, and by the time *Gemini 12* reentered the Earth's atmosphere four months later NASA knew that the moon could be conquered. Project Gemini had proved that a man could live and work in space long enough to reach the moon and return; that spacecraft could maneuver, rendezvous, and link up in orbit; and that the complexity and precision of space operations could be handled. In eighteen months, NASA had flown ten manned misions. During the same period, not one Russian had entered space. With the splashdown of *Gemini 12* in November 1966, the moon, it seemed, would belong to America.

Just ten weeks later, however, the triumph turned to despair. At 6:31 P.M. on January 27, 1967, a disaster occurred that almost wiped out the moon program in a single instant. The crew of *Apollo 1,* Gus Grissom, veteran of *Mercury* and *Gemini,* Ed White, the space walker of *Gemini 4,* and rookie astronaut Roger Chaffee, were asphyxiated during a fire at takeoff. The accident stunned the nation, but not everyone at NASA was surprised. A scathing assessment of the quality of the work being done on the *Apollo* spacecraft had been delivered to the industry contractor a year earlier by Apollo program director General Samuel Phillips.

In his report, General Phillips had pointed out a number of deficiencies in the spacecraft. These had gone uncorrected, according to Richard Lewis, because of the

pressure to maintain a schedule, dictated by political pressure to beat the Russians to the moon. Christopher Kraft agrees that the race with the Russians had been a concern but insists that NASA did not compromise standards to meet deadlines. "The race with the Russians," says Kraft, "certainly put us in a position of pressure, but I don't think it was pressure that caused us to do anything wrong, or . . . things that we didn't think were reasonable."

The investigation into the fire's origin triggered a major reevaluation of the *Apollo* spacecraft, and the review was sobering. The most appalling discovery was that although supposed to be fireproof, the spacecraft was full of highly flammable paper, nylon, and plastic. Corrections in design were made. The disaster, says Kraft in retrospect, "was, I think, a major milestone in the program. It's unfortunate that it had to be brought about by the deaths of three very fine young men, and we'll always regret that that situation happened and feel bad about it. But it probably made a tremendous contribution to the success of the lunar landing program."

In November 1967, with the accident hardly forgotten, Americans began testing the Saturn V rocket, the launch built to take *Apollo* to the moon. A year later, the Russians, who had been lagging further and further behind, seemed about to take a last-minute gamble. The front page of *Pravda* proclaimed Russia's intention of shooting a man around the moon and back. According to Kraft, NASA did not consider this a great accomplishment but feared that to the rest of the world, it would look as if the Russians had gotten to the moon first even if they did not land on it. The Americans, not to be outdone, changed the plan for the first manned flight of a Saturn V.

The plan had been to fly the lunar excursion module (LEM) on the Saturn V and test it in Earth orbit, but work on the LEM had fallen behind schedule. With the LEM not ready, and the Russians threatening, NASA rethought the mission: the first Saturn V to carry men would take them not just around the Earth, but around the moon as well. On

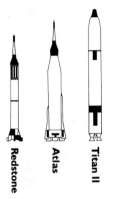

Redstone   **Atlas**   **Titan II**

The launching of the *Apollo 15* space vehicle on July 26, 1971, on a lunar landing mission. The 363-foot-tall Saturn 5 launch vehicle dwarfs the rockets used in earlier space missions *(below).*

Robert H. Goddard (1882-1945), inventor of the liquid fuel rocket. Goddard's pioneer experiments in rocketry in the 1930s were ignored by the U.S. military, but were closely followed by young German rocket engineers such as Wernher von Braun.

When the astronauts stepped onto the lunar surface, they entered a museum world in which the secrets of the origin of the planets were preserved. Because the moon has neither atmosphere nor plate tectonics, it has changed relatively little over eons of time.

This four-billion-year-old piece of the moon's original crust, dubbed "Genesis Rock," was picked up by astronaut Dave Scott at the *Apollo 15* landing site. The Earth's original crust may also have consisted of this feldspar-rich mineral.

Christmas Eve 1968, *Apollo 8* astronauts Jim Lovell, Frank Borman, and Bill Anders took turns reading from the Book of Genesis as they circled the moon.

It was a tremendously risky mission, rushed ahead of schedule, but it had worked. If it had failed, if the crew had been lost, then perhaps the whole lunar landing program might have been lost, too, or at least postponed for a number of years. But it had worked, and Mission Control was jubilant. Now any attempt by the Russians simply to send a manned spacecraft looping around the moon was irrelevant. The space race had effectively ended with *Apollo 8,* and this victory was confirmed by *Apollo 11.*

But before *Apollo 11* could land on the moon, the lunar module had to be tested. Several months overdue, the LEM made its debut in Earth orbit with *Apollo 9.* Two months later, in May 1969, a second lunar module was put in orbit around the moon and was taken down by astronauts Tom Stafford and Gene Cernan to within 50,000 feet of the lunar surface. They skimmed over the landing site, where, in another two months, man would first set foot on another world.

On July 16, 1969, astronauts Neil Armstrong and Buzz Aldrin left Michael Collins

behind in the *Apollo 11* as they descended to the lunar surface in the lunar module *Eagle.* Six hours later, Neil Armstrong opened the hatch for the final descent to the lunar surface. As the world watched on television, air-to-ground communications recorded, "O.K., Houston, I'm on the porch. I'm at the foot of the ladder, and I'm going to step off the LEM now. That's one small step for man, one giant leap for mankind."

In Houston, at Mission Control, it was the supreme moment; the dream was fulfilled. On the moon, the lunar module bore a plaque declaring, "We came in peace for all mankind." But American taxpayers had footed the bill, and the flag came first.

With the goal of a man on the moon accomplished, says Lewis, later *Apollo* missions 12, 14, 15, 16, and 17, all lunar landings, were anticlimactic. Public interest had waned. And although, by the end, twelve astronauts had walked on the moon, 850 pounds of rocks had been collected, and 59 miles of lunar surface had been traversed, the scientific harvest of the Apollo program was little noticed. Even though *Apollo* helped settle some of the questions about the origin and evolution of the solar system, what was memorable about the mission in the minds of many was not the scientific work done on the moon, but the fact that man had been there.

"I think we may all agree that the motivation for going to the moon was political and semimilitary, but the effects of it were surprising and much longer-lasting than the mere competitive aspect had been," says Lewis. "The journeys to the moon gave us a new perspective about the Earth, enabled us to visualize ourselves living on a planet afloat in space, a visualization which is particular to the twentieth century and which I believe had never existed before in human consciousness."

On December 17, 1972, man left the moon for the last time. The moon program was over; NASA's goal was achieved. Now the question was, "What next?" The answer: *Skylab,* the space shuttle, colonies in space, solar power satellites—the frontier was there for the taking.

15

17

14

11

12

16

The *Apollo* program landed six teams of astronauts on the moon between 1969 and 1972. The early, more famous landings, *Apollo 11* and *Apollo 12 (center right and bottom left),* went to relatively safe, flat sites in the dark-colored lunar "seas." But by *Apollo 17 (top right),* the astronauts were threading their spaceship between mountains to set down in a rugged lunar box canyon.

(Right) Astronaut Alan Bean sets up a remote experimental station on the moon during the *Apollo 12* mission in November 1969. The blue glow that envelops Bean is caused by water vapor venting from his life-support backpack. *(Below)* The command module in which the *Apollo* astronauts will return to Earth orbits sixty miles over the moon's Sea of Fertility.

A trip to the moon and back. NASA's ingenious and fuel-saving plan for getting to the moon called for leaving behind various stages of the spacecraft after their purpose had been served. At the end of the trip, all that was left of the mighty stack of rockets that left Earth was a tiny three-man capsule floating in the ocean. The Apollo mission to circle the moon and return to Earth was anticipated 100 years earlier by the French writer Jules Verne. In his novel *From the Earth to the Moon* Verne writes of American space travelers who take off from Florida, fly around the moon, and land in the Pacific Ocean.

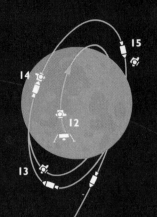

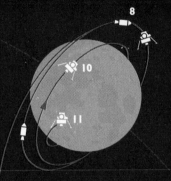

Going to the Moon and Coming Back

1 Liftoff from Florida
2 Second stage ignites
3 Third stage ignites
4 Earth orbit
5 Leave Earth orbit
6 Combined service, command, and lunar modules separate from third stage
7 Midcourse correction
8 In lunar orbit, the two modules separate
9 Command module stays in orbit
10 Lunar module descends
11 Touchdown on the moon

12 Liftoff, leaving descent stage behind
13 Rendezvous with command module
14 Docking and crew transfer
15 Lunar module jettisoned
16 Midcourse correction
17 Command module separates from service module
18 Re-enter Earth's atmosphere; communications blackout
19 Deploy chutes
20 Splashdown in Pacific

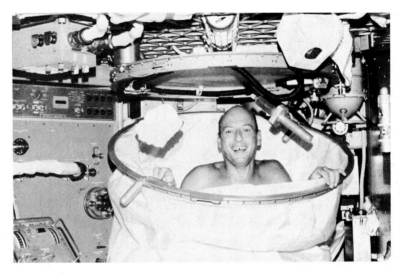

In May 1973, the thirteenth and last Saturn V to fly lifted off with its third stage filled not with fuel to reach the moon, but with a laboratory and living quarters. The Russians, losing the moon race, had turned to space stations in Earth orbit. Using leftovers from *Apollo,* NASA followed suit. With an interior volume as big as a comfortable house, *Skylab* was fitted with bedrooms, bathroom, shower, dining room, and workshops as well as a complete solar observatory. For the first time, space had become truly habitable, and human fitness in space could be tested experimentally. *Skylab*'s first crew stayed a month, its second for two months, and by the end of the third mission, lasting eighty-four days, NASA doctors were convinced that human beings were very adaptable creatures. *Skylab* showed that man can survive long periods of time in space. It also gave a glimpse of the future: people working in space at tasks that cannot be done on Earth.

Meanwhile, the Cold War that had spurred *Apollo* melted into detente. The space race, too, came to a symbolic end in July 1975, when the last remaining *Apollo* spacecraft docked in orbit with a Russian *Soyuz* spacecraft. Cosmonauts and astronauts shook hands in space. The old space race may have ended in caviar and vodka, but in racing for the moon, both Russians and Americans had discovered the economic opportunities of space—that space offers more than an adventure playground for acting out national rivalries or making expensive geopolitical gestures. The time had come to look to space to answer such Earthly problems as shortages in materials, metals, and energy resources.

But a new approach was needed. *Apollo* and its predecessors had been designed to get man into space and accomplish specific tasks—to open the new frontier. New spacecraft had to be designed to make exploring and exploiting that new frontier economically feasible. According to John Yardly, head of NASA's Office of Space Flight, "It became apparent that, as we were landing on the moon, if we were going to use

space in any gross way, we needed a better means to do this. The Saturn V launches cost two to five million dollars each, so the idea of a space shuttle came into being."

In September 1976, the theme from the popular television program "Star Trek" heralded the space shuttle's debut at the Rockwell plant in Downey, California. On hand were two "Star Trek" regulars, Dr. McCoy and Mr. Spock. "Star Trek" fans, in a letter-writing campaign, had succeeded in having the first shuttle named the *Enterprise.* After the Vietnam era, in which science and technology were blamed for the evils to which they were put, NASA, sensing a new and younger constituency for space, was happy to go along.

Charles Conrad showers aboard *Skylab* on the first of the three missions to test long-term living in space, 1973.

The Russian *Soyuz* spacecraft as seen from the window of *Apollo* during the joint Russian-American space flight in 1975.

The *Enterprise* was the prototype of five space planes or orbiters built in an extraordinary shape for an extraordinary task: to be launched into space like a rocket, to operate in orbit like a spacecraft, and then to return to Earth and land like a glider. Each of the five orbiters would be expected to make more than 100 flights into space and back. (This number was later changed to four.)

The shuttle's unusual design, the result of the cost compromises that have characterized the program, make it a peculiar hybrid: part reusable, part throwaway. The shuttle is taken to a height of twenty-seven miles with the help of two solid strap-on rockets. When exhausted, these are dropped and parachuted into the ocean, recovered, cleaned up, and used again. The shuttle's own engines continue. The fuel for these engines is carried in an external tank that is half a football field in length. The tank is jettisoned just short of orbit and burns up on reentry. The orbiter continues into space with small auxiliary engines.

NASA claims that the shuttle will bring down the cost of space flight dramatically, but that will be true only if there is enough work to keep the fleet of five orbiters busy. The shuttle's main job will be deploying satellites that today are launched by conventional rockets. The shuttle is also going to be used to put *Spacelab* into orbit. *Spacelab,* being built by the European Space Agency, is designed to fit into the sixty-foot-long, fifteen-foot-wide shuttle cargo hold. It will provide a shirt-sleeve environment for working in space. NASA also hopes to convince U.S. industry that space is the place to be.

The construction of large structures in space, such as solar power satellites, is another possible function of the shuttle. Solar power satellites, brainchild of Peter Glazer of Arthur D. Little, Inc., are panels of solar cells several square miles in area that capture the full-time sunlight of space and turn it into electricity. This energy would then be beamed down to Earth as microwave radiation. On Earth, the microwaves would be converted back into electricity.

NASA's enthusiasm for solar power satellites is not surprising. Here is a task that could more than match the capacity of the shuttle. It builds directly on NASA's experience with space, gives the agency a role in helping to solve a very practical problem, and could set the sort of dramatically visible goal that NASA has not had since the days of *Apollo*.

According to Glazer, the United States alone would need about 112 satellites operating by the year 2025 to supply about 25 percent of our total electrical power needs. That is a lot of power, so a lot of satellites would have to be built. A new generation of heavy launch vehicles would be necessary to put into high orbit the thousands of tons of materials required to build even one solar power satellite. No one knows the environmental consequences of the rocket fumes from hundreds, perhaps thousands, of such launches each year. And no one knows, either, the cost of such an immense project. The expense of simply getting all the materials needed to build solar power satellites could have killed the idea if it were not for Gerry O'Neill, a physics professor at Princeton University.

O'Neill believes that, in the long run, a planetary surface such as Earth is not the right place for an expanding technological civilization. Space, he has suggested, has many advantages. Energy in the form of sunlight is free, continuous, uninterrupted by night and weather, and essentially everlasting. The weightlessness of space means that huge structures could be built. Inside them, comfortable Earthlike habitats could be created. O'Neill envisions enormous inside-out worlds: spheres or cylinders over a mile across, rotating to create artificial gravity on their inner surfaces, powered by the ever-present sun. The scale of these space settlements, with masses of millions of tons each, would dwarf even a solar power satellite. Bringing all that material up from Earth, for building either solar power satellites or space settlements, would be clearly out of the question. O'Neill has proposed instead

that the moon and asteroids be mined and their materials be catapulted by a mass driver into space, where they can be captured, processed, and used for building.

NASA is taking O'Neill and his ideas seriously, although initial efforts would more likely be aimed at building a laboratory in space. After all, the lunar landings of Apollo showed that the moon was made up of about 30 percent metals such as aluminum and iron, 20 percent silicon, 40 percent oxygen—most of what would be needed to build in space. O'Neill sees the shuttle as playing a central role in his scheme to begin exploiting the resources of space. Within a few years the shuttle could be taking up modular units one at a time for the assembly of mass drivers piece by piece in low Earth orbit. These mass drivers could then roam freely in near-Earth space, fetching asteroids, sending supplies and people out to the moon or high Earth orbit to build solar power satellites and habitats for construction workers.

According to O'Neill, a program of producing a few satellite power stations would not stop there. The manufacturing capability could be upgraded until all the world's needs for low-cost clean electrical power were met. And, he adds, "By the time we get that far, we would certainly need to have many tens of thousands of people in space, and I'm sure that they would by then be living in rather large and comfortable space habitats."

Artist's concept of the interior of a huge space colony. Stationed a quarter of a million miles from Earth and constructed almost entirely of ore mined from the moon, the colony would contain a population of 10,000 people. They would live and work in an Earthlike environment inside a vast wheel more than a mile in diameter.

***(Opposite)* The space shuttle *Columbia* takes off during its first year of testing in 1981.**

"Happy valleys in the sky" are what O'Neill's critics call his ideas for the habitation of space. They suspect that he plans simply to abandon Earth and its problems altogether. But O'Neill and his supporters argue that turning to space is one of the few options we have left to solve the problems of Earth.

"People are beginning to get this new cosmic consciousness that the space program first introduced to us, and I think we're on the verge of a second Copernican revolution," says Brian O'Leary, an astronomer and former astronaut. "Now that the Earth is no longer the center for gravity space, you can move around lots of material in space and draw a new kind of wealth from the energy and materials available in space."

Henry Kolm, the man who designed a model mass driver at the Massachusetts Institute of Technology, believes that the best way to look after Earth is to find new space to live in, new raw materials, and new sources of energy. He likens the need to colonize space to the colonization of the Americas by European countries: "I think the logic is just as compelling in the case of space colonization. . . . It's inevitable, in terms of human population expansion, to move into the universe. Gerry O'Neill predicts that within two hundred years, there will be more people living in space than on Earth, and I fully believe it."

Space has begun to interest people again—especially the young, as huge crowds at the Air and Space Museum in Washington have shown. For a whole generation, space is neither exotic nor in the realm of the military-industrial complex. It is simply the Earth's backyard, an environment waiting to be explored and conquered. The first launch of the space shuttle *Columbia* on April 12, 1981, marked the beginning of this new era in the space age. The shuttle is the first space vehicle designed and intended for reuse, and NASA expects to get much use out of it and at least three other shuttles, named *Challenger, Discovery,* and *Atlantis.* There are plans for each to make 100 round trips into space over the next two decades. Thus, while the dramatic power of *Columbia*'s liftoff and the breathtaking beauty of its gliding return to Earth have enraptured millions of Americans during its first flights, such sights will soon become commonplace. The focus of the manned space program will shift from space travel and from the opening of a new frontier to man's use of space. Scientific discovery has been a major component of our space effort from the beginning, but this shift in emphasis holds the promise of seemingly limitless opportunities for new experiments; for major breakthroughs in our knowledge of our planet, our solar system, and our universe; for new outlets for our human curiosity.

The first four missions, including one in November 1981, one in March, and another in June–July 1982, amounted to a shakedown cruise. Despite some frustrating delays and mechanical problems, a tendency for some astronauts to get motion sickness, and one tragedy (two workers were killed in a launch pad accident about a month before *Columbia*'s first voyage), the shuttle has proved to be a remarkable spacecraft, both to get itself and its occupants into space and back and to serve as a base of operations for a variety of scientific, industrial, and military projects.

The scientific goals of the shuttle are many and varied. With the help of observations from the spacecraft, scientists hope to

Artist's concept of a lunar mining station. Ore is mined in the pit at the lower right and then loaded onto the electromagnetic guideway running diagonally across the scene. From there it is accelerated to lunar escape velocity and launched on its way to a space station for processing into metals and glass for construction, and oxygen for breathing.

make major improvements in our mapping and forecasting of weather and in detecting (through the sensing of infrared and other sources of radiation on Earth) important new mineral deposits and highly productive areas of marine life. They hope to discover and develop new methods of manufacturing metals, fluids, biological substances, and new types of glass, alloys, drugs, and crystals for electronic devices. They also hope to study the role of gravity on plant growth, hormone function, and the development of blood lymphocytes.

Physicists and astronomers are particularly enthusiastic about NASA's space telescope, which is scheduled for launch in 1985 and which will bring into view objects that are fifty times fainter than those now visible through telescopes on Earth; long-time exposure images of celestial objects will be ten times sharper. These scientists also plan to make extended observations of the sun and study the high-energy wave and particle radiation that is plentiful in space, but that is unable to penetrate the Earth's atmosphere.

Some of these experiments are already under way. In the third mission, for example, astronauts C. Gordon Fullerton and Jack R. Lousma deployed instruments to measure solar flares and solar ultraviolet radiation that could affect the long-term climate conditions on Earth. They successfully tested the shuttle's fifty-foot mechanical arm—the critical piece of equipment that future astronauts will use to "launch" new satellites, retrieve others for repair, and deploy a variety of other experiment packages. They also took with them a container of moths, honeybees, and other insects for an experiment designed by a high school student to determine the effect of weightlessness on insect flight.

Other experiments—including those requiring the space telescope and those to be conducted on board the European Space Agency's *Spacelab,* which the shuttle will carry into space within the next two years— remain in the designs and dreams of the scientists who will conduct them. Ultimately,

scientists hope the United States will support the construction of a space station—to be deployed in a higher orbit than *Skylab*—to serve as a permanent scientific outpost where they will be able to carry out experiments in weightless conditions, beyond the effects of the Earth's atmosphere, without interruption —the forerunner, it would seem, to O'Neill's happy valleys.

The success of the shuttle and the variety of scientific experiments planned for it will depend in part on the extent to which industry comes to use it to deploy satellites and conduct research and development into new industrial processes. Industry and other users will pay a significant portion of the shuttle's costs, leasing space in its cargo hold to carry their payloads aloft—costs that are now borne, for the most part, by the military. Without this income, the ambitious plans for the next two decades and beyond may have to be scaled back.

Crane reaching toward cargo bay of the space shuttle as it orbits in its normal upside-down mode.

A Conversation with Bruce Murray

Bruce Murray has been one of the leaders in planetary exploration for the past two decades. He recently resigned as director of the Jet Propulsion Laboratory (JPL) in Pasadena, California, a post he held for six years. It was from JPL that the Pioneer and Voyager missions to the planets were monitored and controlled. It is no exaggeration to say that he has been one of the driving forces behind the explosion of knowledge surrounding the solar system.

Murray recalls fondly his days as a high school student and the first years of his distinguished career. He says that he did not do very well in math and physics courses, but he demonstrated an early love for the outdoors. As a native of southern California, he belonged to the Santa Monica Geological Society, which afforded him the opportunity to sample the joys of collecting and exploring the neighboring countryside. He was also stimulated by his father's interest in gravity, so he naturally turned to science.

He attended M.I.T., from which he was graduated with a degree in geology. He was interested in geology not only as an intellectual pursuit, but also as an outlet for his deep interest in history. "Geology," he says, "is the ultimate study of the history of the Earth." From that point on, his path into astronomy and planetary science is fascinating.

Murray enjoys the diversity of backgrounds of people in astronomy. Nuclear physicists, communication engineers, biologists, and scientists from many fields have become involved with the field. As a geologist, Bruce Murray is no exception. After leaving M.I.T. with a Ph.D. in geology, Murray joined the air force and worked on the gravity and geodesy of the Earth. Here, he became aware of the value of rockets to space exploration. When he left the service at the end of his term, he did postgraduate work at the California Institute of Technology. Somewhat by chance, he began working with a group that was interested in the infrared radiation emitted by stars and other celestial bodies. With what he calls "a willingness to experiment," Bruce Murray became one of

the first infrared astronomers. By the early 1960s, he had built devices that had measured the heat radiation of stars.

He soon began extensive ground-based observation of the planets, including infrared studies of the solar system. At the same time, the group that would later take the first planetary photographs by satellite was forming. Murray joined the team and has been involved in planetary exploration ever since.

Bruce Murray realized that the space age was upon him, and he recognized the need to alter his goals. The rapidly developing technology of space flight was offering new scientific challenges, and in 1967 he left infrared astronomy to devote himself full time to space studies.

Some of Murray's explorations have revolutionized our understanding of the solar system. It is difficult for most people to imagine seeing something for the first time that was previously unknown, but Bruce Murray has done just that from a front-row seat. He speaks earnestly with a sense of wonder at "looking for the first time at God's handiwork. It's a fantastic experience." He is referring, in part, to the Mariner missions, which shed new light on Mars and Venus. He says that the goal is to determine why a planet looks the way it does. What do the surface features mean? Murray likens the process of discovery to that of a historian who is suddenly admitted to a library that has been locked for centuries. Imagine the wonders to be discovered!

Bruce Murray explored that library. He recalls the excitement of the *Mariner* fly-bys of Mars as the tiny spacecraft sent images of the red planet that altered his way of thinking. It was discovered that the polar ice caps of Mars are made of dry ice, as originally proposed by Murray and his colleague Robert Leighton. He took charge of some aspects of the *Mariner 9* orbiting flight, and he was the team leader of the Mercury-Venus probe, *Mariner 10*.

By 1970 he was planning what has been called the "grand tour." This is the proposed encounter of all of the outer planets by one satellite. Murray's vision of the grand tour is being mostly fulfilled by *Voyager II,* which has visited Jupiter and Saturn and is on its way to Uranus and Neptune.

Yet, as great as the successes of the Voyager missions were, including a massive public involvement through the electronic media, he feels that the highlights of his career are personal successes that received far less publicity. He mentions as important to him his experience with *Mariner 4,* which sent back photographs revealing the presence of craters on Mars. This showed scientists that the red planet is not at all like Earth, and Murray realized that Mars never had any oceans. He explains that most craters were formed early in the life of the solar system, and that if they are still present, the planet must not have erosion processes such as oceans or rainfall. This changed the accepted notions of the planet's past.

He also considers the *Mariner 10* mission to have been a high point in his career. The exploration of Venus was exciting because it was "a planet about which we knew nothing," covered as it is with a thick layer of clouds that obscures the surface. Murray had spent all his time promoting the mission and engaged in technical aspects of the satellite, such as choosing cameras. He feels that the rewards of his work and the joys of seeing things for the first time were great.

Bruce Murray cites his experience with *Mariner 10* as an example of a complete project pursued with dedication and courage. He feels that these are two qualities a person must have to be a good scientist. The determination, however, must be properly motivated. It must come from within. A person must be almost compulsive and work with single-mindedness in order to assess his work and adjust to changes brought about by his own successes and failures. It is this drive from within that keeps the scientist going at all times.

In addition, a scientist must be able to focus his attention and energy on significant problems. This transcends hard work and getting the right answers. Murray thinks back to many of his students who, as bright as they were, did not have the ability to focus their attention. He wishes there were some way to teach this quality, but realizes that it must be learned by oneself.

Since a personal dedication has driven Bruce Murray in his career, he is unimpressed with the trappings of success. He has sought rewards inside himself and found them. He warns about the dangers of selecting projects for notoriety or any reason other than really wanting to find out the secrets of nature. The most important thing is to do what one wants to do. Perhaps the recognition will follow, but the rewards, like the dedication and drive, come from within.

Bruce Murray is a scientist who has in effect pursued many careers: geologist, infrared astronomer, planetary explorer, and administrator as the director of JPL. He has enjoyed the challenge of new fields and new projects because they have been what *he* has wanted to do. This inner, powerful will is what gives him the freedom of thought that has propelled him to the forefront of modern science and brought him the reputation and notoriety he never sought.

Stalking the Crab Nebula

I n the year 1054, Chinese astrologers observed the sudden appearance of a star that shone so brightly it was visible in daylight for twenty-three days. Today, in the same part of the sky, this object is barely visible, even through the most powerful telescopes. But it is as fascinating and intriguing an object to today's astronomers as it must have been to the people of the Sung dynasty, who interpreted its appearance as a promising omen.

Although no one is certain, many scientists today believe that those astrologers were the lucky eyewitnesses to a major astronomical event: a supernova, the explosion and total disintegration of an aging star. This particular supernova is believed to have resulted in the creation of a spiraling collection of gas and debris that we call the Crab Nebula.

The Crab Nebula was the subject of some curiosity during the eighteenth and nineteenth centuries, but for lack of adequate telescopes and because astronomers were more avid about planets, asteroids, and comets, it was largely ignored. In fact, Charles Messier, an eighteenth-century French astronomer who cataloged about one hundred celestial objects that could be mistaken for comets, advised his readers: "Don't look here. It's not a comet. It won't move."

Today's astronomers, however, are looking in earnest. When the first modern radio telescope began monitoring the heavens, astronomers found to their surprise that the Crab Nebula was one of the brightest sources of radio waves in the sky. This discovery was so significant that it prompted widespread interest in the fledgling field of radio astronomy and spurred the development and construction of more sophisticated generations of telescopes. It also focused considerable scientific attention on the Crab itself.

Scientists quickly determined not only that its signal was strong, but that it was too strong to emanate solely from a cloud of debris that had been cooling for nine hundred years. They speculated that some unknown force was behind it. Theorists reexamined

their ideas about what happens when a star runs out of fuel.

When our sun's fuel supply is finally exhausted some five billion years from now, it is expected to collapse on itself, crushing its atoms together with enormous gravitational force until it becomes a white dwarf star about the size of the Earth. Astronomers have detected such white dwarfs elsewhere in the sky. But scientists have also proposed that such pressures could build in some stars to the point where the atoms themselves collapse; in that case a star four to eight times the mass of the sun would be crushed into what is called a neutron star just seven miles across. Such a star would be incredibly dense (its gravitational pull would be such that our heads would weigh as much as a hundred ocean liners). A neutron star at the center of the Crab Nebula, it was suggested, could be the source of its power.

Thus, the hunt for the explanation of this source of radio waves became a search for the first neutron star as well. And it turned up another surprising characteristic of the signal: it pulses. This mysterious object seemed to be an entirely new kind of star, and as researchers subsequently found evidence of other pulsating signals, they gave them the name pulsars. Eventually, scientists determined that the Crab's signal has a rate of thirty pulses per second and that it is gradually slowing down. With this information, scientists were able to perform calculations demonstrating that the Crab's pulsar conforms to the theoretical properties of a neutron star.

Thus, the source behind the Crab's signal seems to be a rotating neutron star that scientists believe is surrounded by a magnetic field and gaseous plasma, both of which provide a good source of radio waves. High-energy electrons in the plasma spiral around the magnetic field and emit highly directional radio waves akin to the beam from a flashlight. As the neutron star rotates, the beam of waves sweeps like a lighthouse beacon, creating the pulsating effect. Subsequent investigations have proven the existence of X-rays and gamma rays as well, so that the

A supernova explodes in a distant galaxy. The top photo shows a single star exploding to outshine billions of others in the same galaxy. Months later (below), the star has gone out. The supernova that produced the Crab Nebula in 1054 would have looked like this to an observer in another galaxy.

The Crab Nebula, which scientists believe to be the remnant of a supernova, the explosion and disintegration of an aging star that occurred in 1054 A.D. It is one of the brightest sources of radio waves in the sky.

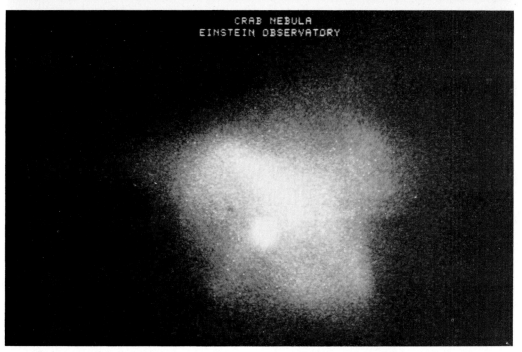

CRAB NEBULA
EINSTEIN OBSERVATORY

The Crab Nebula, viewed in X-ray light that clearly shows the dominating pulsar, a rotating neutron star that is the source of powerful radio waves emanating from the Crab. The bright, diffuse region above the pulsar corresponds to the region that is brightest in visible light.

Crab is now known to emit radiation from all bands of the spectrum.

The Crab Nebula has thus become a Rosetta stone for astronomers and has helped alter scientific thinking about the evolution of stars. Further research into supernovas, some scientists think, may yield clues to the origins of life itself. Filaments of material in the Crab Nebula—formerly the inside of the star—contain atoms of carbon, oxygen, sulfur, and other elements. If they condense into new planetary systems, some scientists theorize, they could be the seeds for the formation of life in a distant part of the universe far into the future.

The Search for Life

In 1976, a spacecraft prepared to land on Mars, its instruments at the ready to make all the measurements we accept as usual in space exploration. But *Viking*'s unique mission was to land on the Martian surface and look for life, the first attempt to find life on another planet. What would the lander be looking for? Trees? Elephants? Space monsters?

The experiments *Viking* was charged with were designed not to find space monsters but to analyze three ounces of Martian soil for microscopic organisms. Microbes cover the Earth's surface. The hope was that the Martian surface, too, had microbes. In fact, the Martian experiments were formulated entirely on our knowledge of the origin and development of the only life we know: life on Earth.

The history of life on Earth can be run backward on a time clock starting from the present day at zero to more than three billion years ago. Geologists track down the origin of life by tracing out the animal forms captured as fossils in older and older rocks. Before two million years ago, the human species is not found. Earlier than that, all other mammals disappear. Then reptiles go. Then fishes . . . plants . . . insects . . . shellfish . . . corals. The earliest sea animals are found just 600 million years ago, less than a tenth of the Earth's age. And then, by grinding down slices of ancient South African rock until they are paper-thin, scientists have found the oldest known fossils: microscopic shapes that are almost certainly primitive single-cell organisms.

The origin of life on the Earth is sandwiched in geologic time between the oldest known fossils and molten lava. Imagine the fiery Earth soon after its formation from a spinning cloud of interstellar gas and dust four and a half billion years ago. Gradually the surface cooled. Collisions with interplanetary debris became fewer. Hydrogen gas built up in the atmosphere, and water vapor condensed. Storm clouds began to gather, and rain poured down. Somewhere on the Earth's surface, life started. Somewhere, in a warm, drying pool, on a beach, or on the slopes of a volcano, the materials of life, amino acids and nucleic acids, were forming·and beginning to group together. Perhaps the same course of events once took place on Mars.

When the space probe *Mariner IX* approached Mars in November 1971, it relayed back to Earth more than 7,000 detailed photographs. They provided surprise after surprise for the scientists waiting in the Jet Propulsion Laboratory control room in Pas-

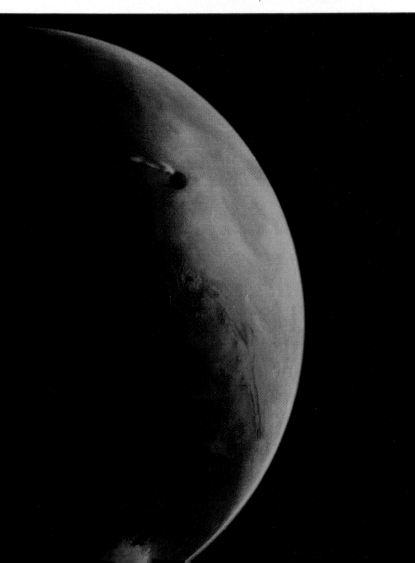

Mars, as the second *Viking* approached the planet in August 1976. In the north, morning clouds trail from the volcano Ascraeus Mons.

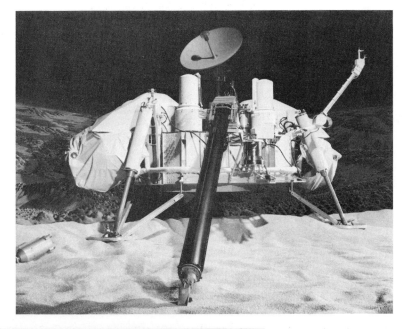

adena, California. The first surprise was the presence of recently active volcanoes. Surprise number two: a vast rift valley, running 3,000 miles around the planet. The geologists were delighted; but the biologists were looking for water, essential for life on Earth, essential in all scientific experiments reconstructing the origin of life, and essential in every living cell. *Mariner IX* did not let them down. There were no canals as predicted by the nineteenth-century astronomer Percival Lowell, but there was something just as exciting: sinuous valleys with tributaries that looked as if they could have been produced by running water.

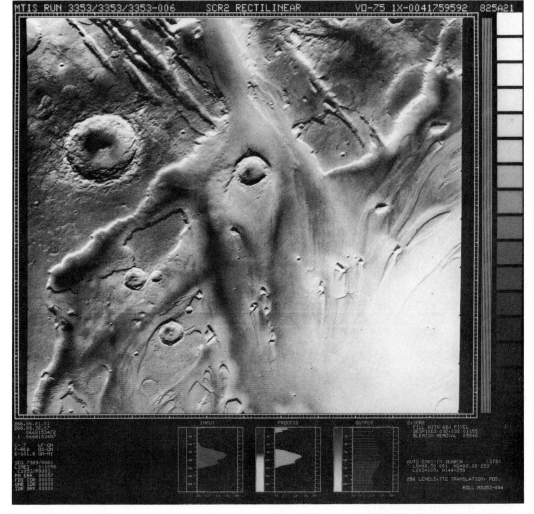

(Above) *Viking* **lander. The telescoping surface sampler is mounted on the front and can turn to deposit soil in an automatic analyzer inside the spacecraft. Cameras are housed in the two cylinders on either side of the disk antenna.**

(Left) **The mouth of the dried-up Kasei river system on Mars.** *Viking I* **landed 180 miles downstream from here in hopes of sampling a part of the planet that had once been wet enough to support life.**

The surface of Mars. The shovel-like sampler arm of *Viking II* extends from the bottom, and a recently dug trench scars the middle ground.

Life on Mars—at least at one time—looked possible. For the Mars observed by *Mariner IX* appeared to be an Ice Age Mars, a Mars that may have had a more hospitable climate in its distant past. In 1974, as *Mariner IX*'s reports instigated *Viking*'s hands-on mission, hopes were high that life would indeed be found on our neighboring planet. In the space program, as in all science, optimism is as essential to life as water.

Norman Horowitz of the California Institute of Technology believed that in the early stages of Martian history, there was a more favorable epoch in which an abundance of water and other atmospheric gases on the planet made the origin of life possible. But, he said, "as conditions in the planetary environment gradually became more and more severe, that life mutated and evolved in such a way as to enable it to survive under conditions as they became harsher and harsher, and if this occurred, then there's a possibility that survivors of that origin of life are still living on the planet."

One of the experiments to be conducted by *Viking* was developed in Norman Horowitz's laboratory. First the soil sample is

sealed in a chamber. The idea was to match experimental conditions with Horowitz's theory of Martian life. Next, carbon dioxide gas, labled with a trace of radioactivity, is added. If there's life on Mars, it may have evolved to use only what's available now: very little water, carbon dioxide from the atmosphere, and Martian sunlight. After incubating for five days, a radioed command from Earth would heat up the chamber. The radioactive carbon dioxide and microorganisms will be driven off and the gas pumped out. But microbes, if there are any, will be locked into a special absorbent column. At that point, a new command heats up the column, decomposing the Martian life. If any radioactive gas has been incorporated into their bodies during the past five days, it will be detected by a miniature Geiger counter. The experiment will have found life on Mars.

There was, however, another theory for Martian life—hibernation. Carl Sagan of Cornell University, perhaps the most vocal advocate of extraterrestrial optimism, said, "If it's true that Mars has oscillated between warm, wet and cold, dry environments, then there may be warm, wet kinds of organisms that are in a dormant state, awaiting the coming of the long summer. It may be that by dropping them in liquid water, they will think the waters of the climatic change have come, and they will do their stuff, whatever that is. The *Viking* may have a very pleasant surprise in detecting such organisms that are just waiting for naturally occurring liquid water to arise on Mars."

Vance Oyama, a soil biologist at NASA's Ames Research Center, developed the second *Viking* experiment to test the hibernation theory. Oyama's idea was that life on Mars, like that on Earth, may require water and rich nutrients. The method will be to add to a soil sample a water-based nutrient, packed with nearly everything an Earth microbe could possibly want to eat, including amino acids and trace elements. Under incubation, Earth organisms give off gases as they eat the food. Oyama hoped that Martian organisms would do the same. There was even a third experiment, using conditions midway between carbon dioxide only and rich nutrients. Why did NASA send this many?

The chief biologist for the *Viking* mission, Gerry Soffen, explained: "At the time we simply did not know enough about the planet and where to go, so the idea was to cover this broad spectrum of possibilities, hoping that one of them will hook. It's a little like going fishing with a variety of bait. What makes good bait in one place doesn't necessarily make good bait in another. You use frogs' eggs in one and you may use caviar in another."

A section of the enormous Mariner Valley on Mars. Color pictures on *Viking* were obtained by successively photographing through blue, green, and red filters. The tables at the bottom give various technical information, such as latitude and longitude, sun position, and the dimensions of the scene (in this case 726 × 950 kilometers, an area the size of Colorado).

The inhospitable surface of Venus, photographed by a hardy Russian spacecraft in 1982. Being about the same size as Earth, Venus is often called Earth's twin, but the resemblance stops there. For some reason not yet clear, Venus developed an atmosphere (mostly carbon dioxide) that is 100 times more dense than Earth's. The perpetually cloudy skies trap the sun's heat, pushing the surface temperature up to 900° F., hot enough to melt lead.

The relative sizes of the sun and planets. The inner planets—Mercury, Venus, Earth, and Mars—are composed of rocky materials, while Jupiter and the planets beyond (with the exception of icy Pluto) consist largely of gases. Before the space age, Venus and Mars were considered the most likely candidates to harbor extraterrestrial life.

Going fishing for microbes on Mars is an expensive and difficult business. In many ways, the mission was a huge gamble, with the stakes nearly a billion dollars. These theories of Martian life could well have been wrong. But if this billion-dollar gamble were to succeed, what could we gain? Gerry Soffen believed that if the biological experiments were to give a positive result, extraterrestrial life would be likely to become one of the most significant discoveries of the twentieth century. "I doubt whether there will be a single experiment that one could ever do again during the twentieth century that would possibly come up to the discovery of indigenous life on Mars. It would indicate to us a fantastic possibility that the universe is literally teeming with life; that we are not alone; that the universe has many other civilizations."

But, to the disappointment of many who had pinned their hopes on *Viking*, it appears we *are* alone, at least as far as we've ventured since Alan Shepard's first five-minute ride in space in 1961. The two *Viking* orbiters with detachable capsules, launched in the summer of 1975, completed an eleven-month journey, touching down on Mars in two separate locales. While cameras scanned the landscape for large-scale biological life, the ten-foot mechanical arms scooped up soil and delivered it to the landers for automated biochemical analyses. The peoples of the world held their breath, and then came the disappointing results—not even the simplest organism was detected.

The biological experiments did show, however, an exotic soil chemistry. It is sometimes said that perhaps the Martian chemistry of life is so different that the biochemical analyses could not have discovered it. Or possibly that Martian life exists only below the surface. Or even that we are not correctly interpreting the data. Perhaps we will unearth life on Mars yet. For while both *Viking* orbiters have been turned off and the *Viking II* lander stopped working in 1980, the *Viking I* lander is expected to continue transmitting signals—and searching for life—until 1994.

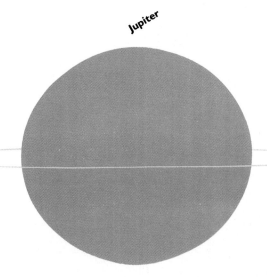

Sun Mercury Venus Earth Mars Jupiter

The Milky Way is only one of the millions of galaxies that form the universe. It is composed of more than 100 billion stars, of which our sun is only one. The Earth may be the only inhabited planet in our solar system, but the odds seem overwhelming that life exists elsewhere in the cosmos. Shown here is the Andromeda Galaxy, 2.3 million light-years distant and yet the closest galaxy to our own. When the denizens of the planets in the Andromeda Galaxy look in our direction, they must see a view much like this one.

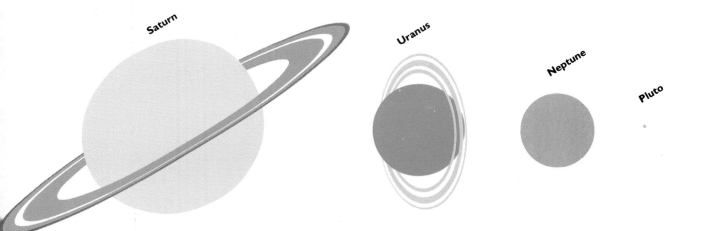

Saturn

Uranus

Neptune

Pluto

Jupiter and Beyond

The trajectories through the solar system of the *Voyager I* and *Voyager II* spacecraft. The mission of the *Voyagers* was to fly beyond Mars and the asteroid belt to study the giant gaseous outer planets–Jupiter, Saturn, Uranus, and Neptune. *(Facing page)* The one-ton *Voyager* spacecraft. Normal and high-resolution cameras hang off the boom at the right.

Once every 175 years, the planets line up in such a way that a spacecraft from Earth can visit each of the outer planets, Jupiter, Saturn, Uranus, and Neptune, in one grand tour. That special opportunity came in 1977, and the United States launched two unmanned spacecraft, one to make the whole tour, the other to leave the solar system after Saturn. Their mission, to view worlds that had never been seen in detail before, would carry them billions of miles from Earth and would take over a decade to complete. This was the Voyager Project.

Voyager I and *Voyager II* were launched atop Titan-Centaur rockets one month apart in the fall of 1977. As they cruise through space they are powered by nuclear batteries driven by the heat of plutonium. Three on-board computers sense spacecraft orientation and follow commands that have been transmitted up to as much as a year earlier. A large antenna beams radio signals to Earth. These signals are captured by three large receiving dishes and transmitted to Mission Control at the Jet Propulsion Laboratory in Pasadena, California.

The *Voyager* spacecraft are small space observatories designed to help scientists understand how the Earth fits into our solar system and into our universe. The Earth is one of nine planets circling the sun. Together with the moon, our place in the solar system is among the inner planets close to the sun. These planets, Mercury, Venus, and Mars, are small, dense, and rocky; the others, Saturn, Neptune, Uranus, Jupiter, and Pluto (called the outer planets), are giant and gaseous. It is hoped that the *Voyager* observations of the planets of the outer solar system will illuminate, says Hal Masursky of the U.S. Geological Survey, "the general principles that govern the kinds of things that have happened in the past on the Earth, and that will happen in the future."

Scientific data on planet atmospheres, magnetic fields, satellites, and rings are collected by ten different *Voyager* instruments including wide- and narrow-angle color video cameras, cosmic ray detectors, magnetome-

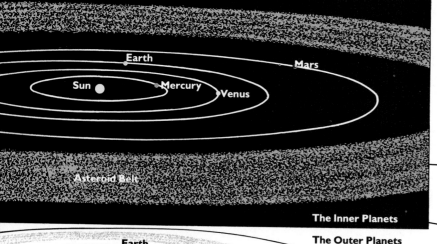

Earth

Sun   Mercury   Venus

Mars

Asteroid Belt

The Inner Planets

The Outer Planets

Earth
Voyager I: September 5, 1977
Voyager II: August 20, 1977

Mars

Jupiter
Voyager I: March 5, 1979
Voyager II: July 9, 1979

Saturn
Voyager II: August 27, 1981

Saturn
Voyager I: November 13, 1980

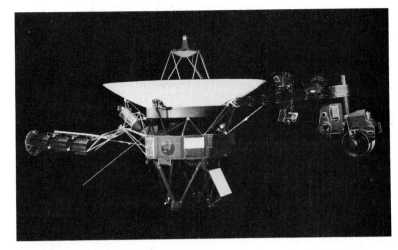

ters, infrared spectrometers, radiometers, and devices for detecting and recording ultraviolet radiation and radio emissions. The data, which come back in digital form—as a binary code of dots and dashes—are transmitted at high speed and reconstructed to form pictures. Each picture is made up of 64,000 dots or pixels, and each dot is numbered to indicate how bright or dark it is. Various scenes, each photographed through a different colored filter, can be assembled into a single picture of remarkable clarity. By computer-manipulating the numbers, scientists can enhance and modify the pictures. And by sequencing many of the pictures together, color movies can be created.

Eighteen months out and traveling ten times faster than a rifle bullet, *Voyager I* encountered Jupiter. In Western mythology, Jupiter is the father of the gods. For Galileo, the planet provided a proof of the Copernican system. For later observers, it has been—and remains—an object of awesome majesty and mystery.

Jupiter, as seen by *Voyager I*, is a huge banded globe of subtle blue, white, orange, and brown hues with a large red spot. Eleven Earths could fit side by side across its diameter, and more than a thousand Earths

could fit into its volume. Aside from the sun, Jupiter contains twice as much material as all the rest of the solar system, but it is not solid. Jupiter is mostly made of hydrogen and helium gas. Underneath an immensely deep ocean of compressed hydrogen, which behaves like liquid metal, there may be a small rocky core somewhat larger than the Earth. From *Voyager I*, scientists gained new information about Jupiter's weather and moons. This information, by extension and analogy, gave them a better understanding of the Earth's history.

Pluto

The *Voyagers* will cross into interstellar space in the mid-1990s. This occurs at the point where the solar wind of charged particles is overcome by the interstellar magnetic field, and it is here that the solar system ends.

Neptune
Voyager II: August 24, 1989

Uranus
Voyager II: January 24, 1986

Jupiter is a cloud-belted world of rapid jet streams and complex cloud forms. In this view made by *Voyager I*, Jupiter's inner satellite, Io, passes in front of the giant planet.

Jupiter's weather takes place in an atmosphere rich in the materials thought to have made up the early solar system, especially such compounds as methane and ammonia. Could the conditions in this atmosphere be the same as those of the early Earth when life began over three billion years ago? Cyril Ponnamperuma, director of the Laboratory of Chemical Evolution at the University of Maryland, has discovered that organic matter can be synthesized under the conditions that exist today in Jupiter's atmosphere. By mixing methane, a source of carbon, with ammonia, a source of nitrogen, and passing a spark—representing lightning in Jupiter's atmosphere—through the mixture, organic matter can be formed.

"There is no question that building blocks of life can be made many ways," says Ponnamperuma. "And if we can do it in a laboratory, presumably it must happen elsewhere; it must happen in planetary atmospheres, on planetary surfaces, and so on. But we must have some mechanism by which these molecules can be brought together. They have to interact. And I believe that one single necessary ingredient for such an interaction is liquid water. So if we can find liquid water somewhere in the universe, I think we may be able to conclude that there is also life there."

At certain depths Jupiter's atmosphere does hold water. Since the organic building blocks are abundant, and there is electrical energy, shouldn't the chances be good for life? The problem, according to geneticist and life scientist Norman Horowitz, is that Jupiter contains a heat source deep inside its great atmosphere. Jupiter actually produces twice as much heat as it receives from the sun, causing its atmosphere to be in constant convection; gases are forced from the deep interior up into the upper atmosphere, then circulate back down again.

According to Horowitz, these flow patterns mean that any organic compounds formed in the upper atmosphere of Jupiter would be carried by convection into the deep interior, where they would be destroyed by temperatures as high as 10,000 degrees absolute. As a result, the kinds of complex organic substances that scientists think are necessary for life have no chance to evolve on Jupiter.

During its brief encounter with Jupiter, *Voyager I* was not able to prove Horowitz right or wrong. But *Voyager's* observations on Jupiter's moons, particularly the four satellites first seen by Galileo in 1610, have shed new light on the processes that led to the formation of the Earth.

Callisto, the outermost of the Galilean satellites, was the first moon to fall under *Voyager's* scrutiny. It is about twice the size of the Earth's moon and is half ice. A rocky core lies under a frozen ocean hundreds of miles deep. Its surface is intensely cratered. In some places, gigantic impacts seem to have turned the ice to slush before it refroze.

From Callisto, *Voyager* flew inward to the largest moon in the Jupiter system, Ganymede, which is also half composed of ice. Much of Ganymede's surface looks as if whole chunks of Callisto have been grafted onto it. But the rest of Ganymede's surface is quite different and unique: it is covered with grooves—strange parallel lines twisting over the surface and superimposed on one an-

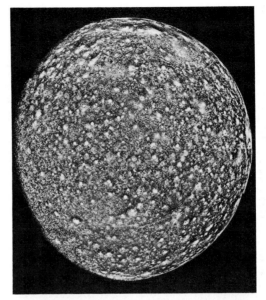

Callisto *(above)* **from a half-million miles and** *(left)* **in a high-resolution view of its heavily cratered surface.**

other. Like earthquake fault lines caused by the tensional pulling apart of the Earth's crust, the grooves may have been formed as ice froze and stretched apart.

From Ganymede, *Voyager* flew still farther inward, to Europa. Europa is similar in size, density, and mass to the Earth's moon. A frozen ocean slightly deeper than the Earth's water body surrounds a large rocky core. But unlike Ganymede and Callisto, Europa is almost perfectly smooth, like a billiard ball. Its only distinguishing feature is that its surface is covered with a complex network of cracks, resembling the ice of Antarctica when viewed from space, which appears to have been caused by the expansion of Europa as the ocean froze.

Scientists believe that Callisto, Ganymede, and Europa illustrate three different stages in the natural history of a planet or moon. The solar system condensed from a huge disk of primordial matter swirling around the sun about four and a half billion years ago. The planets and moon emerged from the disk by gathering up smaller objects in the solar system—meteors, asteroids, and comets—through an intense bombardment that lasted millions of year. The bombardment left their surfaces scarred with impact craters. As eons passed, part or all of the cratered surface was covered over or erased by new land forms. Even this new surface became lightly cratered by the remaining debris. Meteor Crater, Arizona, is one of the rare impact craters on Earth, left over from this later period, which has not been eroded away or covered up by sediments.

The surface differences among Callisto, Ganymede, and Europa are thought to be a consequence of when their oceans froze relative to the duration of bombardments. There are virtually no impact craters on Europa because the meteorites splashed down into a liquid sea that subsequently froze and cracked. On Ganymede, the ocean froze at about the same time the bombardment was lessening, erasing some of the original cratered surface with grooves. But on Callisto, the ocean must have frozen

Ganymede, Jupiter's largest moon, to scale with the other Galilean satellites in this column. *(Left)* **Close-up view of craters and oddly grooved terrain on Ganymede.**

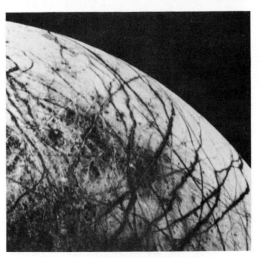

Europa from a half-million miles. *(Left)* **Detail of intersecting streaks on Europa. These are probably filled-in fractures in the satellite's thick icy crust.**

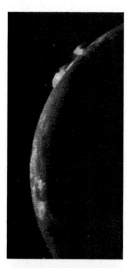

Io (above) from a half-million miles, and (top left) displaying two erupting volcanoes. (Top right) Detail of one of eight active volcanoes observed on Io by Voyager I.

Thin ring of particles orbiting Jupiter, photographed from the planet's night side by Voyager II.

before the end of the bombardment, which left on its surface a testimony to the violent birth of the solar system.

If Jupiter's moons did develop this way, scientists reason, then Europa could support life. Its early ocean must have been liquid for millions of years, possibly long enough for life to have begun. If there is liquid water today underneath its completely frozen surface, life that may have once started might still remain—living in total darkness like the newly discovered sea worms on Earth.

From Europa, *Voyager* traveled to the innermost of the Galilean satellites, Io. Io is the size of the Earth's moon, but its cratered surface is unlike anything ever seen before. It was the stage for one of the most unexpected discoveries of the *Voyager* mission: the observation of several simultaneous volcanic eruptions. Io's mottled yellow, red, and blackish surface turned out to be even more volcanically active than the Earth's.

"We found out that not only was Io's surface being modified by volcanic processes,

it was being done literally in front of our eyes," says Torrence Johnson of the imaging science team (the graphics specialists) at the Jet Propulsion Laboratory. "There were active volcanos going off at the time we were there with *Voyager I*—at least seven or eight, perhaps nine clearly identified explosive volcanic eruptions throwing materials hundreds of kilometers above Io's surface. We estimate that this continual resurfacing of Io amounts to spreading a layer of volcanic material about a millimeter's thickness over the entire planet every year. Now, at that rate, if it continued over geologic time, it would amount essentially to turning Io inside out."

Io is volcanic because its interior is alternately pulled and stretched by the immense gravity of Jupiter's tides, much as our moon raises tides on Earth. The heat of this constant deformation has melted its interior so that Io remains a thin crust of sulfur and rock sitting on a ball of molten lava. Its volcanoes are powered by sulfur dioxide boiling under the surface at − 300 degrees Centigrade. This sulfur, spewing out of the intensely hot volcanoes, may account for Io's vivid colors.

Altogether, *Voyager I* caught sight of fifteen of Jupiter's moons, three of which had not been known previously. The most unexpected discovery of all, however, was that Jupiter, like Saturn, has a ring. The ring may be 30 kilometers thick and 7,000 kilometers wide. The size of the particles composing the ring is not known, but judging from the ring's brightness, the particles are believed to be small. The Voyager Project scientists hadn't expected to find a ring in Jupiter. All their calculations indicated that such a ring should have dissipated long ago. Yet it exists, with some constant replenishment. Still, Jupiter's rings can hardly compare to the rings of Saturn—the next stop on *Voyager I*'s "grand tour."

In approaching Jupiter and its inner moon, Io, the spacecraft trajectory had bent through more than a right angle. The encounter was designed to boost *Voyager*'s speed to 35,000 miles an hour, though it slowed Jupi-

ter in its orbit by about one foot per trillion years. One year later, after traveling another half billion miles, *Voyager I* spied Saturn.

Voyager images showed that Saturn is distinctly oval. The planet is flattened by the speed with which it spins; a Saturnian day is only ten hours long. Saturn lies ten times farther from the sun than Earth. Like Jupiter, it radiates twice as much heat as it receives. More than 650 Earths could fit into the volume of the planet, but it is made up mostly of gas, not rock.

Saturn's atmosphere is blanketed in a thick layer of smog, so the imaging team had to construct complicated computer programs that would reveal the detail underneath. They succeeded, and a computer-enhanced image of a previously unseen Saturn emerged below the haze. Saturn's atmosphere is patterned with spots and vortexes trailing for tens of thousands of miles. Equatorial winds blow up to 1,000 miles per hour. Beneath this turbulent atmosphere is an ocean of liquid nitrogen more than 30,000 miles deep. Deep down, this ocean is so compressed that it behaves like liquid metal.

At the very center is a rocky core, a solid planet rather larger than the Earth. Girdling the planet are the rings.

Although the rings were probably first noted by Galileo and later by the Dutch astronomer Christian Huygens in the 1600s, it was centuries before it was understood that there were several rings and that they must be made up of particles. Composed of billions of small, independently orbiting chunks of ice, the entire ring system is almost 40,000 miles across and, surprisingly, less than a mile thick.

The names for Saturn's rings were worked out long before any spacecraft arrived to view them. The names are based on the two most noticeable divisions in the rings: the gap known as Cassini's division and the boundary with the faint inner "crepe" ring. In the interest of simplicity, the rings are called A (outer), B (central), and C (the inner crepe ring). But as the resolution has improved on the rings, the naming has been extended in the alphabet and has become absurdly complicated. Even so, it underestimates the rings' structural complexity. The ring system is so intricate and full of

A false-color view of Saturn obtained by *Voyager I*. Saturn's cloud tops are more nearly the color of butterscotch. They have been enhanced here to bring out distinct banding such as that displayed by Jupiter.

The crescent Saturn photographed by *Voyager II*.

detail that it has been compared to the grooves on a gigantic phonograph record.

Close as *Voyager* flew to the rings, it was not close enough for the spacecraft cameras to pick out the individual boulders of the rings and determine their size. But by catching the Earth in the gap between the rings and the planet, and then transmitting the spacecraft radio beam back to Earth through the rings and listening to the interference, the spacecraft was programmed to find this out. The particles in the outer or A ring are, on average, thirty feet across. Some are as large as square city blocks. In the inner or C ring, the average particles are about the size of automobiles. In the central or B ring, the boulders are jostling so close together that, on first analysis, the spacecraft radio beam appeared not even to have penetrated through. But as *Voyager* moved closer to Saturn, the team became aware of curious irregularities in the ring.

The B ring looked as if it had spokes, dark lines radiating out from the planet and patterning the ring material. The baffled imaging team decided to make a movie from the *Voyager* images. From a different view, the spokes not only looked very unspoke-like and irregular, they also seemed to have changed color. Seen from the other side of Saturn, they were not black, but white. This was the vital clue to what the spokes actually are, for the property of changing brightness with the direction of light is shown only by minute objects, like specks of dust caught in a

Bright-colored spokes cutting across Saturn's rings.

sunbeam. The spokes are thought to be clouds of tiny particles, possibly composed of dust or frost from ice, levitated above the boulders of the rings.

Formed with Saturn more than four billion years ago, the rings may provide a clue to conditions in that ancient time when the solar system was born. Over time, some process seems to have steadily marshaled the myriad swarms of independently orbiting boulders of ice into the complicated ring structures. Before *Voyager*, scientists theorized that the gravitational influence of Saturn's moons, orbiting far outside the rings, had displaced the icy boulders over billions of years and given them their precisely spaced divisions. According to the resonance theory, if a ring boulder orbits Saturn at twice the rate of a moon, for example, every second orbit the boulder will be bumped out slightly by the gravity of the moon. The effect of this happening over and over again for eons would be to pull the ring boulder into a different orbit from its neighbors'. Its elliptical orbit would cause it to collide with nearby boulders, clearing a gap in the rings by slowing them all into different orbits. Long before *Voyager I* ever got to Saturn, the resonance theory had predicted exactly where all the divisions in the rings would be found.

The only problem with the resonance theory, which so elegantly linked the rings with the moons in a kind of divine harmony of the spheres, is that it turned out not to work in most cases. Not only were many divisions within Saturn's rings found by *Voyager I* to be elsewhere than where the resonance theory placed them, but there were also many more divisions than the resonance theory could account for. Since resonances with known moons did not completely explain the divisions in Saturn's rings, additional mechanisms had to be discovered.

A clue to one mechanism was photographed by *Voyager I* at the very edge of the rings; here was a moon that had never been seen before. According to Bradford Smith, head of the imaging team, the gravitational interaction of this moon, only about the size of Kansas City, has kept the outer edge of the rings in place for billions of years. *Voyager I* also discovered two moons orbiting just inside and outside the F ring. These moons seem to act as shepherds, fencing the F ring into a narrow line. Scientists wonder if other small shepherd moons might be embedded among the boulders of the rings, invisible to *Voyager*'s cameras.

The resonance theory, so bad at predicting where gaps were in the rings, was better at predicting the locations of concentrations of material. Instead of clearing gaps in the rings, the gravitational influences of Saturn's moons appear to be slowly concentrating the boulders of the rings together at the regularly spaced resonance locations. This unexpected discovery may help explain how the Earth and other planets came to be at their present locations in the solar system.

The sun and planets formed from a huge cloud of gas and dust. About four and a half billion years ago, this cloud began to collapse. The sun was to form at the center. The dust and gas that were to become the planets settled into a great disk of material swirling around the sun much as Saturn's rings do. If Jupiter and Saturn formed first out of this immense ring, the resonances caused by their huge collecting masses may have determined where the inner planets would coalesce. Later, the sun heated up, and a hot wind of particles destroyed the remnants of the primeval ring and pushed the debris far into space, erasing the evidence of the birth

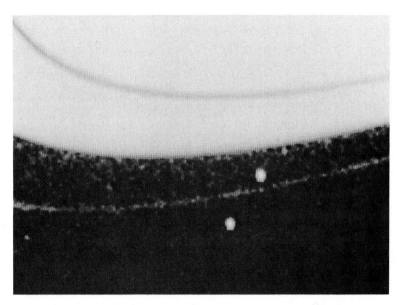

Shepherd moons herding in Saturn's F ring.

of our solar system. Perhaps we owe our existence to Jupiter and Saturn; if our planet were not almost exactly at its present distance from the sun, life as we know it may never have evolved.

Besides its rings, Saturn has more than fifteen moons; *Voyager* photographed thirteen of them. The closest large moon to Saturn is Mimas, named after a mythological companion of the god Saturn. Roughly the size of the state of Iowa, Mimas appears to be mostly made of ice. Its surface is richly cratered and cracked as a result of huge impacts. On one side of Mimas, *Voyager* photographed an enormous crater believed to have been caused by the impact of a comet traveling at more than ten kilometers per

Saturn's small moon, Mimas, showing a sixty-mile-wide impact crater that must have come close to destroying the satellite.

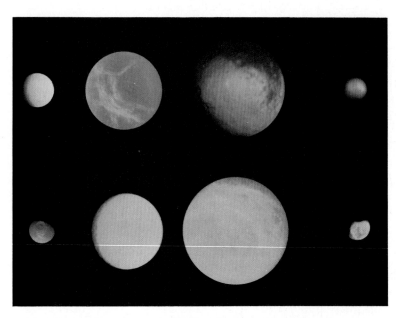

Eight of Saturn's satellites to scale. *Top row, left to right:* **Enceladus, Dione, Iapetus, Phoebe.** *Bottom row:* **Mimas, Tethys, Rhea, Hyperion. On the facing page Titan is shown at the same scale.**

the visible part of a fault running right through Tethys.

The moon Dione is about 700 miles across. At first sight, it seemed similar to Saturn's other icy moons—a heavily cratered sphere of ice with cracks spreading across its surface for hundreds of miles. But it seems that at some point in Dione's history, new material escaped from the moon's interior, modifying the color of portions of its surface. For one side of Dione is painted with strange patterns, christened "wispy terrain," that had never been seen before anywhere in the solar system. Besides its unusual coloring, Dione exhibits an orbital curiosity. It has a minute companion, Dione B, another satellite of Saturn in exactly the same orbit, but slightly ahead. Dione and Dione B are gravitationally locked together by a technicality of gravitational theory known as Trojan points. Unfortunately, Dione B remained tantalizingly out of range of *Voyager I*'s cameras.

By the time the spacecraft reached the moon Rhea, it had been accelerated to 45,000 miles per hour by Saturn's gravitational boost. Worried that pictures of Rhea would be blurred by *Voyager*'s velocity, scientists programmed the on-board computers to pan the spacecraft's cameras repeatedly. The technique worked, and the imaging team was able to study the icy surface of Rhea, nearly a thousand miles in diameter.

It was just as well that the pictures were sharp. While examining the moon's cratered surface, Lawrence Soderblom, the geologist on the *Voyager* team, found an important clue to the origin of the material out of which Rhea and possibly the entire Saturnian system were formed. The photographs showed that portions of Rhea had only small craters compared with the larger craters elsewhere on its surface. According to Soderblom, there was only one good way to explain how one part of Rhea could be relatively free of cratering: volcanism. Rhea's volcanism, he says, is "probably some form of slush and mush that flows out in the form of a slurry. But it had to flow out on the surface and blanket all the large craters. Then crater-

second. The shock was just below the threshold that, if exceeded, would have broken Mimas into fragments. Resembling a huge eye staring into space, this gigantic crater gives Mimas an eerie appearance.

Voyager I moved on to view the moons Enceladus, Tethys, and Dione. *Voyager* did not take high-resolution pictures of Enceladus because its trajectory carried it too far away. This enigmatic moon is roughly 300 miles in diameter and is apparently made of slushy ice.

Tethys is about 650 miles in diameter. Black-and-white spacecraft images reveal that one side of the moon is marked with what may be a very large impact crater, as Mimas is. On the other side is a chasm more than twice the size of the Grand Canyon. Like its sister moons, Tethys is thought to be made of ice. But the ice of Tethys is extremely cold. Unlike ice at 0 degrees Centigrade, which is relatively soft, the ice of Tethys, frozen to −200 degrees Centigrade, is very brittle. Tethys is cold and brittle enough to have been broken up by comet impact, but it would have reassembled under its own gravity into a fractured ball of icy debris. Indeed, there is evidence that such a fragmentation and re-collection occurred, for the huge chasm running across the moon is probably

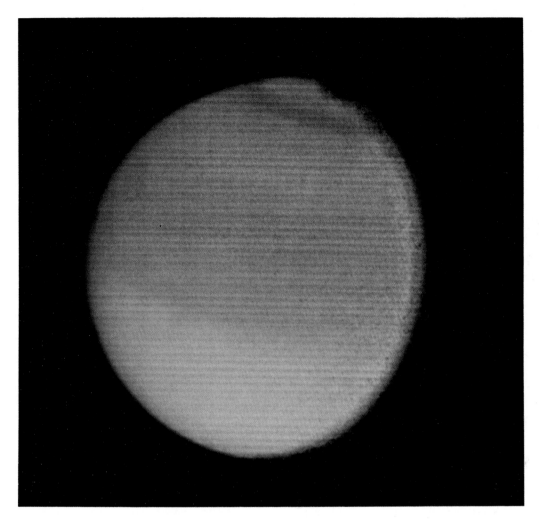

Titan, the largest of Saturn's moons. Titan falls within the same size range (diameter 2,000 to 4,000 miles) as Mercury, Mars, Earth's moon, and the four Galilean satellites of Jupiter. Of these, Titan has by far the densest atmosphere, as can be seen by the absence of any surface detail in this high-resolution image.

ing had to continue because this area that's been resurfaced still has a dense population of craters, but no large ones."

The hypothesis is that the surface of Rhea was formed in two separate bombardments. The first bombardment included objects that were large enough to make big craters that evenly peppered the entire surface of the moon. Then parts of Rhea cracked, and some process akin to volcanism caused new material in the form of slush to be forced out from the interior. This resurfaced a whole area of Rhea, erasing the original record of cratering. Then a second phase of bombardment started. This time, there were no large objects to produce big craters, so the resurfaced area was patterned only with small impact craters. Now, thousands of millions of years later, Rhea's surface bears a record of the two epochs of cratering that formed it.

The main objective of the entire *Voyager I* mission was a close encounter with Titan. Bigger than the planet Mercury, this giant moon of Saturn is almost a planet in its own right. Titan's reddish-brown atmosphere was known to contain the gas methane. And it was believed to be warmer at the surface than Saturn's other moons—about − 160 degrees Fahrenheit—and to hold liquid nitrogen boiling or erupting in hot springs or geysers. For years, astronomers had hoped that alien life might be found on Titan.

As *Voyager I* flew 2,400 miles over Titan, the imaging team had planned to photograph Titan's surface through gaps in the cloud deck. With this in mind, the spacecraft was programmed to cover the entire surface with high-resolution images. Unfortunately, the haze in the atmosphere was too thick for surface details to be photographed. In orange light, scientists found that Titan looked

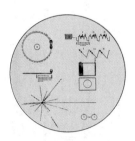

On board both *Voyagers* are identical gold-covered copper phonograph records that contain pictures, music, and messages from Earth. Scientists hope that they may be intercepted, deciphered, and answered someday by intelligent beings dwelling on some distant planet. Shown above is the etched cover of the record that gives playing instructions in binary arithmetic and elementary units of time.

like a fuzzy tennis ball, but at shorter wavelengths—in ultraviolet light—a few mysterious details emerged: a dark solar cap, a northern hemisphere that is darker than the southern, and an observable border between the two hemispheres right at Titan's equator.

Titan's surface secrets, whatever they are, remained guarded under a seemingly impenetrable layer of haze. But the *Voyager* team had had the foresight to equip the spacecraft with an infrared spectrometer, an instrument designed to look through cloud layers by measuring the heat emitted by Titan's gases. Since every gas can be identified by its unique infrared signature, readings from Titan revealed the presence of methane, ethane, acetylene, and hydrogen cyanide in the atmosphere. The existence of hydrogen cyanide was the least expected and most exciting discovery, according to Toby Owens of the imaging team, because it played a very important role in getting life started on Earth.

"You can polymerize it to make adenine, which is one of the bases of DNA, the famous double helix, which gets us all going," Owens says. "You can make amino acids if water is present, which is true on Earth but not on Titan. So . . . we're not saying that because we have found hydrogen cyanide in Titan's spectrum, we think life is beginning on Titan. But we do think that some of the chemistry on the Earth before life began is probably happening on Titan today. And if we could study it in more detail, we could learn something about pre-life chemistry on the Earth."

Some source of energy is needed in Titan's atmosphere, however, to ensure that chemical processes can take place. But what is this source of energy that bonds chemicals together on Titan and may have fused the complex organic molecules that precede life? *Voyager I* gave scientists a hint. A protective magnetosphere encloses all of Saturn's inner moons except Titan. This is the region of space around Saturn that is dominated by the planet's magnetism. A wind of highly energetic particles from the sun is constantly bumping the protective magnetosphere in

and out. There energetic particles of the solar wind may spark Titan's exotic chemistry, for it is in Titan's upper atmosphere that the chemistry appears to be happening.

According to Owens, the substances that are forming high in Titan's atmosphere are some kind of organic material, possibly polymers of acetylene and hydrogen cyanide. He expects that these small particles may aggregate into larger ones and rain out of the atmosphere like a slowly settling fog. But as this organic matter gets closer to the planet's surface, it passes through a region where methane may condense in the atmosphere.

"So we may have clouds of methane and methane rain on Titan," Owens says, "because in some respects, methane plays the same role in that atmosphere that water vapor does on Earth. So standing on the surface . . . we have these organic materials floating down—manna from heaven, as it were—and accumulating on the surface. We might expect pools of the liquid methane with some of this gunk dissolved in them and the shores lined with it. And the solid places would not be rock as we find on Earth or on the inner planets, but a combination of . . . ices covered by this organic material. It would be an extremely alien environment, totally different from anything that we're accustomed to."

In 5,000 million years the sun will run out of hydrogen fuel and expand into a much larger star, a red giant. It will enlarge to engulf the Earth, which will cease to exist. But Titan, farther out, will warm up, and the huge amount of water that is now trapped as ice will begin to turn into an ocean. If conditions stabilized, says Owens, it could stay an ocean for some time. It might also just totally evaporate, leaving a small rocky core, rather like a large satellite. Sadly for any life that might have started on Titan, there will be only a few million years of warmth before the sun contracts to a tiny vestigial white dwarf. Then Titan will refreeze.

In March 1981, a new mathematical analysis of the *Voyager I* data was completed. It revealed that most of Titan's atmosphere,

On January 24, 1986, *Voyager II* will fly by Uranus (shown here in a computer simulation). *Voyager*'s photographs and instrument readings will be virtually the only information that we possess about the third largest planet. Three and a half years later, *Voyager II* will sail by Neptune at a distance of 3 billion miles from Earth.

which is thicker than the Earth's, is composed of nitrogen. Evidence is growing that Titan has not pools but a huge ocean of liquid methane up to a half-mile thick. At the poles, this liquid natural gas may have even frozen into methane ice. Perhaps if a spacecraft is ever sent to land on Titan, it will have to be a submarine.

In the geography of space Titan is in our neighborhood. It not only provides a vision of our own planet just before life began, but also suggests that if Titan is so near to life and so close to us, somewhere out there is a warmer Titan with life. But *Voyager I*'s mission has been completed. The spacecraft is heading up and out of the solar system. Its next encounter is in 41,000 years' time, with a star in Ursa Minor known as AC + 793888.

Voyager I's twin, *Voyager II*, following one year behind on a different trajectory, encountered Saturn in August 1981. It repeated the observations of *Voyager I* from different distances and angles. *Voyager II* is programmed to visit still more distant planets: Uranus in January 1986 and Neptune in August 1989. Then it, too, will leave our solar system. But both *Voyagers* will continue on a journey, one that could possibly last a billion years. Perhaps in all that time, they may yet again come to rest upon a friendly shore.

"When Columbus stood on the shore of Europe and looked across the Atlantic," says Cyril Ponnamperuma, "everybody thought that he was on a foolish quest. Modern space explorers are like Columbus: they are looking out there. Who knows what might come out of all this? One single word from a civilization beyond. . . . Eventually that kind of communication might be of great importance to us here on Earth."

Given our present technology, we cannot follow the interstellar tour of the *Voyagers*. But with the hope that the spacecraft may one day be intercepted by intelligent life, the *Voyagers* are each carrying a record of our species—twelve-inch gold-plated phonograph records containing two hours of sound, 118 pictures of life on Earth, and greetings from the peoples of the Earth in sixty languages, including this salutation from Kurt Waldheim:

"As the Secretary General of the United Nations, I send greetings on behalf of the people of our planet. We step out of our solar system into the universe seeking only peace and friendship, to teach if we are called upon, to be taught if we are fortunate."

The Sunspot Connection

The worst drought in United States history made a dust bowl of the wheat-producing states in the Midwest in the mid-1930s. Similar droughts hit the same region 20 years earlier and about 20 years later, causing some scientists to wonder whether the droughts were related to the 20- to 22-year cycle of sunspot activity.

In the 1930s, huge clouds of black, choking dust swirled across America's Great Plains after months of rainless weather turned fertile wheat fields into a dust bowl. One-fifth of the nation's land mass was affected. It was the worst drought in U.S. history.

But it wasn't the only one. Twenty years earlier, the same area had suffered a similar spell of dry weather, and twenty years later, the drought returned again. In fact, by the mid-1950s, the western plains had had eight successive droughts twenty to twenty-two years apart, an apparent pattern that scientists in the past decade have pondered with increasing curiosity. Some researchers anticipated another drought in that region in the mid-1970s, and not only out of habit. They speculated that the pattern was linked to another cycle: the appearance and disappearance of spots on the surface of the sun.

The spring of 1976 started dry, threatening America's wheat crop and bearing out scientific expectations. But just before summer, the rains came to the plains, ending

fears of devastation. Still, there was a drought of record proportions beginning that year and extending into 1977; it occurred in California. For scientists, the timing was right but the geography was slightly off, enough to render their observations inconclusive. The question remained: do sunspots directly affect the Earth's climate? In the view of some scientists, it is one of the most important questions an astronomer can ask.

"There's only one thing that society really needs from the sky," says Jack Eddy, a solar physicist at the High Altitude Observatory in Boulder, Colorado. "How does the sun change? That is, how does it change in ways that affect the Earth? We haven't been very good at answering that question."

Until recently, the question itself was regarded as heresy. Indeed, when Charles Greeley Abbot, a pioneer physicist who spent most of his 101 years making meticulous measurements of the sun's energy, died in 1973, his view that the sun's output was cyclic rather than constant was largely disregarded by the scientific community. After all, he undermined thousands of years of mythology about the star that gives the Earth life, and even among scientists such myths die hard.

The sun was seen as a constant, unchanging star, so predictable and reliable that Stonehenge, the massive stone calendar on England's Salisbury Plain, marks the sunrise on summer solstice as accurately today as it did when it was built 4,000 years ago. When Galileo and other astronomers looked at the sun through the newly invented telescope in the seventeenth century, they were astonished to find that it was blemished, spotted with a kind of solar acne. So shocking was this observation that Galileo was reluctant to publish it for fear of reprisals from the church.

In the mid-nineteenth century, astronomers keeping watch on those spots discovered that they seemed to come and go on a regular basis. Sometimes the sun appeared virtually covered with spots—sunspot maximum. Sometimes it was as smooth and perfect as mythology had imagined—sunspot minimum. Eventually, scientists determined

that the length of the cycle between two maximums was eleven years.

Sunspots remain an important scientific curiosity, and while the church no longer has a stake in the sun's purity, the spots are nonetheless the subject of intense scientific speculation and debate. For if the sun, which is so fundamental to life on Earth, undergoes constant and periodic change, it is hardly unreasonable to expect such change to have an important impact on our planet. And any connection we might be able to make between the sun's cycle and weather and climate patterns on Earth might enable us to plan for and minimize the effects of such events as the great drought of the mid-1930s.

Sunspots are dark blotches on the sun's surface that usually extend for thousands of miles and are visible because they are cooler than the regular surface of the sun. When sunspots are present, scientists have determined, the sun's surface is especially active. Huge prominences of gas leap tens of thousands of miles into space. Flares explode across its surface with the force of a billion hydrogen bombs. Both flares and prominences are much more common at sunspot maximum, and they seem not only to make the sun's surface boil, but also to make the corona—the sun's atmosphere—appear more turbulent all over, thus creating a symmetrical halo. Thus, not only does the sun change its spots every eleven years; it changes all over.

Scientists have used massive telescopes, satellites, and outer space experiments on *Skylab* and other space missions to help them understand these changes. One outcome is the discovery of the solar wind, a constant outward flow of particles from the sun's corona extending far out into space. This wind is extraordinarily thin—a billion-billionth of the density of the Earth's atmosphere—but scientists believe that unusual disturbances in this wind result in showers of solar particles in the regions near the North Pole and the South Pole. These showers produce one of nature's most spectacular displays: aurora borealis, the northern lights. Some physicists have sought to connect strong gusts in the so-

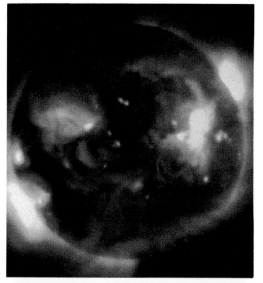

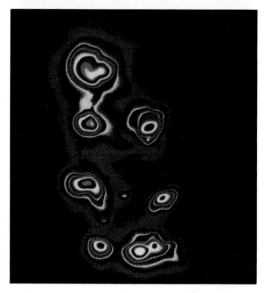

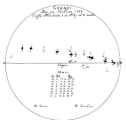

Italian astronomer Galileo Galilei (1564–1642) (*below*) used one of the first telescopes to see the sunspots he recorded in his notebook (*above*). Today we use massive telescopes, X-ray and computer-enhanced photography, and satellites to capture the views of sunspots shown at left, center. The top and bottom photographs were made by the solar telescope on *Skylab*.

Solar flares erupting from the sun's surface on this computer-enhanced photograph (above) may influence weather changes on Earth, scientists believe. Other scientists search for patterns or clues to the sun's influence on our climate in the rings of trees, which provide a kind of annual weather record: wide rings indicate a good growth year; thin rings a bad one.

lar wind to the development of large storms on Earth. While they have found some interesting statistical evidence, they have yet to explain how something as delicate as the solar wind can generate the kind of energy on Earth necessary to create a storm.

Solar flares, which generate much more energy than solar wind, also emit particles that rain down onto our atmosphere, creating nitrogen oxides that eat ozone molecules at a remarkable rate. Since a thinner ozone layer lets more ultraviolet radiation through to the lower atmosphere, scientists speculate that the flares make weather changes possible.

But neither flares nor solar wind explains the apparent twenty-two-year drought cycle in the American West. Evidence supporting these cycles themselves—both on Earth and on the sun—includes some puzzling and intriguing inconsistencies. To find a correlation between the two cycles, scientists

have gone beyond the reports of droughts to one of the most reliable records of weather history available—tree rings.

The late Andrew Ellicott Douglass, founder of the Tree Ring Laboratory at the University of Arizona and the first scientist to make the connection between tree rings and climate (thick rings, good growing year; thin rings, poor year), spent much of his life gathering data and looking for evidence in the rings of an eleven-year cycle, to match the sunspot cycle. He believed he found such evidence, but more modern, computer-assisted analysis and the accumulation of additional information have raised doubts. However, another researcher, who used tree rings to study drought in the western plains, found unexpected confirmation in nearly three centuries of data of a twenty-two-year cycle—exactly twice the normal sunspot period. Why would a drought occur at every *other* sunspot minimum? Scientists don't know. And while the statistics on the twenty-two-year cycle represent a strong correlation between sunspots and weather, scientists have yet to find a physical explanation for the connection.

Meanwhile, other research has raised questions about the regularity of the sun's cycle. Despite minor fluctuations in the length of sunspot cycles, solar scientists have come to assume that the sun is regular. This was only an assumption, but it was never really questioned until Jack Eddy of the High Altitude Observatory read about a nineteenth-century English astronomer named E. W. Maunder who had claimed that for about seventy years, between 1645 and 1715, the sun had completely lost its spots. Eddy thought the notion ridiculous and set out to disprove it. But the more he dug into historical records, the more support he found for Maunder's contention. He collected tree rings from the seventy-year period and measured the amount of carbon 14, the radioactive form of carbon that scientists use to determine the age of scientific and archeological artifacts, contained in the rings. He theorized that if sunspot activity were particularly low, more cosmic rays would penetrate the atmosphere, causing the production of higher amounts of radiocarbon. Indeed, he found abnormally high amounts of the element in the rings.

Eddy was convinced that Maunder's report was accurate. Interestingly, during that same seventy years, there was a prolonged cold spell so unusual that it is often referred to as the Little Ice Age.

Eddy has also reanalyzed the once-discredited records of Charles Greeley Abbot and found evidence that everyone else, including Abbot, has missed. Eddy thinks that Abbot's measurements of the output of solar energy over the thirty-five-year period between 1920 and 1955 yield no strong evidence of an eleven- or twenty-two-year cycle, but they do point to something else. "I see a gradual increase in the value of the solar output during that time," he said. "That increase . . . is about one-half of one percent per century. That's almost exactly what is required to explain the gradual warming of the Earth during the same period of time."

The Earth's gradual warming trend this century has never been explained, and it's never really occurred to anyone that the sun may simply be getting hotter. In addition, Eddy also found in Abbot's data that as the solar constant, or solar energy output, has increased, so has the number of sunspots at

each peak of the sunspot cycle. The whole sun has been getting much more active as the sun's radiation has seemed to increase. Eddy's speculation about this increased output was confirmed unequivocally in 1981 with new data from the solar maximum mission satellite. That doesn't specifically explain the cycle of drought, but it does suggest that sunspots may account for some long-term changes in the Earth's climate.

The sunspot-weather connection, then, may be as simple as this: as sunspots increase, the sun gets hotter and the Earth's climate is slightly warmer; as the sunspots decrease, the sun and the Earth get cooler. The effect may draw itself out over hundreds of years; it may also have an eleven- or twenty-two-year cycle. The question is, then, what is the sun doing now? Is it getting warmer or cooler? We don't know. At the moment, the sunspot mystery remains unsolved.

E. Walter Maunder, (above), a nineteenth-century English astronomer, maintained that the sun had completely lost its spots for seventy years between 1645 and 1715. During those same years, there was a prolonged cold spell that is often referred to as the Little Ice Age. Artists of the period often used cold winter scenes, such as *Skating on a Frozen Lake (left)* by P. H. Avercamp (1585–1663), as their subjects.

Charles Greeley Abbot (1872–1973), former director of the Astrophysical Observatory of the Smithsonian Institution, spent much of his life making observations of the sun. He believed that its output was cyclic rather than constant, a view largely disregarded by the rest of the scientific community as recently as the year of his death, but now the basis of much solar research.

Is the Sun Getting Hotter?

The chart below, compiled by Jack Eddy, a solar physicist *(right)*, records the annual mean sunspot number from 1610 to 1980 and shows the "Maunder Minimum" (1645–1715), which corresponds to the Little Ice Age. Eddy did extensive research into data compiled by both Maunder and Abbot to produce this record. It does not support a 20- to 22-year sunspot cycle, but Eddy believes it does suggest something else: the sun may be getting hotter. The record shows a small, gradual increase in solar energy output—about one-half of 1 per-cent per century—an amount consistent with the gradual warming of the Earth during the same period. As the sun's energy output has increased, Eddy found, so has the number of sunspots at each peak of the sunspot cycle.

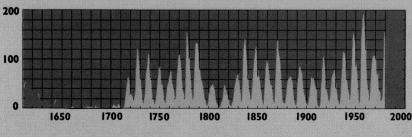

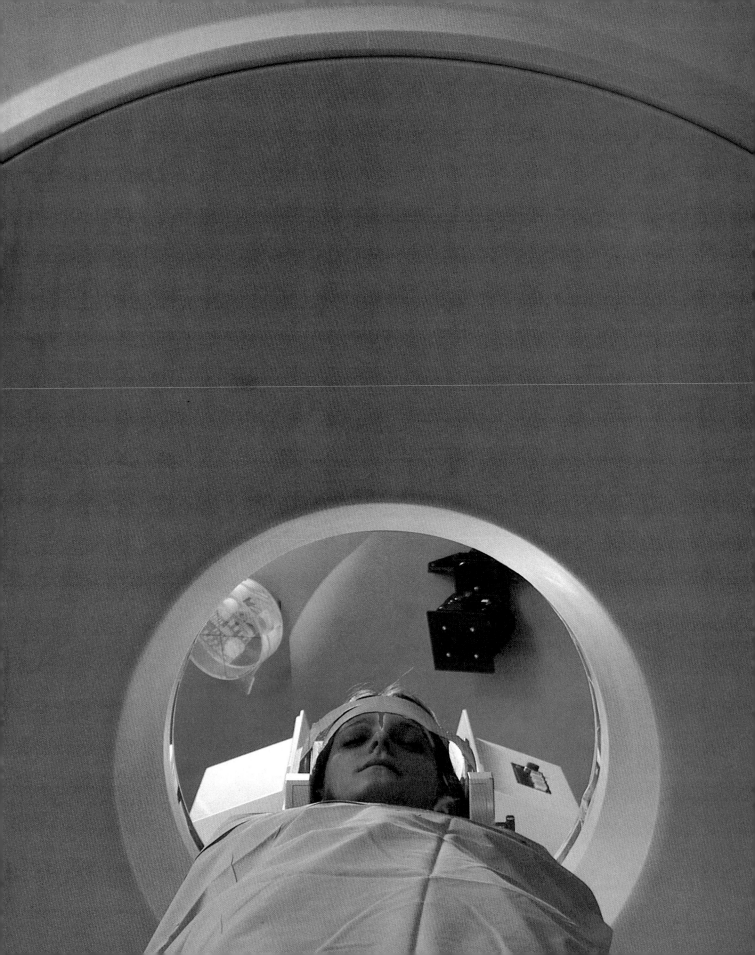

Science is more than discovery for discovery's sake. Although we often stand in awe of our origins and we find space exploration thrilling, it is probably the application of science to our everyday lives that affects us the most. New discoveries and techniques can ensure better health, provide safer environments, and perhaps even prolong life itself. Ultimately we rely on science in a very personal way; we depend on it to heal ourselves and our planet.

In pursuing the practical aspects of their field, scientists find themselves playing two different roles. First, they are detectives hunting for clues that will solve the mysteries of disease. Often they have no idea who, or what, the villain is; all they may have to work with is a partial, muddied footprint. And just like detectives on a case, scientists find that luck plays an important part in the serendipitous discoveries that signal major medical breakthroughs. At the same time, they have become the guardians of our planet, defining and protecting the precarious balance of nature's delicate ecosystem. Their concerns with health go beyond conquering disease to preserving–and in some cases restoring–the integrity of all life on Earth. As modern society becomes ever more complex, their tasks and responsibilities grow awesome.

A woman is examined
with Computerized
Axial Tomography (or
CAT scan), a vast re-
finement of the normal
X-ray. CAT scanners

The Keys to Paradise

It is not by coincidence that medical crisis often precedes medical discovery: emergency catalyzes innovation, and Mother Necessity is indeed the parent of wily invention. When, a decade ago, biochemical researchers first began to question the structure and mechanisms of the class of drugs called opiates, none dreamed that their eventual answers would disentangle some of the deepest knots in the human brain.

By the late 1960s, drug addiction in America had risen to near-epidemic proportions. Wars of one sort were rampaging through the physical world; wars of another were tyrannizing the psychological. Many Americans found release from unbearable outer and inner chaos in the opiates heroin and morphine. Both drugs are derivatives of the juice of the poppy plant; both provide a sense of euphoria and a promise of paradise for those haunted by visions of fallen lands.

Eden has never been a state people own for long, though—and this paradise is particularly costly. The same drugs that induce euphoria also depress respiration and blood pressure; the same analgesics that deaden pain cause impotence and constipation. Heroin and morphine act upon the central nervous system; in the brain and spinal cord, their heavenly emotional effects exact devilish physiologic prices. When the problem took on crisis proportions, federal funds were finally redirected, and intense research into opiate action was begun.

Morphine exists in two forms. Like right and left hands, the forms mirror but do not duplicate each other: they are nonsuperimposable. One form exerts the traditional euphoric and analgesic effects associated with opiates. The other is absolutely inert. Since morphine acts first upon the brain, researchers began tracing its effects from the top downward. Somewhere in the brain cells, receptors had to exist that were capable of combining with morphine molecules; and—if they existed—these morphine receptors molded to the shape of the drug precisely, as a lock molds precisely to a key. In this case, the morphine key opened emotional doors easily. But where was its complementary lock?

Tissue culture of nerve cells from a human brain.

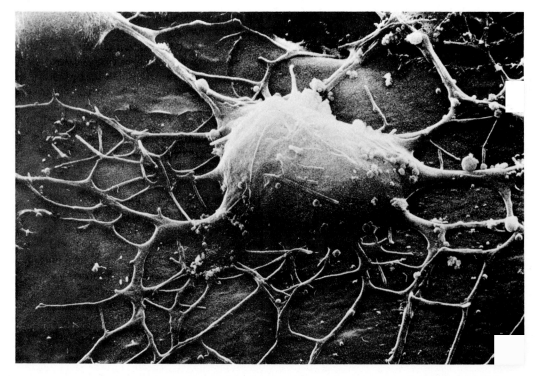

All nerve cells share a fundamental structure: they have a central cell body, a set of stringy dendrites that receive impulses, and a single elongated axon that transmits these impulses to the next cell. Two nerve cells meet and forward signals at their synapse, a microscopic junction between the end of one cell's axon and the beginning of another's dendrite. Could the receptors be housed in these borders? Did specifically shaped molecules bind and stimulate the action of morphine here?

Theory is a mental challenge: it requires a clear head and an armchair. Locating a single molecule within a cell, though, is somewhat similar to locating a single keyhole within an entire city. The hunt calls for techniques of nearly incomprehensible sensitivity, while the hunters themselves must be imperturbable enough to withstand scent after false scent, alley after blocked alley.

Under Dr. Lars Terenius, a research team searching for steroid receptors at the University of Uppsala in Sweden had, by the late 1960s, perfected a technique called radioactive labeling. When the search for opiate receptors reached international shores, Terenius and his staff turned their hound into a slightly different wind. Following a complex sequence of homogenization and centrifugation, the team separated thousands of rat brains into extracts, each composed of an isolated component of the original organ. Some extracts, containing nerve cells, were further purified for their fragments most likely to house morphine receptors. These fragments were shaken with radioactive molecules, tags whose visibility guaranteed that their activity in the brain would not go unnoticed.

The extracts were thoroughly washed, to flush out any loose molecules. All radioactivity then measured was from molecules bound too tightly to be rinsed away: molecules bound to the elusive morphine receptors. American researchers had reasoned that if the receptors existed, they could be located. Terenius reversed the premise: if the receptors were located, they had to exist. Activity affirmed existence. I think, therefore I am, said the philosopher Descartes; molecules bind, therefore receptors exist, answered the biochemist Terenius.

The best research is both original and redundant—discovered by one team, confirmed independently by another. While Terenius was quantifying radioactivity in Sweden, an autonomous group at Johns Hopkins University in Baltimore had arrived at the same labeling technique in the same hunt. Although the Baltimore team netted and published the first set of technically perfect data, the Swedish team announced its findings almost simultaneously, as did a third team working in New York. All the evidence confirmed a single conclusion: morphine receptors abound in the brain.

Science, though, is not a process simply of problem and conclusion; it is the marriage of one good question to another. If receptors served as the keyholes for synthetic morphine, did natural keys also exist? Nature holds no truck with biological decadence; she permits no cells the luxury of life without practical function. If receptors were pre-

Endorphins fit receptors in the brain in the same way a key fits a lock. Specifically shaped molecules in the brain, in other words, bind and stimulate the action of certain chemicals. Receptors are strings of amino acids folded into a unique geometrical template of loops and crannies to which only a precise complement—a chemical with the identical amino acid pattern—can attach. If the key fits, the lock will open, setting in motion the drug's effects.

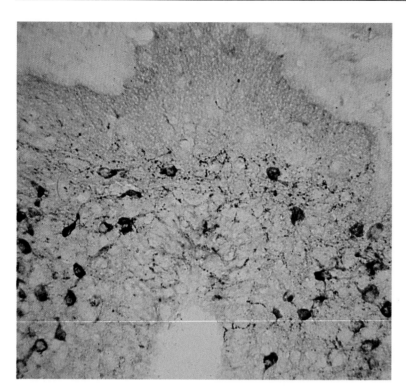

Presence of beta endorphin can be demonstrated in the arcuate nucleus cell group of the rat hypothalamus by a technique that uses antibodies to tag the B-endorphin-containing cells with the enzyme horseradish peroxidase, which catalyzes the polymerization of the brown pigment diaminobenzidine. This photograph shows the preparation under a light microscope. (Each cell is approximately 25 micrometers across.) Brown pigment marks many nerve cells, which appear to be the only cells in the rat brain that contain B-endorphin. The space at the top of the field is the third ventricle, a fluid-filled chamber inside the brain.

existent in the brain, then, by her pragmatic laws, they bound morphinelike substances also preexistent in the brain. Where were these chemical keys, and what was their purpose?

Any *Physician's Desk Reference* contains at lease one page ragged from use: it describes Naloxone, the drug used as an antidote to morphine. A single injection reverses the effects of morphine overdose and breaks the opiate coma. It also breaks the opiate euphoria and dampens its analgesic effect. Pain returns almost immediately after treatment, and pain thresholds are reduced dramatically. Naloxone acts as a copy of the morphine key; it competes with and displaces it in brain cells. Somehow, its affinity for the opiate receptors is more powerful than the opiate molecules themselves; as an imitation, it offers a more persuasive deal than the original. If opiates are inhibited by Naloxone, then can't we assume that a chemical whose effects are reversed by Naloxone is an opiate?

In Aberdeen, Scotland, John Hughes was among a research team disassembling the puzzles of addictive behavior. Hughes had found animal tissue in general a responsive testing ground for drugs, and one tissue, in particular, responsive to morphine and Na-

loxone. The vas deferens is a small duct in males that conveys sperm from the testes to the sperm sac. Hughes attached one end of a mouse deferens to an electric strain gauge, but kept the other fixed. When he applied an electric shock, the tissue contracted. Morphine reduced the contractions. Naloxone stimulated them again.

Hughes used extracts from thousands of pig brains, purifying and repurifying the extracts by separating out molecules by weight. The extracts were tested, directly and indirectly, for their morphinelike properties: directly by their effect upon the vas deferens's response to shock, and indirectly by the tissue's subsequent response to Naloxone. When one small bottle of extract was summarily retested on its way to the garbage can, it provoked a tissue response quite opposite from the one of its original trial. As if under the influence of morphine, the vas deferens twitched less when stimulated, then renewed its contractions when given Naloxone. In true storybook style—science enjoys its share of rescue fantasies come true—the natural chemical key was salvaged in its crudest form. Hughes called the extract "enkephalin," meaning "in the head." Rightly, though, it belonged to a different part of the body: here, at last, was the shoe that fit.

The extract had been identified; yet it was still chemically anonymous. It took another year of rigorous purification to confirm that enkephalin was not, as the Scottish team had assumed, a single compound. It seemed, instead, to be a peptide, a string of complexly folded amino acids. Within any peptide, the sequence of acids specifies its final shape, and this shape in turn directs the peptide's activity. The science of molecular interaction is as much architecture as chemistry. Amino acids fold into a geometrical template of loops and crannies to which only a precise complement can attach. If the key can't fit, the lock won't budge. Biochemistry is ultimately that simple. Enkephalin was a peptide, but until its amino acids were named and numbered, it was a peptide misunderstood at best.

Wise scientists are never ashamed to wear the same wardrobe twice; often, success follows as much from the recycling of old and effective investigative approaches as from the invention of new. Howard Moss, a British researcher, responded to the anonymity of enkephalin with a familiar analytic technique—mass spectrometry. In a spectrometer, the components of an unknown chemical sample are separated by weight to produce a spectrum recognizable to specialists. Individual elements and groups of elements can be interpreted and identified. Measuring the last few grams of available enkephalin at the University of Cambridge, Moss uncovered a surprising pattern: the spectrometer was actually analyzing two peptides, not one. Each pentapeptide contained five amino acids; and though chemically distinct from morphine, the two peptides, assembled, took a nearly identical shape in space. They fit the morphine receptor as accurately as any opiate; they turned the same lock in the brain and set in motion the same magical analgesic effects.

The idea was fantastic, as fantastic as the discovery of the morphine receptors. Once again, though, separate research and redundant conclusions turned the fantastic into the credible. In England, scientists at the National Institute for Medical Research (NIMR) were studying a molecule of unusual length. The molecule contained a series of shorter peptides: possibly, the group reasoned, new hormones synthesized by the pituitary. Located directly underneath the brain, the pituitary gland responds to a complicated cycle of needs and signals by releasing its hormonal contents directly into the blood. They circulate through the body, but act upon only specific target organs. When the long molecule was analyzed, something in its order of amino acids seemed familiar; some sequence rang chemical bells. What was that five-acid segment built into the molecule's structural ladder? And why was it beginning to look so familiar?

Nature is too thrifty not to repeat a good idea. In addition to containing the enkephalin sequence, the mystery molecule proved to be one hundred times more potent an analgesic than morphine: the most powerful natural painkiller currently known. At the University of California, a second team had independently isolated the same molecule fifteen years earlier. They had called it, tentatively, beta lipotropin; pondered but could not clarify its functions; and

Endorphins

Endorphins are substances produced by the brain and stored in the pituitary gland that have powerful opiate effects on the body.

ACTH, for example, is a pituitary hormone. When released, it triggers a series of chemical and hormonal changes that help the body adjust to sudden or prolonged stress. Contained within a molecule of ACTH is the 91-amino-acid protein beta lipotropin.

Within this long molecule is a 31-amino-acid peptide, beta endorphin. A contraction for endogenous (produced by the body) morphine, B-endorphin behaves as an opiate in the body and is, in fact, the most powerful natural analgesic yet known.

Hidden within it is a tiny 5-amino-acid fragment, methionine-enkephalin, a pentapeptide with its own analgesic effects.

Met-enkephalin has a sister pentapeptide, leucine enkephalin, not found in B-endorphin. The two enkephalins are distinct molecules with distinct receptors in the body. Met-enkephalin fits the M (mu) receptor, which controls analgesic action; leu-enkephalin fits the S (delta) receptor, which affects emotional behavior.

Jogging beside the Charles River in Cambridge, Massachusetts.

A Runner's Paradise

Joggers and junkies: both claim an intensely pleasurable high from their separate activities. The junkies' high is one effect of exogenous morphine; could joggers receive theirs from endogenous morphine? Both the brain and pituitary gland release endorphins in response to stress and pain, such as that of childbirth. Is there a threshold of athletic stress after which endorphins are also released? One rule of thumb for joggers is: though the first three miles are rough, the remaining distance is almost restful. Is this an effect of time-released natural opiates, or the attitude of masochists who take forty laps to admit it hurts? Evidence indicates that endorphins are indeed the keys to paradise for sufferers of pain as well as for joggers. Patients

plagued with chronic pain have already begun to reap the benefits of this new knowledge of body chemistry through techniques developed to electronically stimulate endorphin release. One treatment recently developed from experiments with rats is to give electrical stimulation to the pain center of the brain itself. When electrodes are

placed in a patient's brain and electrical stimulation has alleviated the patient's pain, a sample of fluid from the brain reveals an eightfold rise of beta endorphin or enkephalin levels. Such artificial stimulation of endogenous opiates may someday allow even the most lamentably bedridden sufferers to enjoy the jogger's painless high.

shelved it—with realism if not finality—in favor of more accessible projects. When enkephalin was identified and named, they realized that their mystery molecule contained within it the same pituitary molecule being unraveled at NIMR. They retained the name B-lipotropin for the long molecule and called the 31 amino acid segment within it "beta endorphin"—a contraction, officially, of "endogenous [naturally produced in the body] morphine." With the title, more than a decade of international work was conferred legitimacy at last.

Since 1975, the year in which Hughes rescued the first sample of enkephalin from an unglorious end, other endogenous morphinelike peptides have been located. The pentapeptide enkephalins are found in the brain, spinal cord, and gut. Their effective life is brief, only a few seconds. Longer peptides, beta endorphin most actively among them, are produced in the brain but stored and released by the pituitary gland. Their life is more prolonged—two to three hours—and their functions are more complicated.

The roles that beta endorphins play in the body are mainly protective. In response to pain, their concentrations in body fluids increase up to eightfold. Pregnant women, one population under particular physiologic stress, have blood levels consistently higher than normal. They also have an extra lobe of the pituitary gland, a storage barn, perhaps, for the additional painkillers they need during labor and delivery. Maternal endorphins, like many chemicals, are capable of crossing the placenta selectively during pregnancy. Research into fetal effects is still speculative, but awfully seductive: is the blissful prenatal state—that psychoanalytic backdrop against which all adult disappointments are measured—nothing more than an overload of endorphins?

Placebo effect is another door whose locks may turn, eventually, in response to the beta endorphin key. When patients recovering from minor surgery are given an innocuous saline solution advertised as an analgesic, one in three claims a reduction of

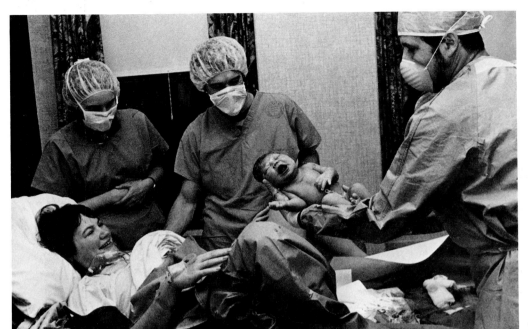

Pregnant women have blood levels of beta endorphins consistently higher than normal— and the capability to store extra doses of these natural painkillers needed during labor and delivery.

pain. Naloxone reverses the soothing effect of genuine analgesics, of course; but it also reverses the effect of some placebos. If an imaginary cure produces real chemicals, placebo and Naloxone must bind the same chemical receptor in order to compete for its fit; if the psychology of expectation affects the physiology of response in a legitimate and quantifiable way, then their clinical implications must be compelling. Medicine suddenly becomes more of a mutual business. When physicians adjust analgesic dosages to a patient's concept of his suffering, regardless of the objective nature of his injury, will treatment become more or less successful?

Mental illness is a third house in which jammed entrances are opening to now familiar keys. Rats injected with large doses of beta endorphin freeze promptly into the rigid posture characteristic of catatonia. A single injection of Naloxone relaxes their stance and returns their mobility. Cerebrospinal fluid taken from schizophrenic patients also contains elevated endorphin levels. And in several cases, patients suffering from auditory hallucinations claimed a spontaneous disappearance of bells and voices when given Naloxone. The music by which they lived their secret lives vanished; someone, as one relieved patient explained, had turned the radio off.

Subsequent testing has shown a connection between schizophrenia and Naloxone considerably less direct than the one between catatonia and Naloxone. If the mechanisms of delusion are more sophisticated chemically than those of paralysis, then Naloxone may need modification in order to target certain sites of the brain selectively. Nonetheless, the marriage of psychiatric and biochemical questions seems off to a fine start.

Finally, obsessive behaviors like alcoholism, overeating, and—coming full circle—drug addiction may also be rooted in an endorphin imbalance. If the blood of addicts is deficient in endorphins, the dependencies may be nothing more than the overshooting of an otherwise normal, compensatory instinct.

The history of enkephalins and endorphins has been one of doors opening upon more doors, opening upon more doors that finally lead to the first threshold again. Exogenous opiates led researchers to endogenous ones; the endogenous opiates led, in turn, to a deeper understanding of the exogenous. Discovery has unsnarled some of the tangles of pain and its cessation, pleasure and its enhancement, behavior and its deviations. One wonders, sometimes, if Necessity is ever surprised by the circularity of her own inventions. Yet one also knows she has little time for such feelings; she is as busy with conception as scientists are with delivery.

Best of Both Worlds

In our culture, we are accustomed to sophisticated prescription drugs containing a variety of inscrutable chemical ingredients. Few of us realize that many of the drugs we use today originally came from forests or gardens instead of large pharmaceutical laboratories. Valerian tea is a sleep inducer from the valerian root; digitalis, used in cases of congestive heart failure, comes from the foxglove; and oral contraceptive drugs are extracted from the black yam.

Chief Fagbenro of Lagos, a high priest of Ifa and a prominent traditional healer.

The revolution taking place in Nigeria today is one of culture mix rather than clash; experimental establishment of a dual health care system that draws from tradition and modern science and offers the people of the most populous nation in black Africa the best of both worlds.

Under British colonial rule, the practice of traditional medicine was discouraged or sometimes forbidden in Nigeria. Traditional healers were called "witch doctors" by the colonizers, who viewed their medical practices as inferior. Since independence, the Nigerian government has decided to give these healers official recognition. Although discouraged, the practice of traditional medicine never died out during the colonial era. The government's current position is a recognition of reality—that healers serve a great majority of the Nigerian people. Nigeria's experiment is a chance to see whether developing countries can make use of a great resource: traditional doctors who have inherited centuries of folkloric knowledge about medicinal drugs.

Traditional healers practice a rich and ancient art based on an oral literature. It is a medical lore that uses herbs and roots from which are derived basic drugs, as in the West, but that has another vital element: the spiritual. It is not uncommon in Nigeria to follow up a hospital treatment with treatment by a priest, for the equally important cure of the soul.

The members of Nigeria's National Herbalist Association are pooling their knowledge about the medicines they use and the dosages they prescribe. Under colonial rule, these healers practiced in great secrecy. Now they are eager to share and compare their knowledge—and to cooperate in the modern research being carried out to systematize it. Chief Fagbenro, a healer and an Ifa priest, is an excellent advertisement for his own cures. At seventy years old, he has had to go to a doctor of modern medicine only once, when one of his legs had to be amputated. He's quite willing to admit the value of orthodox surgery, but prefers to live by the ancient traditions. He believes that traditional medicine has the advantage of using nature's own cures to conquer human ailments. "But," he points out, "all medical practices have their merit."

The new collaboration between the herbalist and the modern doctor has uncovered previously unrecognized benefits of traditional cures. The fagara root, for example, has long been used in Nigeria to clean the teeth. While testing the root chemically, Western-trained scientists discovered that, brewed and drunk like tea, it also appears to

A Western-trained psychiatrist, Dr. Olatawura of the University Teaching Hospital in Ibadan, greets his traditional counterpart, Chief Salami Foworade, a healer specializing in treatment of mental illness, at the latter's clinic in Ibadan. Traditional and Western medical knowledge are shared in Nigeria in a brave experimental coalition, encouraged by the Nigerian government and the World Health Organization.

Herbs and roots form the basis of a tradition of healing that has developed over the centuries in Nigeria and continues to refine its methods of treatment in collaboration with Western knowledge of many of the drugs derived from these very plants.

combat the genetic blood disease sickle-cell anemia.

Scientific methods are also classifying and verifying the healing properties of other herbal remedies. The oldenlandia root, used by traditional healers to accelerate labor contractions in pregnant women, has been tested in the laboratory by university-trained scientists, who have used it to produce similar contractions in pregnant rats. A tea brewed from the leaves of the neem tree combats malarial fevers. Socopa, known in the West as redberry, helps cure jaundice. Alukrese, a creeping plant, is used to prepare the Nigerian equivalent of iodine. Its leaves, when crushed and applied to an open wound, stop blood flow and kill bacteria. And oruwa—whose botanical name is sincona—is mixed with water to form a healing potion that has been found to combat yellow fever.

In the marriage of traditional and Western medicine, neither partner reigns superior. Where one fails, the other succeeds. One of the strengths of traditional medicine is its spiritual emphasis. Western doctors have recently begun to acknowledge the importance of psychological factors in maintaining health. But the Nigerians have understood this connection for ages. Traditional Nigerian healers view physical illness as an outward manifestation of a metaphysical problem.

While traditional medical practitioners like Chief Fagbenro are helping Western scientists catalogue an international pharmacopoeia, Western medicine makes contributions to the promotion and maintenance of public health in Nigeria in areas previously unappreciated by traditional practice. Concepts of patient hygiene and public sanitation to provide a cleaner environment are crucial to Nigeria's public health policy. Although taken for granted in industrial nations, the keeping of written records and the immunization of children against common diseases are becoming as important to Nigerian mothers as thanking the gods in song.

The Barefoot Doctors

In the mid-1950s, lethal viral epidemics were routine occurrences throughout the West. Polio and influenza were killing thousands. World health organizations tallied the fallen helplessly: antibiotics were powerless against the invasion. Meanwhile, in an area of China twenty miles square, one out of four people was dying from cancer of the esophagus—the tube connecting the throat with the stomach. The incidence of this cancer was higher there than anywhere in the world. No one understood why.

In 1959, a small interdisciplinary team of scientists gathered in the valley of Lin Xian, China. Esophageal cancer had been a constant among the peasants here for more than 2,000 years. Yet thirty miles from Lin Xian its incidence dropped by half, and 300 miles farther it was down to normal levels.

Half the scientists were local, half from the Cancer Institute in Peking. Among them

noted the foods they favored and the ways they prepared them. They looked at the crops they grew and how they stored them. They measured chemical levels in the soil and nitrite levels in the water. They even took a thermometer to the soup, following up an old folk rumor that its excessive temperature caused esophageal burning, which in turn caused esophageal cancer.

The scientists intended to organize their clues into four categories: chemical carcinogens in cigarettes or pesticides; environmental carcinogens such as ultraviolet light or viruses; local behavior that brought the peasants into contact with either type of carcinogen; and physiologic conditions that prevented their internal defenses from working. If they could identify cancer-related factors individually, they could also block them.

The peasants survived on a diet of maize, wheat, and turnips, supplemented by delicacies they made themselves. They prepared persimmon cakes by wrapping the fruits in husks of sharp, prickly wheat before drying them solid. They pickled vegetables into a sauerkraut of greens without salt or vinegar, using a process of natural acidification that took weeks. They steamed their maize bread for hours and, the team discovered, in so doing concentrated nitrite in the water left in the bottom of the steamer—water then used for their broth. Perhaps, after all, there was truth in the folk rumors.

But in 1966 research came to a sudden halt. The little red book of Chairman Mao became, in that year, a guide for all action, and his Cultural Revolution severed all contacts with the outside world. Universities were closed, Western books burned. The project in Lin Xian, considered too refined and narrow, was ended. Its primary lab was turned into a sick ward, and its team members were dispersed throughout China. One leader was sent to the laundries, where she worked for two years. Another, a pathologist studying the incidence of cancer in the valley's chickens, was permitted to continue his routine work, but was prohibited from his research. He read on the sly at night.

A Lin Xian woman preparing corn according to traditional methods, which turned out to be a part of the cause of esophageal cancer that afflicted residents of this valley for more than 2,000 years.

were specialists in virology, epidemiology, pathology, and surgery. Beginning with a single question—why was there so much of such a rare cancer here?—the team agreed to concentrate on external clues rather than on internal origins of the disease.

The team moved into the villages to watch the way the peasants lived. They

Finally, in the early 1970s, the upheaval began to subside. The task of relocating the scientists was taken up, and the team was reassembled from laundries and libraries. Among those who returned was the pathologist, Dr. Liu Fu-sen. Before his research was discontinued, Dr. Liu had discovered a number of chickens in Lin Xian with cancer of the gullet—the chicken's equivalent to an esophagus. He screened 14,000 chickens after returning to the project. In those areas with the highest human cancer rates, the rate in chickens soared correspondingly. Where the peasants' cancer was low, the birds remained uninfected. When 50,000 peasants suffering from the cancer were evicted from the valley to a distant district, they were given new birds. In contrast to the cancer-free countryside around them, these birds alone developed tumors. Clearly, transmission of the cancer had a direction and an order—and the chicken, in this case, did not come first.

The research team thus directed attention to a group of carcinogens called nitrosamines, all of which affect the esophagus specifically. Nitrosamines are Y-shaped; the bottom stem is built from nitrite, the two ascending branches from amines, which are products of decaying proteins. Was it possible, the team wondered, for the stomach to synthesize these chemicals during its normal digestive rounds?

One team member fed nitrites and amines separately to pigs marked for a local slaughterhouse. After they were butchered, he removed and analyzed the partially digested food. The discrete fragments had been linked into complete nitrosamines. Given the right ingredients and the right environment, the body could indeed function as its own black alchemist, manufacturing poisons from nutrients. With so little specific knowledge about the process, the most practical line of defense was to prohibit these ingredients from reaching the stomach. Once again, the Lin Xian team returned to find clues to the peasants' life-style. What were they eating, and how could they be dissuaded after 2,000 years of eating it?

Testing the urine of the valley's women, one team member found high levels of nitrite and, mysteriously, low levels of vitamin C. After a week of supplementary vitamins, she again measured chemical levels in the urine: the nitrite had decreased by a third. Lack of vitamin C was somehow disrupting the body's defenses against nitrites. Another member tested the valley's soil and found it deficient in the trace element molybdenum. Without molybdenum, plants concentrate nitrite in their leaves and simultaneously synthesize less vitamin C. Unable to detect the high levels of nitrite in their vegetables, and unable to protect against it in their bodies, the people of Lin Xian were left at double risk.

The source of amines, nitrite's partner in crime, lay no farther than in the local staff of life. When steamed bread made from maize flour molded after several days, it was considered a delicacy much like bleu cheese. Scrapings of the fungi caused amine

Dr. Liu Fu-sen, a pathologist on the medical team that investigated the incidence of cancer in the valley of Lin Xian in China.

Cancer Facts

In a healthy body, no one group of cells is dominant over any other. But when cells undergo a change that leaves them insensitive to normal growth restrictions, they begin to divide uncontrollably at the expense of neighbor cells. Cancer is this wildly proliferative cell growth. More than one hundred types exist; the majority are epithelial cancers, for cells of the skin and the linings of body cavities are the most easily worn out and the most continuously replaced. Tumors, or cell masses, can be benign; rapidly spreading or malignant; or metastatic, in which case cells leave their original site to grow at a distant one.

Cancers arise in response to drugs, radiation, hormones, viruses, environmental carcinogens like pesticides or cigarettes, and chemical carcinogens in food and drink. Some types of cancer are sex-linked; most are diseases of old age. Treatments are broadly organized into two areas: those that stimulate the body's natural defense system, and those that attack without enlisting the aid of natural resources. Certain drugs—interferon, hormones, and adjuvants (mixtures of heat-killed bacteria or detoxified viruses)—encourage and enhance the immune system's own mechanisms. Chemotherapy, radiation, and surgery operate independently of them.

Less quantifiable approaches include laetrile, a derivative of apricot pits; drastic "cleansing" diets; and psychological therapies in which patients visualize and conquer their cancers on mental battlefields.

Professor Shen Cheung, a pathologist, at work on the barefoot doctors' anticancer campaign in Lin Xian. To save the lives of the valley's residents, the campaign asked them to change a way of life rooted in ancient tradition.

levels in laboratory bread to soar seventeenfold. Rats on a diet of moldy bread with nitrite added developed cancer, and the links fell into shape—a Y, to be exact. To the team's surprise, though—if surprise was still included in their repertoire—rats fed certain types of mold alone also developed the cancer.

The Lin Xian scientists began to examine individual tumors removed from the peasants. Almost every tumor had live fungi growing on it. They took biopsies from the esophagi of uninfected peasants. Twice as many peasants with hyperplasia, a precancerous form of tissue growth, had fungi on their tissues as healthy people. Did the fungi induce the hyperplasia? Did the hyperplasia induce the fungi? Was this yet another chicken-and-egg problem, with neither chickens nor eggs nor answers?

Fifteen years had passed. The causes of the Lin Xian cancer remained imprecisely understood. At least one chemical carcinogen, nitrosamine, had been identified. At least

some of its sources were clear: low molybdenum in the soil, mold in the food, nitrites in the water. Their interrelationships, and the means by which they induced tumor growth, however, were still unclear. But it was 1974, and time to supplement investigation with the strong stuff of prevention.

A campaign was launched. Using paramedical "barefoot doctors" as their foot-soldiers, the team began to educate and mobilize 70,000 villagers in the area of highest cancer incidence. On bicycles and in local garb, the barefoot doctors moved through the villages, approaching the villagers one by one. The message they carried was a bitter pill: keep away from your persimmon cakes, your moldy maize bread, your pickled cabbage—all the delicacies that link you to your ancestors; eat only fresh vegetables, drink only purified water, boil off your thrifty soup stocks. They brought difficult suggestions and demonstrated difficult techniques: guard your food in large drying areas, for the mold you encourage is not to be trusted; build your barns on waterproof platforms, for your methods of storage are not to be trusted; add molybdenum to your seeds before leaving them to the soil, for the soil itself, your soil, is not to be trusted. The barefoot doctors committed themselves to a task simple and impossible: they began to change an entire community's way of life.

Household by household, field by field, barefoot doctors discouraged familiar habits and replaced them with routines less familiar, but also less deadly. Education was their first assignment. But by now enough was known about the onset of the cancer to make early detection their second. Deterioration begins with a hyperplasia, a thickening of the inside of the esophagus; unchecked, it develops into a more serious hyperplasia, and a more serious thickening. Up to this stage, the condition is reversible, but the symptoms are undetectable. Finally, inevitably, the hyperplasia crosses into cancer and invades specific muscles. Then comes the difficulty in swallowing. Then comes the impossibility of it. And then, within months, comes death.

The barefoot doctors began carrying a new parcel with them on their rounds, a packet whose contents could test for cancer in its first, symptomless stages. Each villager was asked to swallow an abrasive balloon, right down to his or her stomach. It was inflated internally and, when pulled out again, snagged a scraping from the esophagus as it passed through. The scraping was stained, examined, identified, and diagnosed by a nearby hospital.

Balloon tests were given in all corners of the valley. Those peasants found to have the hyperplasia were treated with preventive chemotherapy. The first drug tried was *kang-i-e-san*, an herbal mixture used in traditional Chinese medicine. When it was tested on patients with severe hyperplasia, 75 percent regressed to normal; with a Western drug, Tylerone, 50 percent regressed. Given no drugs at all, only 30 percent of hyperplasial patients survived.

For those villagers whose hyperplasia had crossed irreversibly into cancer, only one cure existed: an eight-hour operation, and an incision halfway round the body. The chest is opened, lungs and heart are pushed to one side, and a rib is removed. The esophagus lies deep, just in front of the backbone, and must be rerouted to the other side of the main artery leading from the heart. That part containing a tumor is removed, and the stomach is stapled to the remaining section. The operation is long, expensive—$200 in Chinese currency—but relatively successful: 42 percent of those who undergo it survive for five years. It is not the first time in history that a single rib has been sacrificed for life.

Chinese Folk Medicine

The barefoot doctors of the People's Republic of China utilize some therapeutic techniques that most Westerners find unorthodox, to say the least. Yet these herbal remedies and folk cures have thousands of years of civilized doctoring behind them, and the Chinese are not about to toss their tradition of healing aside in favor of penicillin and bed rest.

Chinese herbal remedies fall into eight broad categories. There are the herbs that induce perspiration to "sweat" an ailment out, those that stimulate vomiting, others that act as purgatives, and still others with neutralizing or "smoothing-out" effects. Certain preparations act as "heating" agents to dispel chills, and others "cool" the body and blood to combat fever. "Deflection" herbs improve blood, energy, and digestive circulation by redirecting the disrupted processes; the "tonic" herbs act as general supplements to a patient's blood, energy, yin, and yang.

Acupuncture and acupressure are well known in the West, but other folk cures are less familiar. The skin scrape—scraping a patient's extremities with an old copper coin dipped in wine—is used against heat stroke and indigestion. Bloodletting, the therapy that finished off George Washington, is still used to treat diarrhea, vomiting, strokes, and abscesses. Perhaps the most extraordinary technique is moxibustion, consisting of the application of flaming moxa leaves directly to the patient's skin to cure mumps, nosebleed, vaginal bleeding, convulsions, and rheumatism. All this may seem a bit strange to us, but who are we to quibble with the experience of millennia?

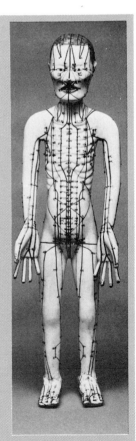

The Question of Interferon

Viruses are primitive organisms. Nothing more than short stretches of nucleic acids wrapped in overcoats of protein, they walk a semantic line between living and nonliving. They are unable to reproduce themselves. Instead, they practice a more sinister method: they inject themselves into nonviral cells, take over their hosts' reproductive machinery, and direct it toward their own duplication. Afterward, they burst the cells open in order to release their progeny, and the new viral forces move out in search of fresh hosts. Antibiotics, which fight bacteria on their cell surfaces, are unable to reach these intracellular highjackers.

In London, a researcher was trying to unravel the reasons—other than undemonstrable luck—that prevented people from catching more than one virus at one time. Infecting chick cell membranes with flu virus, Alick Isaacs was able to create a microscopic battlefield in his test tubes. Somehow, the chick cells responded to invasion by producing a potent chemical. The chemical was unable to save them from the effects of the virus, but when it was transferred to uninfected chick cells, they became flu-resistant. Isaacs recognized this naturally synthesized chemical as a viral warrior. He called it interferon.

Dr. Alick Isaacs, discoverer of interferon.

For several years after interferon's discovery, researchers set to work with the vengeance of hounds on the trail. Every scent was new, every result publishable. In one trial, interferon controlled a herpes virus also known to cause eye ulcers and blindness. In another, it triumphed over a form of the common cold. For $2,000, a small tube of nasal spray could clear out all symptoms of rhinovirus before the first sniffle could be sounded.

Chick cells were the only affordable and so the only viable source of experimental interferon. But chick interferon was ineffective in humans and rodents. Mammalian cells themselves make only the minutest amounts of interferon. Isolating and extracting it was like looking for a needle in a haystack. Slowly, the hounds lost their vigor. Unshakable expectation slid into professional skepticism and, from there, downhill to simple disappointment. After Isaacs's death in 1967, "misinterpreton" research—as the disenchanted of some scientific communities had it—was shelved in a corner of the laboratory along with other dusty talismans.

With time, advances in biochemistry, cell biology, and genetics began to cohere; more precise purification of chemicals had become possible, and more types of cells in which to study interferon production were available. The firecracker question that had prompted interferon's first burst of interest—how much can it do?—was replaced with a less explosive one: how does it do it? Emphasis shifted from the miraculous to the mechanical—a prudent exchange, given that miracles had been the downfall of interferon once before.

The mechanism of interferon, though, is almost as dramatic as its results. After infiltrating a host cell, but before exploding it apart, a virus triggers some chemical alarm within the cell nucleus. The dying cell responds by making interferon, which diffuses out and toward all neighboring cells. Binding the surface of each individually, it transmits a warning: uninvited company coming up the front path. The cells, forewarned, are able to pull up their mats, pull down their shades, and produce

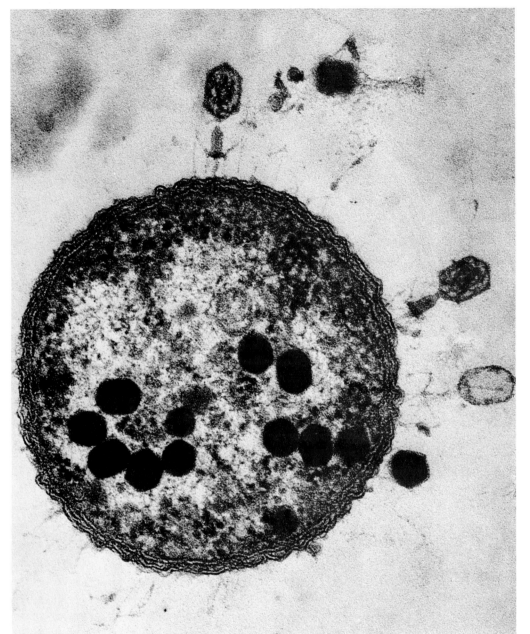

T2 bacteriophage invading _E. coli B._ Activated by interferon, phagocytes like T2 recognize and engulf foreign or worn-out cells by the processes of "cell eating" and "cell drinking." In some cases, research suggests, interferon may cause phagocytes to turn their appetite to cancer cells and thus shrink or eliminate tumors.

yet another protein, an antiviral protein that destroys viruses before they can cross the cellular hearth. Interferon, synthesized by infected cells within hours of infection, catalyzes all the body's dormant natural defenses. The sacrifice of that original cell, in a fine example of genetic altruism, ensures survival for the rest.

In the midst of this renewed study of interferon, an American scientist happened upon a startling observation. Treating mice for a virus that also causes cancer, Ion Gresser

observed that interferon not only prevented viral replication; it shrank the tumor as well. Furthermore, it managed to activate the scavenger cells of the body, the phagocytes that recognize and engulf foreign or worn-out cells by the processes of "cell eating" and "cell drinking." Stimulated by interferon, these scavengers were able to turn their appetites to the cancer. Interferon had transformed the tumor to make its surface appear foreign, and the body responded as if it were a virus. In the United States, where

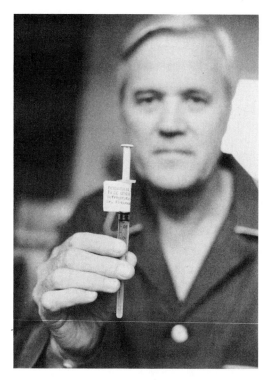

A cancer patient holds $2,000 worth of interferon in one small syringe. With the recent advancements in gene-splicing techniques, this precious drug may eventually be available in large quantities at reasonable costs.

1,000 people die daily from cancer, such dramatic news elevated interferon once again to its unsought image of wonder drug.

Researchers now had enough of a theory to test, more than enough patients on whom to test it, and plenty of refined equipment with which to quantify the results. Only one element was missing. The amount of human interferon necessary for large-scale clinical trials was impossible to obtain. It took Kari Cantell, a Finnish researcher, almost ten years to supply it. Working with blood donated to the Helsinki Red Cross, Cantell perfected a brew of human leukocytes and viruses, incubated together under optimum conditions. He set up an interferon kitchen, and the first experiments on human cancer were done in Stockholm, Sweden, using interferon from his recipe.

In 1971, a young boy suffering from osteogenic sarcoma—a rapidly spreading bone cancer—began to receive a series of interferon shots. The prognosis for bone cancer is poor. But the boy survived two and a half years of injections, remained well, and lost none of his limbs. More trials were run. One child had had fifteen operations by the age of three, each to remove papillae (small tumors) that were clogging up his throat. After each routing, the papillae, like weeds, rose again.

Within weeks after beginning interferon shots, the tumors disappeared. They did not return.

Survival rates for tested children were 100 percent. For adults they were only half as good—still, the results were extremely encouraging. Some patients, the young particularly, felt no side effects. Some adults felt an achiness or exhaustion. But no complaints were heard. These were payments exacted for breath and heartbeat. By all personal standards, the trial results lay somewhere between medical success and divine intervention.

Scientific communities were less generous with their approval. Traditionally, research protocol is as significant as its results; the correctness of a procedure validates or invalidates its findings. The Stockholm trials broke all golden rules of empiricism: studies were not double blind; no placebos were given; there was no untreated control group. The doctor in charge, a clinician, had refused to withhold interferon from any patients he felt might benefit. His ethical priorities clashed openly with the demands of research for testing, retesting, and correlation—for comparative findings, above all else.

In the 1970s, research on interferon finally began to move out of European and into American labs. Private drug companies and the American Cancer Society (ACS) had greeted it with initial restraint, in their optimism and in their funding. Those Stockholm tales crossing the sea—something about survival as dramatic as resurrection—had loosened neither their reservations nor their pursestrings. American immunologists and oncologists, on the other hand, were becoming believers. Privately, some made arrangements to procure interferon for their own patients; publicly, many began to call for financial support. In 1975, the ACS pledged $2 million toward interferon research, its largest monetary commitment ever.

It was the beginning of an era of competition in the finest free-market style. Production of human interferon guaranteed unparalled profits for the company that could synthesize it most swiftly and least expensively. Some of the methods developed were more effective than Cantell's, but more dangerous: Burroughs-Wellcome, a British drug manufacturer, purified its interferon in enormous vats of lymphoma cells. Although the cancer had long since exterminated its original host, the cells themselves were capable of eternal life. Six hundred liters of lymphoma cells per tank, cared for judiciously, gave thanks by producing interferon perpetually.

Another technique of interferon production, even more controversial than using live cancer cells, involved recombinant DNA. As a virus multiplies, it triggers a signal deep within its host's nucleus that switches on one or more specific genes. These are the genes that direct interferon synthesis. Within the nucleus, these nucleic acids make blueprints of themselves; each blueprint complements its maker as a right hand mirrors a left. Messengers bearing these blueprints diffuse into the rest of the cell, to the sites of protein synthesis, and release their instructions.

By isolating and treating them with a chemical converter, scientists were able to trace the blueprints back to their master architects: they could identify the original interferon genes. The genes were spliced out of their cell and inserted onto a circular loop of bacterial DNA called a plasmid. When the plasmid was transferred into another bacterium, the bacterium duplicated according to its instructions; it became a virtual interferon factory.

Science was actually mimicking the object of its original studies. Interferon is, after all, a response to viruses; and viruses do, after all, help themselves, uninvited, to the machinery of their hosts. Now interferon genes were being coupled into bacteria and encouraged to direct the bacterial machinery. In conjunction with the plasmid, they created a totally new form of life, a new DNA strand.

Creation still calls forth hazards, both ethical and practical, however. Can this technique survive a scaling up to commercial

proportions? Several biotechnology firms, including Boston's Biogen and San Francisco's Genentech, are manufacturing large-scale amounts of interferon, for use mainly in clinical trials. They have tested its properties as an antiviral and antitumor treatment and based on their test results are confident that in the future interferon will live up to its promise.

The ACS is still cautious. Dr. Frank Rauscher, the ACS research chief, states that "before we advocate interferon use for the public, and before the public gets too enthusiastic, we need much more information—much more."

Genentech laboratory in San Francisco. The company manufactures interferon on a large scale, using recombinant DNA techniques, for use mainly in clinical trials.

Blueprints in the Bloodstream

Human beings inherit susceptibility to disease in much the same way we inherit curly hair or brown eyes—genetically. Just as we are not born with equal chances of having fair skin or a big nose, we do not enter this world with equal chances of falling ill. The genetic dice are loaded from birth, or even before. Scientists now believe that many of the diseases we contract are the result of a complex interplay between genes and the environment. How did scientists come to the conclusion that illness, like good looks, can run in the family?

The body's fight against disease depends on the white blood cells, which patrol the bloodstream as part of the immune system's complex set of defenses. The immune system recognizes invading cells through the surface of our cells, which carry more than 6,000 markers known as antigens. These antigens are like chemical fingerprints

Bone marrow with red blood cells (green in this photo) and white blood cells magnified 630 times.

on our cells—each individual's antigens are distinguishable from those of others—and new knowledge of them is leading to breakthroughs in the prevention of disease.

Rose Payne led the way with her discovery that there are different types of white blood cells, just as there are different types of red blood cells. Jan van Rood, a researcher in the Netherlands studying antibodies in pregnant women, found that mothers form antibodies against their own babies. A child developing in the womb represents a massive invasion of the mother's body with foreign cells. Each type of antibody, van Rood found, fits a specific type of antigen on the baby's cells. The scene shifted next to Stanford, where white cell typing began in earnest. Scientists soon called the system HLA, Human Leukocyte Antigens (leukocytes are white blood cells).

HLA research has already paid big dividends in medicine. Blood transfusions are safer, and organ transplants now have a greater chance of success. In the future, scientists may be able to analyze an individual's HLA types, predict the diseases he or she is most likely to catch in the course of a lifetime, and then protect him or her from them by vaccination. Doctors and researchers all over the world are collaborating in this effort, which could revolutionize medicine, and have formed an HLA clan, emphasizing the importance of cooperation in the competitive field of scientific research.

Dr. Frank Lilly of the Sloan-Kettering Institute for Cancer Research in New York City noticed that mice with a certain tissue type were vulnerable to a form of leukemia caused by a virus. If they had that HLA type, they died when exposed to the virus. If not, they survived. Already, Dr. Derek Brewerton, an English researcher, had found a very high correlation between ankylosing spondylitis (a form of rheumatism) and the HLA antigen B27. A patient born with the B27 marker is more than 300 times more likely to develop this disease than one born without it. But many people with the B27 marker remain healthy all their lives. There had to be

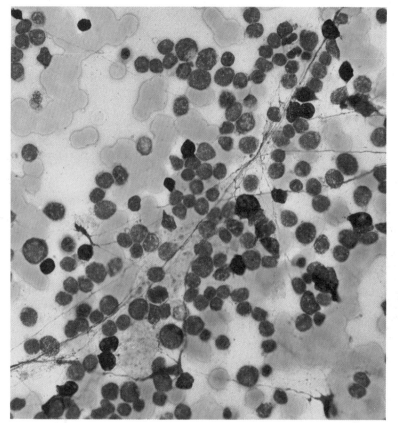

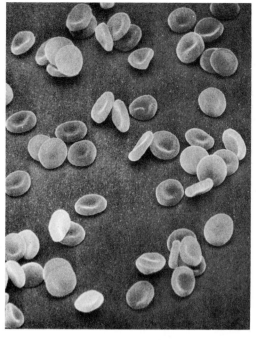

Normal red blood corpuscles photographed through a scanning electron microscope, magnified 6,750 times.

some other factor. Lilly's mice provided one clue, but stronger evidence that something in the environment can trigger a disease in susceptible people came from an unexpected source.

A report by the ship's doctor on the U.S.S. *Little Rock* in 1962 led tissue-typing sleuths to yet another discovery. Six hundred crew members suffered from dysentery aboard the ship. Four weeks later, ten of these sailors had developed a painful form of arthritis. Ten years after that, a trio of Stanford University immunologists, James F. Fries, Andrei Calin, and June Calin, painstakingly tracked down half of the arthritis-stricken sailors. When these five men were tissue-typed, four of them were found to be B27-positive. The dysentery-B27 link provided strong evidence that at least some of our reactions to our environment are governed by our genetic inheritance—and that there are various viral triggers that instigate something in the bloodstream that, in turn, causes a person to develop arthritis.

Genetic markers are also emerging for other diseases, such as diabetes, multiple sclerosis, and rheumatoid arthritis. What hope does this hold for the future? Doctors can already identify susceptible patients with HLA typing. And in families where one child contracts a disease thought to be partly genetic in origin, a physician can try to pinpoint a similar risk in his or her siblings. The next step will be to hunt for triggering viruses, develop vaccinations against possible infective agents, or use blocking antibodies to prevent people with these genetic susceptibilities from developing diseases just because they come across a precipitating infection.

Blood

An antigen is any substance that provokes formation of antibodies, the body's natural response to foreign invasion. All red blood cell surfaces are marked with the antigens A, B, or O; these genetically inherited markers determine human blood types. A body will make an antibody to any blood type not its own; A blood types make AB against B blood types, and vice versa. AB types make no antibody against either A or B, since they carry both markers on their cell surfaces; as "universal recipients," they can accept any blood type. O types have no antigen on their cell surfaces and are "universal donors"; since they carry no provocative markers, they will be accepted by any other blood type. If, during transfusion, the blood types of donor and recipient are not matched, the recipient's antibodies will attack the foreign antigens. Until almost 1900, patients were as often killed by mismatched transfusions as by blood loss.

The RH factor is another blood antigen. If a mother who is missing this antigen has a child who carries it, she will produce antibodies against the embryo's blood; these antibodies cross the placenta and doom any subsequent children to death in utero or in early infancy. The effect can be supressed by an injection of the synthetic chemical RhoGAM (trade name for Rh. [D] immune globulin) immediately after a first pregnancy.

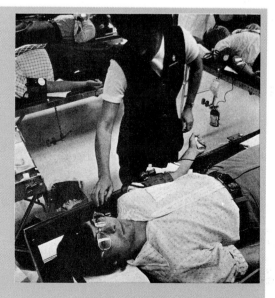

A Conversation with Rosalyn Yalow

I believe in little science—the Mom and Pop shop," says Rosalyn Yalow. "I have always run a very small laboratory. We keep ahead of the game because we can always run into areas where there are very few people. Whenever it gets too crowded and people move into the same thing, we move on."

Yalow and her colleagues have been running successfully since 1947, when she started work at the Veterans Administration Hospital in the Bronx, New York. Along the way, the onetime nuclear physicist and electrical engineer has worked with physicians, developed and trained countless scientists in the use of a powerful analytic tool, and picked up a Nobel Prize—the ultimate scientific accolade.

The basis of Rosalyn Yalow's remarkable scientific career is radioimmunoassay (RIA). The technique, devised by Yalow and Solomon Berson in the 1950s, allows physicians to measure very precisely the amount of a variety of specific substances, such as hormones, drugs, viruses, and other objects of medical interest, in a patient's bodily fluids. The test involves the use of a precisely measured amount of a tailor-made radioactive compound, chosen according to the nature of the substance being assayed. The radioactively "labeled" compound competes with the natural substance to link up with biological materials known as antigens. Precise measurement of the amount of radioactivity emitted gives an exact value of the amount, in the blood or urine, of the substance under assay. Thousands of hospitals, clinics, and laboratories around the world now use RIA for tasks as diverse as treating diabetes, diagnosing thyroid disease, reducing the risk of transmitting hepatitis in transfused blood, and identifying tuberculosis and other infectious diseases.

Yalow's training itself seemed tailor-made for the achievement of devising RIA. "I decided to become a scientist when I was about eight," she remembers, still with more than a trace of her native Bronx accent. "I was always good at math, but not at languages and music, and I was a lousy athlete." High

school, where a math department head and a chemistry teacher took her under their wings, strengthened Yalow's resolve, and when she entered nearby Hunter College in January 1937, she decided to take advantage of its excellent physics department and major in that subject.

Women physicists were a rare breed at the time, but fortune favored the local girl. By the time she graduated, the military draft had started, and Yalow was able to land an assistantship at the University of Illinois. After World War II, she spent one and a half years in industry as an electrical engineer and then returned to Hunter to teach. But it was research that excited her, and in 1947 she started what was to become a career-long association with the VA Hospital in the Bronx.

Yalow's specialty opened the way to her team's achievements. Radioisotopes—specific radioactive forms of different elements—were brand-new in medicine; the first was released from Oak Ridge in 1947. Thus the expertise of a nuclear physicist was essential to apply them to the world of the physician. Yalow's task was to develop diagnostic and treatment methods for veterans using the exotic new isotopes.

The research needed more than a good physicist. In 1950, Yalow linked up with Berson, a physician with impeccable credentials who had just completed his residency in internal medicine. Their talents mingled in a synergistic way. "During the twenty-two years of our partnership, I learned a good deal about medicine and Berson showed a remarkable talent for physics," Yalow recalled. "In these days when we hear so much about interdisciplinary and multidisciplinary research, we must recognize that such collaborative work is most effective when each member of the team makes a commitment to at least on-the-job training in the discipline of the other. Hence, our ability to talk the same hybrid language was a major factor in our success as a research team."

The eager pair's major success came when they first developed RIA, to determine the amount of insulin in the bloodstreams of diabetics. The work showed for the first time that diabetes does not necessarily involve an absolute deficiency of insulin. And it pointed the way to analytical techniques for determining a laboratoryful of hormones, viruses, and other substances in patients. Yalow has continued to extend RIA, moving into fresh applications whenever the current ones attract too much attention from other researchers. She has also taught her rivals, would-be rivals, and colleagues the ins and outs of RIA—for a consideration. "We devised it, we sold it," she says laconically.

Her collaboration with Berson ended tragically in 1972, when her friend and partner died at the early age of fifty-four. Shortly after, a grief-stricken Yalow brought in Eugene Straus, a gastroenterologist, as her associate, and the tiny VA team continued its incredibly successful string of research results. Straus contributed strongly to adapting RIA in such a way that it could be used to diagnose tuberculosis, a disease that still strikes millions of people in the Third World and shows signs of reemerging in the United States.

Berson's death threatened to cause Yalow distress in more than one way. "I think Sol and I expected to get the Nobel Prize for RIA," she muses, "but after he died the scuttlebutt was that I wouldn't get it alone. Never before had one member of a close team won the prize after the other had died." The Nobel committee in Stockholm, in its inscrutable way, decided not to respect its own precedents. In 1977, it awarded Yalow half of the Nobel Prize in physiology and medicine (the other half went jointly to Andrew Schally and Roger Guillemin, who had used RIA extensively in their ultimately successful effort to isolate and characterize the substances that determine how the brain controls the body's hormonal system). Nevertheless, Yalow recounts, "some people said that I won the prize twice."

Yalow gloried in the award. Photographs taken at the Nobel ceremony show her resplendent in a long blue silk dress, beaming with radiant pleasure. "This gave me my place in history," she says simply. "All the

other awards are imitations."

The prize didn't change her approach to her work. Nearly every weekday she spends the hours from seven A.M. to past midday at the VA Hospital's Solomon A. Berson Research Laboratory and then puts in an afternoon of activity at the nearby Montefiore Hospital. Weekend mornings, away from the jangle of telephones, she reserves for talks with her assistants. She and her team continue to run for the open spaces in her research field. Having developed a means of using RIA to diagnose tuberculosis, the group is drifting inexorably into the general field of infectious diseases. "We aren't experts yet," Yalow says cheerfully, "but we're moving into it."

Life in a Test Tube

"The doctor came into my hospital room," Judy Carr recalled, "and he said, 'The good news is you're alive. The bad news is you will never have a child.'"

But Judy Carr did have a child—after nine childless years. Hers was the first baby in America to be conceived outside the mother's body, or "in vitro" (literally, "in glass"). "Test-tube baby" is the colloquialism. In vitro fertilization, though, actually takes place on a Petri dish. A test tube is not used in the process.

In 1978, after ten years of British research, Louise Brown became the world's first so-called test-tube baby. She took to life with a healthy, yowling vengeance. Since then, in Britain and Australia more than twenty similar babies have been born.

Normally a woman can become pregnant when, once a month, an egg or ovum is released from an ovary into a fallopian tube. The ovum is sheathed in a follicle produced from the ovary. At conception, a sperm fertilizes the ovum within the fallopian tube. The fertilized ovum migrates downward, attaches

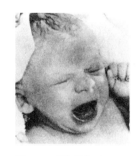

The world's first test-tube baby, Louise Joy Brown (above), in a still from a film made shortly after her birth by Caesarean section at Oldham General Hospital, near Manchester, England, July 25, 1978. (Right) Louise at age 22 months, in May 1980. Louise was conceived in a test tube and later implanted in her mother's womb. A small but healthy 5 pounds 12 ounces at birth, she has thrived in her first years of life with her parents, John and Lesley Brown, in Bristol, England.

itself to the wall of the womb (the uterus), and grows—after nine months of increasingly complex attention to detail—into a baby.

Judy had previously become pregnant in the normal way, but the natural sequence of events was short-circuited—her pregnancies lodged in the fallopian tube itself and not in the womb. After three of these ectopic pregnancies, both her tubes were removed. Thus ended her ability to conceive a child normally.

Howard and Georgeanna Jones are a husband-and-wife team; he's the gynecologist, she's the endocrinologist. For thirty-five years at Johns Hopkins Hospital in Baltimore, the Joneses built a reputation for their work on human infertility. Opting for what they thought would be a more relaxed life in semiretirement, they moved to Norfolk, Virginia. They arrived on the birthday of Louise Brown. A local interview with the Joneses, who had worked with the British medical team, led to a funding offer to bring a test-tube-baby clinic to Virginia. Sailing caps were shipped to sea, and sterile garb was donned once again.

Controversy threatened their work from the start. A routine hearing to seek state approval turned into an arena of public concern; the Joneses were accused of tampering with nature. In twelve months they had moved from Maryland to Virginia, from the quietude of private research to the noisy transparency of a fishbowl; an entire nation was tapping at them through the glass.

Research began, in 1979, with meticulous imitation. They followed the British techniques slavishly, recording each operation on videotape. Using a laparoscope, a medical telescope through which the surgeon can see into the abdomen, they attempted to aspirate a single ovum by drawing off fluid from the ripe follicle. Somewhere in the fluid, they hoped, was the ovum. In the laboratory beside the operating room, they became increasingly expert at finding it.

But being governed by the woman's monthly cycle had drawbacks. Doctors had to wait for the patient's ovary to produce an ovum, which it might do at any hour of the day or night. After thirteen months and nineteen patients, there were still no pregnancies. Morale among the team was low.

Five hundred miles away in Boston, Judy and her husband, Roger, were having an informal but historic conversation with Judy's doctor.

"We had a talk about what I was going to do now, and whether we would be happy without having children. I said, 'You know, we're happy. But I don't want to be forty years old and wish we'd done something.' At that point my doctor brought up the topic of in vitro fertilization."

Normal conception through intercourse is notoriously inefficient. A woman has only a one-in-four chance of becoming pregnant each month. The Joneses decided to modify their process. Following the lead of their predecessors, they began giving their patients a fertility drug, to stimulate the production of more than one ovum at a time.

Their next case was Judy Carr, age twenty-eight. As usual, the team used a technique called ultrasound to visualize and monitor the follicular growth in Judy's ovary. Several weeks later they were able to recover a healthy ovum. After eight hours in an incubator, it was inundated with a million and a half mobile sperm from Judy's husband. Sperm take several minutes to reach their goal. When the ovum is penetrated by one sperm, it promptly walls itself off chemically from all other suitors. This is the moment of conception.

The Carrs waited two days. Their egg had been fertilized.

Returning it to the womb was simple only in principle. The fertilized egg, the embryo, is bathed in a catheter by air and fluid above and below it. The catheter is slipped into a sterile sheath, taken to the operating room where patient and doctors are waiting, and inserted through the vagina. As if it had entered normally from the fallopian tubes, the embryo should ascend toward the womb and embed itself there. Too often, though, the transfer fails.

Dr. Georgeanna Seegar Jones, endocrinologist, and Dr. Howard Jones, Jr., gynecologist, at the Eastern Virginia Medical School in Norfolk, where they performed the first successful in vitro fertilization in the U.S., in May 1981. Elizabeth Jordan Carr was the result, born to Judy and Roger Carr of Westminster, Mass., on December 28, 1981.

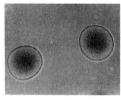 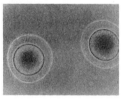 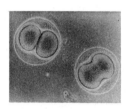

Nine stages in the first day of the development of an embryo—in this case, of a sea urchin.

(*Left*) **The unfertilized egg.** (*Above*) **Five minutes after fertilization.**

First division at 3 hours.

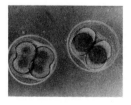 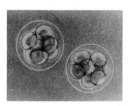 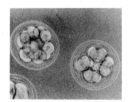

Second division at 5 hours.

Eight-cell stage at 6½ hours.

Sixteen-cell stage at 8½ hours.

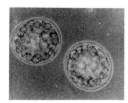  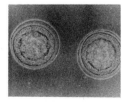

Early blastula at 15 hours

Mid-blastula at 20 hours

Cilia formation begins at 24 hours

Judy recalls: "The doctors explained that the people in the lab had to be given all the time in the world, that this couldn't be rushed, and although I might be very uncomfortable—which I was, upside down with all my weight on my neck—once they were ready to go, they would. And when they were, they ran out of the lab, and I couldn't see the rest of it, but it took only about thirty seconds. Afterwards they congratulated each other— and then it hit me that this isn't something they do all the time."

Judy's pregnancy took. The baby was scheduled to be born Christmas week. "After the press release had been made," she said, "I picked up one of the papers and saw the headline: 'Barren Woman Impregnated.' All we thought was 'This is our baby . . . that's what's the big deal.' "

During 1981, six additional pregnancies gave the team cause to celebrate. They now believe their patients have a one-in-five chance of success, figures close to those claimed by the other successful teams in Britain and Australia. But thousands want to join the program, and because older women have the greatest chance for abnormal births, there's a ceiling age of thirty-five. This means, for some couples, a race against body and clock.

Each treatment costs $3,000, so those who go through it three or four times also shoulder hefty financial responsibilities. Those couples prefer to call it a realignment of priorities. In a hotel a mile away from the

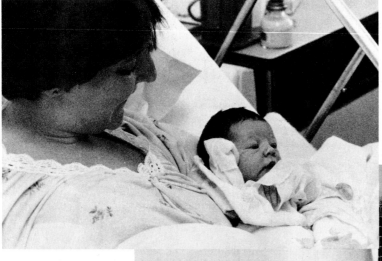

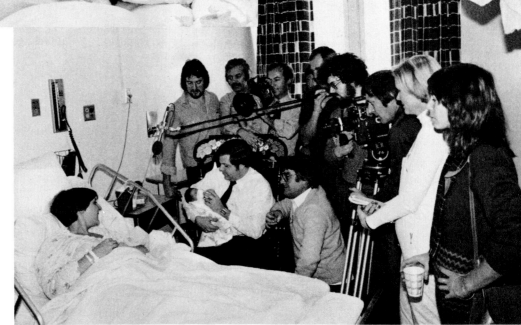

(Above) **Judy Carr holds her three-day-old daughter, Elizabeth, the United States' first test-tube baby, as the Carr family met the press for the first time, December 31, 1981.** *(Right)* **Roger Carr holds his daughter.**

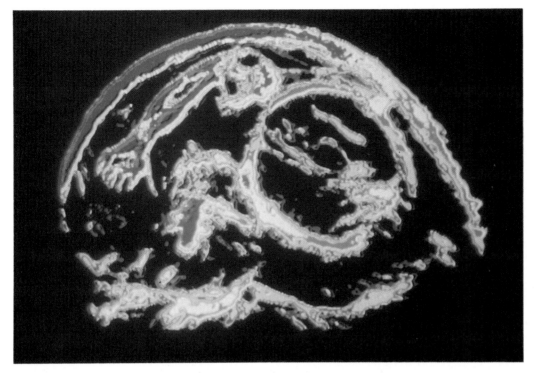

A "sound" picture of a human fetus in the womb. Ultrasonography is the medical version of sonar, used to study infant position and development in the uterus. Sound waves are used because X-rays are potentially harmful. Computers are able to reassemble the complex sound echoes into pictures from any angle; this photo is a cross-section through the mother's body, looking down on the head of the baby.

hospital, they wait for news of transfer, news of pregnancies. As they wait, they hope.

The team at Norfolk tries obsessively to improve every stage of the process. They use mouse eggs to test the medium in which human egg and sperm will be brought together. They look for abnormalities in the growing mouse embryos caused by the fertilization process. One criterion they use is whether the egg divides properly. A strict developmental timetable exists, whether division occurs within the mother or on a Petri dish.

Experiments have yet to be done using human embryos. Such work would make clearer the reasons one embryo "takes" and produces a baby while another, identical under the microscope, does not. It would demystify the state of embryonic chromosomes at the critical two-cell and four-cell stages. The degree and nature of abnormalities could be compared precisely. The opportunities are endless. So is the opposition.

According to the proposed Right to Life Act, a human being exists "from conception." Various commissions, at various times, have defined the onset of life differently; some have declared it to be at fertilization; others, when the heart first beats. In 1979 a report from the Ethics Advisory Board stated that in vitro fertilization was controversial and should not receive federal funding. The report concluded, however, that it should not be stopped for ethical reasons.

The day before the Carr baby was due, an ultrasound test revealed that the fetus weighed only about five pounds. Because small babies stand a greater risk of abnormality at birth, the team quietly prepared for a terrible disappointment.

Judy Carr was taken to the operating room at 7:30 A.M. on December 28. The operation was to be a Caesarean section because of the baby's size and because an infant undergoes a small additional risk in normal childbirth. Judy was anesthetized locally by spinal injection. Fully conscious, she chatted with the team throughout the operation.

At 7:46 A.M., Elizabeth Jordan Carr entered the world. At the same time the world around her entered a new era.

"The label 'test-tube baby' doesn't worry me," says her mother. "I think the newspapers and the telegrams and the offers will die down. Then she'll just be our Elizabeth. Everything is filled with medical aspects. And yet there's still something here that is truly a miracle. Really, a miracle has taken place."

An 1832 woodcut of homunculus, the dwarf man, in a scene from Goethe's *Faust* shows an alchemist brewing up a child in his chemical retort.

The Secrets of Sleep

The Earth's daily clock, measured in a single revolution, is twenty-four hours. The human clock, however, is actually about twenty-five hours. That's what scientists who study sleep have determined from human subjects who live for several weeks in observation chambers with no sense of day or night. Sleep researchers have come up with other surprising discoveries as well.

We spend about one-third of our lives asleep, a fact that suggests sleeping, like eating and breathing, is a fundamental life process. Yet some people almost never sleep, getting by on as little as fifteen minutes a day. And more than seventy years of research into sleep deprivation, in which people have been kept awake for three to ten days, has yielded only one certain finding: sleep loss makes a person sleepy and that's about all; it causes no lasting ill effects. Too much sleep, however, may be bad for you.

These findings challenge some long-held views of sleep, and they raise questions about its fundamental purpose in our lives. In fact, scientists don't know just why sleep is necessary.

"We get sleepy, and when we sleep, that sleepiness is reversed," Dr. Howard Roffwarg of the University of Texas in Dallas explains. "We know sleep has a function, because we feel it has a function. We can't put our finger on it, but it must, at least in some way, direct or indirect, have to do with rest and restitution."

Other scientists think sleep is more the result of evolutionary habit than of actual need. Animals sleep for some parts of the day perhaps because it is the best thing for them to do: it keeps them quiet and hidden from predators; it's a survival tactic. Before the advent of electricity, humans had to spend at least some of each day in darkness and had little reason to question the reason or need for sleep. But the development of the electroencephalograph and the resulting discovery in 1937 of dramatic differences in brain activity between sleep and wakefulness opened the way for scientific inquiry into the subject.

Scientists divide sleep into five stages: one through four and REM (rapid eye movement) sleep. During a normal night's sleep, a person moves gradually through the first four stages to a deep sleep, then back a stage or two and to REM. The discovery of the REM stage in the 1950s was revolutionary, for it indicated that several times during the night, sleep is not a passive but a highly active state. Indeed, scientists have since learned that we dream during REM sleep and we often burn as many calories then as we do during heavy exercise while awake. Researchers think dreams are crucial to our psychological health and that they play an important neuro-

During a single night, we experience varying degrees of sleep. After a period of deep sleep, we eventually enter the REM stage where most of our dreaming takes place. As the night progresses, our dreams tend to increase in length.

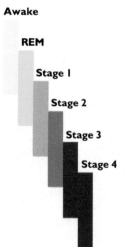

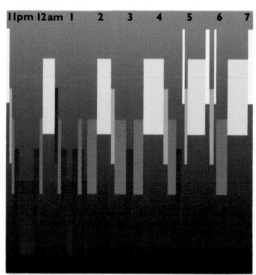

Factor Szzzzzz

In 1982, scientists at Harvard Medical School announced the discovery of a substance in the human body that may lead to a greater understanding of why we sleep. The chemical is so elusive that the researchers had to process several tons of human urine to come up with thirty millionths of a gram of the substance they call Factor S. It is so powerful that about one-trillionth of a gram put test animals into a deep, natural sleep that lasted from five to twelve hours. Although it is possible that Factor S is manufactured in the brain, scientists think it may consist of bacteria and other substances in the intestines and that it may be absorbed by the body, in the way that vitamins are, and used as a sleeping agent.

logical role in helping the brain sort out the barrage of electrical impulses it has to deal with every day.

But REM sleep is more than just dreams. Infants, who sleep about sixteen hours a day, spend fully half that time in REM sleep, while adults, who sleep seven and one-half hours, experience REM only 20 percent of the night. Thus Roffwarg and others speculated, and proved experimentally, that REM sleep is an important prenatal and early postnatal brain stimulant and critical also in the development of the central nervous system. REM sleep also plays a role in the way our eyes coordinate each other's images so that we see clearly and with depth perception. Research subjects awakened at the end of REM sleep had much better depth perception and eye-movement coordination than those awakened at the beginning of REM stages.

Other research indicates that REM function may be connected with a variety of sleep disorders that plague millions of people. These include the widely prevalent insomnia, the inability to sleep; the rarer narcolepsy, the inability to stay awake; and sudden infant death syndrome, also known as crib death. Narcoleptics seem to begin sleep at the REM stage, an indication of a central nervous system defect. Infants are most vulnerable to crib death at age two to four months, at the time they begin to experience less REM sleep. Sleeping pills taken to relieve insomnia become ineffective after prolonged use, in part because they cause a significant decrease in the amount of REM sleep.

None of these developments explains why we sleep in the first place, but pursuit of the answer to that question seems likely to lead to breakthroughs that will bring relief to those with sleeping problems.

An actual time-lapse photography sequence of a night's sleep suggests there is little truth to such claims as "I didn't move after my head hit the pillow." Rather, we shift in bed between ten and twenty times a night. Major posture shifts often follow periods of REM sleep, while periods of little movement are related to deep-sleep stages.

Biochronology

Despite occasional scares produced by widely publicized reports of airplane crashes, the airlines' safety record seems good when compared, in terms of passenger casualty per mile traveled, with the record of other modes of transportation. The industry takes considerable pains to ensure the structural and mechanical integrity of the craft it sends into service. It has less control over the people who fly the planes, however, and four out of five accidents can be attributed to pilot error. The science of biochronology, which studies the daily physiological rhythms of people and other organisms, offers a partial explanation: the rapid movement between time zones produces gross disturbances of normal sleep patterns and the general desynchronization of normally integrated body rhythms. Ordinary long-distance travelers are discomfited

by "jet lag," but in pilots, it can lead to the much more serious disorientation, fatigue, and reduced efficiency that make disastrous accidents possible.

A Long and Healthy Life

The dream of immortality is ages old, as generation after generation has wished for a magic formula for staying young. The dream has always been disappointed, of course, though it has repeatedly been the substance of literary and cinematic imagination. But science has now begun to unravel the biological mechanisms that cause aging, positioning us on the brink of controlling what happens to our bodies over time. Like Methuselah, who lived nearly a millennium, we may find ourselves living longer than we ever imagined.

At the turn of the century, a newborn in the United States could expect to live an average of forty-seven years. A baby who managed to live through infancy would find childhood fraught with dangers. Child labor practices and malnutrition killed and crippled many youngsters. Childhood diseases were epidemic in the years before vaccines were developed. Disease also took the lives or sapped the strength of adults in their middle years. Tuberculosis was common. For an indi-

Seventeenth-century engraving of Methuselah. The Bible says he lived 969 years.

vidual who didn't succumb to illness, an accident in the workplace might prove fatal.

Only as these difficult conditions were overcome, one by one, did life expectancy rise. It took nearly half a century for Americans to achieve an average life expectancy of sixty-eight. Today in the United States, men can expect to reach age seventy and women seventy-seven. In truth, people over seventy-five compose the fastest-growing age group in the country.

Age alone is not a reliable indicator of how old people feel or how effectively they can function. Ronald Reagan was elected president of the United States in 1980 at the age of sixty-nine. John Kelley, at seventy-four, had run in more than fifty Boston marathons. And Katharine Hepburn won her fourth Oscar in 1982 at age seventy-three for her performance in *On Golden Pond*.

Dr. Jack Rowe, chief of gerontology at Boston's Beth Israel Hospital, addressed the question of what being old is. Experiments in separating chronological time from physiological time in some systems of the body have been successful, Dr. Rowe says, and this may revolutionize what we think of as old.

"Any approach to answering the question 'what is aging?' should include some statements about what it isn't," Dr. Rowe states. "Aging is an irreversible, inevitable change that occurs over time. It occurs in all organs and has a major influence on the function of the organism. But perhaps even more important is an understanding of what getting older is not.

"There is a highly prevalent view in America that could be described as ageism. Like sexism or racism," Dr. Rowe explains, "ageism is an innate negative view of the elderly. We have in our minds a view of what it's like to be old, and it's not a very pretty picture. It's a view of an end-stage, irreversibly ill, irretrievable human being, sitting in a corner or a room in a nursing home, staring blankly out the window. There are specific myths that old is dying, that old is sick, that old is poor, that old is senile, that old is sexless and hopeless and helpless. What we have

Johnny Kelley has run in fifty Boston marathons, from the 1930s (*above, wearing number 2*) to today (*right*), and he is still going strong.

CAPE COD
ATHLETIC CLUB

learned about aging and about the elderly over the last couple of decades has helped debunk some of those myths.

"How older individuals feel is really a mixture of two kinds of effects. There is the effect of aging itself, and then there is the effect of any specific diseases the individual might have. The goal in studies of normal aging is to try to understand how a healthy, fit, old individual is different from a young individual. What are the changes that occur with age in how we think, act, and feel; in how our heart works, our kidney, our liver? These changes are important because they form the basis for the influence of age on the presentation of a given disease, the symptoms that people have when they are sick, and the response they have to medications."

Although there are many "young" old individuals—people in their sixties and seventies still vigorously engaged in all life's activities—there are others who suffer from

a variety of illnesses. The tradeoff for eliminating early death has been an increase in chronic illnesses that show up later in life. They are often incurable and incapacitating. Arthritis, cancer, and many diseases we associate with old age can happen to us because our immune systems deteriorate over the years. A look at the way the immune system works makes it clear why age increases the risk of getting these diseases and how they may be treated in the future.

Many old people, like this pair of dancers, are still healthy and active.

This nineteenth-century chart gives a sobering view of aging: "At Ninety every trifling care becomes a burden hard to bear."

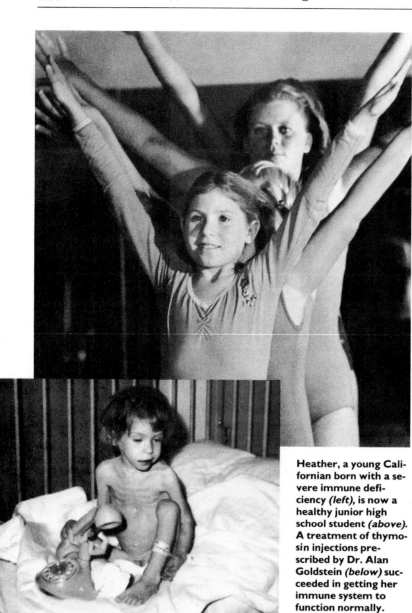

Heather, a young Californian born with a severe immune deficiency *(left)*, is now a healthy junior high school student *(above)*. A treatment of thymosin injections prescribed by Dr. Alan Goldstein *(below)* succeeded in getting her immune system to function normally.

Alan Goldstein of George Washington University hopes his thymosin treatment will strengthen the immune systems of elderly cancer patients against the weakening effects of radiation and chemotherapy.

Defending the body against both internal and external threats are two kinds of white blood cells produced in the bone marrow. The B-cells attack invaders with antibodies. The T-cells—so called because they get their instructions from the thymus, the master gland of the immune system—have a variety of jobs. By passing through the thymus directly or by being activated by thymic hormones, the T-cells learn their immune functions. At puberty, the thymus has begun to shrink. As it shrivels with age, T-cells lose their functions and we become more susceptible to both infectious and autoimmune diseases.

Although immune-system aging begins at puberty for most of us, for a young Californian named Heather it never began. Because of the success of the therapy she received, scientists have made great strides in understanding how the immune system functions as people age.

Alan Goldstein of George Washington University in Washington, D.C., thinks we may soon be able to compensate for our lost T-cell immunity by artificially supplying thymic hormones. His optimism is based on the extraordinary results of his work with Heather, who was born with a severe immune deficiency. Heather's thymus was not functioning, and her immune system was therefore similar in several respects to that of an aging person. The fact that she had no functioning thymus gland really put her at the end point of what occurs naturally as the immune system ages. With no thymic immunity, every viral and fungal infection was a life-threatening situation for her.

Dr. Goldstein treated Heather with thymosin, a thymic hormone he helped discover. "In animals without thymus glands, when we administered thymosin, we could turn on their immune systems," Goldstein says. "We could get their bone marrow cells to mature into functioning white blood cells, functioning T-cells that could then carry out their natural immunity. We reasoned that they could provide Heather with her own natural immunity in a similar way. And the thymosin appears to have prodded her immune system into functioning normally, just as in mice.

"Heather was the very first person to receive thymosin, and her dramatic clinical improvement accelerated the interest in and the potential of using thymic hormones to manipulate the immune system to raise our immunity as we age," he concludes.

Aging cannot be attributed to deterioration of the immune system alone, however. Today there are nearly as many theories of aging as there are researchers, and perfectly plausible hypotheses often seem to contradict one another.

Average Life Expectancy (Years)

Animal	Years
Galapagos Tortoise	100
Swan	70
Elephant	60
Whale	50
Walrus	40
Toad	35
Horse	30
Lion	23
Dog	15
Cat	13
Pacific Salmon	6
Mouse	5

From his work with human cells, Leonard Hayflick, head of gerontology at the University of Florida, became convinced that aging was not something that just happened with the passage of time. He found it was the result of a prearranged genetic plan.

"Those of us who subscribe to a program theory of aging simply think that age changes are programmed, in the way a developing infant is programmed to maturity, in such a way that the program continues beyond maturity to old age," Hayflick says. "There is a purposeful program that dictates changes over time."

Before Hayflick's work growing cells in the lab, it was thought that normal cells in culture could live forever and that there was no intrinsic limit to life. That implied that we could also live forever if the cells in our bodies could be cared for as well as those in the lab dishes. But Hayflick's cell cultures, after going through distinctive phases of life, inevitably died. A series of experiments conducted over a number of years indicated that cultured normal human cells have a clock that determines the number of doublings they are able to undergo.

"I suppose the most pessimistic view of programmed deterioration is to say that our cells have a limited capacity to function and to divide to ensure our deaths," Hayflick comments. "I say that because it's interesting to contemplate that aging as we know it—that is, the extreme manifestations of aging seen in humans—occurs only in humans and in the animals that we choose to protect, like zoo animals and domesticated animals. Aging, in its profound manifestations, does not occur in animals in the wild."

In the wild, animals don't get a chance to get old. They usually either succumb to predators or die of disease or starvation. When animals are allowed to live out their life span, however, there seems to be a maximum age for each species. Mayflies live out their entire adulthood in a single day. The longest-lived cat managed twenty-eight years. The maximum for elephants is sixty-nine—just short of the average life span for people in Western nations. Maximum life span seems to be controlled genetically by some mechanism within the DNA master molecule that makes up genes. But our DNA is constantly being bombarded by ultraviolet light, which causes damage. Normally, enzymes rush in to fix the damage. One theory of aging is that there is a breakdown in our repair system. Damaged DNA then sends incorrect instructions to the cells. Receiving the wrong messages causes malfunctions in the cells' activities, in turn causing genetic diseases that distort normal life processes.

Leonard Hayflick, head of gerontology at the University of Florida, has discovered that cells have a fixed reproduction limit—that aging is a prearranged genetic plan.

Another Look at Blindness

Surveys indicate that next to cancer, Americans fear blindness most of all diseases and disabilities. A half-million of our citizens are totally blind; another 6 million have severely impaired vision. Of the principal causes of blindness, only cataract (a progressive clouding of the lens) can usually be corrected. New medicines and surgical procedures offer limited relief to people afflicted with glaucoma, the injury to the optic nerve caused by increasing pressure from fluid within the eyeball. For those who suffer from diseases of the retina, especially older people and diabetics, there is little hope at present; vision loss is inevitable and irremediable. The incidence of this last cause of blindness (and of other

chronic, degenerative diseases associated with aging) is unfortunately rising, as average longevity is extended by general medical progress.

A dead Pacific salmon after the strenuous journey to its upstream spawning grounds *(above).* Caleb Finch *(below),* of the Andrus Gerontological Center in Los Angeles, is looking for a pacemaker that may regulate the aging process. His study of the female reproductive system, whose functioning seems to be linked with age, prompted his search.

Another clue to the mystery of aging may be found in the lives of Pacific salmon. The journey from the ocean to their upstream spawning grounds requires a tremendous amount of energy. Researchers have found that once these salmon enter fresh water, their pituitary glands swell up and the fish produce large amounts of hormones of all kinds. By the time they reach the spawning grounds, their metabolism has become chaotic from the heightened output of hormones. Many of their organs cannot cope with the stress due to this metabolic upheaval. Not long after spawning, females and males alike succumb to the havoc wreaked throughout their bodies.

The combination in salmon of an enlarged pituitary and accelerated aging has suggested that a similar, if subtler, mechanism causes aging in humans. One researcher believes the pituitary in our brains may be the pacemaker that initiates the aging process. We know the pituitary sends out hormones to trigger growth and the onset of puberty. Perhaps it also sends out "death hormones"

that prevent life-sustaining processes from occurring and so cause us to age.

At the Andrus Gerontological Center in California, Caleb Finch is trying to identify a pacemaker that regulates aging. His search began with a study of the female reproductive system, whose functioning appears to be linked to age, since old women no longer menstruate.

"For a long time, it had been thought that the ovary was the sole cause of reproductive aging," Finch explains, "because at birth, in all mammals, the ovary has acquired its maximum number of egg cells, and then they are gradually used up. What wasn't realized, and what is now very clear, is that there are still a large number of egg cells remaining in the ovary at the time the estrus (or reproductive) cycle ceases. This is true in both experimental animals and humans. The proof that the remaining egg cells in the ovary are still viable is very simple," Finch continues. "One can transplant ovaries from old animals into young animals, and the cycles resume. It was previously assumed that defective eggs might be a signal for the estrus cycle to cease—but these eggs can be shown to have fertility and cannot therefore be considered defective."

The signal to begin estrus goes from the brain to the ovary, which produces a hormone responsible for maintaining fertility. But with each cycle, the hormone accumulates in the brain, aging it and causing damage. Eventually, the brain stops sending the estrus signal.

"My hunch is that in the next twenty to forty years, we'll have a substantial and solid list of brain-aging processes which are controlled or influenced by specific factors," Finch concludes. "In some cases it may be possible to intervene to prevent the preconditions for serious deterioration. It may be possible then to live much more of the human life potential with good health and with major capacity."

A researcher at the University of California at Los Angeles, Roy Walford, is also doing all he can to make sure life will be long and healthy—for himself as well as for others among the middle-aged. He believes the

Life Span of Plants (Years)

Plant	Years
Bristlecone Pine	4900
Sierra Redwood	2300
English Oak	1500
White Pine	450
European Beech	250

Life Expectancy

Greece	700 B.C.	18
Rome	50 B.C.	22
England	1500	33
USA	1790	33.5
USA	1850	40.0
USA	1900	47.3
USA	1910	50.0
USA	1920	54.1
USA	1930	59.7
USA	1940	62.9
USA	1950	68.2
USA	1960	69.7
USA	1970	70.9
USA	1980	73.0

immune system is the pacemaker that controls aging. His experiments are aimed at finding out what controls immunity. Discovering that could make it possible to transcend the current biological limits to life.

He began his investigation by repeating the classic experiment of extending the maximum life span of mice by underfeeding them. The same food was given to three different groups of genetically identical mice, but the amounts differed. One group was fed daily and a second group every other day, while mice in the control group could eat as much food as they wanted, whenever they wanted it.

The mice fed every other day were energetic and healthy, whereas those fed daily were sickly. The control group on unlimited rations all died. "Underfeeding has not only allowed the animals to live their regular maximum life span," Walford states, "but has actually extended the maximum life span for the mouse species. This is true retardation of the aging process because it extends maximum life span. That would be the same as extending human maximum life span, for example, from the present 110 years to, say, 150 years."

Another food-related line of research that interests Walford has to do with the free-radical theory of aging. As our bodies metabolize food and turn it into energy, chemical reactions release molecules of oxygen and other substances called free radicals. These free radicals can bond with the fats that make up cell membranes, thus damaging the cells' ability to function.

Some of the physical changes we call aging are the result of free-radical damage. However, to prevent this, our bodies produce scavenger enzymes called antioxidants. The antioxidants roam the body in search of free radicals and combine with them before any damage is done. But as we get older, we produce fewer and fewer of these helpful enzymes. Walford hopes that diets supplemented with antioxidants will minimize the ravages of age and prevent our bodies from deteriorating.

Even those of us who take no active measures to preserve our bodies may find cause for optimism in Walford's research. He believes that many age changes can be traced to a supergene that could eventually be manipulated.

"It looks as though this supergene regulates immunity, free-radical scavengers, and DNA repair," Walford states, "and these are four of the major theories of aging."

If we manage to extend life but are not successful in preserving vigor, how will that affect the way we spend those extra years? By the year 2000, a 50 percent increase in nursing home residents is expected. If we find a way both to prolong life and to maintain vitality, the dramatic increase in the number of elderly will put a strain on housing, the job market, and health care. Either way, if we are not careful, the dream of immortality could turn into a nightmare.

Roy Walford, of the Department of Pathology at the UCLA School of Medicine, believes the immune system is the pacemaker that controls aging. He supplements his own diet with antioxidants to minimize the ravages of age.

Portrait of a Killer

One American in five suffers from some form of cardiovascular disease, which accounts for half of all deaths each year. Why do heart attack, stroke, and related illnesses afflict so many of our citizens? Medical research points to several factors that place us at risk, most of them associated with our vaunted "life-style."

The stress produced by our rapid pace of life certainly contributes. So does hypertension, or chronic high blood pressure. So does the general lack of exercise in a population dependent on cars and "labor-saving devices." So does cigarette smoking. And so does our diet rich in animal protein and fat, productive of high levels of serum cholesterol, which clogs and hardens blood vessels. It will be interest-

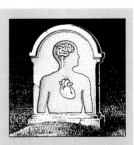

ing to see whether the new national fitness mania will make inroads on the incidence of cardiovascular disease, our number one killer.

A Conversation with James Birren

When James Birren lopes out across the sun-drenched terrain of southern California to jog his customary two miles each day, he does so with a particular understanding of what he is doing. To most of us jogging may be a sweaty quest for a trimmer waistline, an obeisance to a fad, or the meeting of some personal challenge. But Dr. Birren, executive director of the Andrus Gerontology Center at the University of Southern California, has spent his professional life probing how the nervous system and the rest of the body intimately intertwine. He searches for the physical expression of that evanescent quality we call mind and, particularly, how the mind can affect the body as both age.

"We're beginning to get objective evidence that physical activity is useful far beyond just the maintenance of, say, the cardiovascular system," says Birren. "It influences the whole tone of the central nervous system. The effects of exercise are such that the fit older man is probably performing better than the unfit younger man. That's a crucial point," he says, "one that wasn't taken seriously early on."

Birren, who is sixty-four, first became sensitive to the effects of aging during World War II, when he began studies for the Naval Medical Research Institute. The young psychologist was asked to find out why men in the hot, cramped, steamy confines of ships in the Pacific suffered health problems and declining efficiency.

"When we simulated those environments back in Bethesda, Maryland, I was intrigued by the fact that very often age was a bigger variable than some of the other things we were studying. When I found these big age trends in physical strength, I got curious about them, even though that wasn't my field. And I found that the trends in age in physical strength vary considerably with the muscle group. The group that declines the most is the back muscles, and the ones that stay the most constant are the hamstrings. So the problem wasn't so much physiological aging as relative disuse. That, I guess, was a turning point in my research."

In 1946, Birren joined the National Institute of Child Health and Human Development of the National Institutes of Health and continued his studies of performance efficiency in aging. There his studies began to concentrate on mental performance.

"We all know subjectively that at some times we perform better than at others. We have our ups and downs, but it's very difficult to measure them objectively. So I began to try to study the nervous system, its capacity and performance, as analogous to other organs.

"I did a few studies in which I tried to load the nervous system systematically. I asked subjects to add long columns of digits. I started them with two digits and then progressed to twenty-five digits, and I found that when they did long strings of numbers, they lost their place; they were distracted. In a similar study, I asked subjects to list words beginning with the letter *S* and examined the change in that ability with age. I found that there was a change with age in the speed of processing information. Understanding that change became a basic object of my research."

As a result of such studies, Birren has developed theories about the origin of the "slowing" process.

"I think that it has something to do with the arousal of the nervous system itself. There's a system in the brain called the ascending reticular system, and the fibers in this system secrete norepinephrine, which is a stimulant. This system is concerned with wakefulness and sleep—and, I think, arousal. The current hypothesis is that for some reason with advancing age it does not keep the nervous system aroused near its optimal level.

"Now, if we identify the issue as slowness in processing information, we ask the question: how does it change in healthy older men, above age sixty, who are active compared with inactive men? We found that, usually, older men who are physically fit are faster in handling information than are young sedentary males."

Birren himself has gone through a change in fitness that brought his research vividly into personal focus.

"A professor of exercise physiology, Dr. Herbert Reed, gave me the first evidence that I was not as fit as I thought. I was living under the myth that I was physically fit because as a young man I played basketball a lot. But when I did a bicycle ergometer test for him, I couldn't get to the point of exhaustion because my muscles were so out of shape.

"So I started jogging, and I've been doing it for thirteen years. Recently, the same man tested my reflex speed, and he found my reflexes were equal to those of a college-age student.

Mood effects can also profoundly affect physical health, Birren believes, a fact that has been neglected in our society. "I think we're on the brink of a period of great emphasis on behavioral medicine, as the public pendulum swings away from expecting the molecular fix for health problems. Not that the two approaches invalidate each other. They're concurrent truths. But when you think that the molecular fix will take care of all medical problems, you're not going to put that effort into what I would call a hygienic way of life."

Both Birren's studies and experiences point to a profound effect of attitude on the aging process, a phenomenon he illustrates with a personal anecdote.

"We've always thought that we age a little bit every day, but that's not necessarily reflected in incapacity in older people. There's one particular encounter that has affected my life in this respect. I once met a 102-year-old woman who arrived at the center, and I asked her how she came there. She said she came alone on a bus. That was to me a startling response. So I asked her why she had come, and she said, 'I was just curious about what a university gerontology center did.' Now that attitude clashes so much with the stereotype that if you're 100 you're in a bed in a nursing home somewhere. Here's a 102-year-old woman, traveling around Los Angeles on a bus alone, motivated by curiosity.

"Interestingly, I had a letter a few years later that said she died at 104. She didn't go downhill gradually; her whole system seemed to come apart in a period of a month or two, like the proverbial one-hoss shay.

"Thus, I think that a true senescent death might be called a systems death. The various physiological systems no longer have the capacity for self-regulation. Their boundary conditions have been exceeded, and the result is a cascade of failure. That's a very different notion from the idea of aging as dwindling down a little bit every day. Obviously, though, if you have a genetic disease it's a different ball game.

"My research and experience have given me a rather fundamental optimism about getting old, one that's in sharp contrast to the negative projection of young people about aging that you have to be dragging around when you're eighty or ninety or one hundred. With a little bit of luck we can be one of those one-hundred-year-olds traveling around just because we were curious."

For Birren, age sixty-four is indeed a time of continued curiosity. He plans to retire only from administration and to continue active teaching and research. He will also continue to conduct an Institute for Advanced Study in Gerontology, which hosts visiting scholars. And he will continue to enjoy the feel of his daily sprint through the California sunshine.

The Healing Touch

A seven-week-old infant grasps its mother's finger. Tactile contact is our first means of communication with the world after we're born. The lack of a mother's touch, scientists believe, may lead to both physical and psychological problems for a child.

When you were a child, your mother probably used more than Band-Aids to treat your playground cuts and scrapes. She kissed your wound to make it all better, gave you a reassuring hug, and sent you back out to play, feeling as good as new.

Mothers and children seem to know instinctively that touching is the most basic form of comfort and communication. Recently, scientists have begun to research the place of the tactile sense in our lives—and they have discovered the hidden meaning and extraordinary power of what is possibly the most highly developed of our senses.

Dr. Peter Bruggen, consultant psychiatrist to a treatment unit for disturbed adolescents within the National Health Service in England, has used continuous and varied types of tactile contact to help young people work through their problems.

"Tactile contact is the first means of communication between us and other people after we are born," Bruggen says. "It's a peculiarly powerful one. It's also one that is much more honest and clear than words, which are so sophisticated. It's so easy to lie in our words."

To find out the true extent of nonverbal communication through touch, scientists conducted experiments with monkeys to ascertain the importance of touching to the developing infant.

Maternal love and care naturally involve a massive amount of tactile stimulation. What happens if that contact is absent in early life? At the University of Colorado Medical Center, Professor Martin Reite studied the detailed effects of maternal separation, in which deprivation of the mother's touch plays a highly significant part.

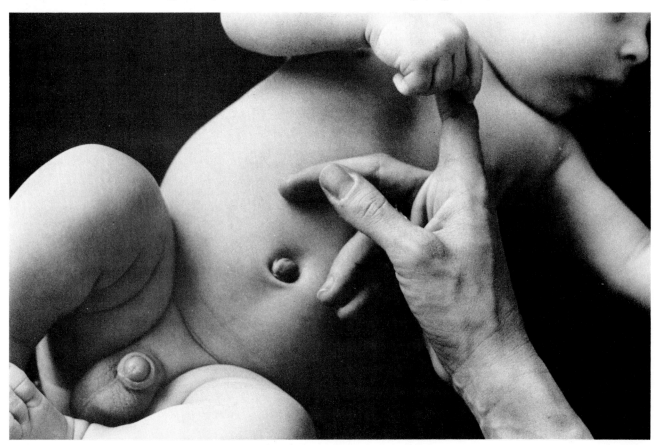

Infant monkeys were separated from their mothers. Before, during, and after the separation, the infant's body functions were monitored through a tiny radio unit linked to sensors on various parts of its body. Gradually, during the separation, severe changes in mood and behavior were matched to changes that were going on in the body.

At first, the baby primate feels helpless and confused when its mother is taken away. Over the first few days it engages in classical primate gestures of protest. The youngster, hunched up in a corner of the cage, displays irreconcilable grief, despite gestures of support from peers. It remains deeply depressed until the return of its mother. Only her unique touch can redeem the situation. After three or four days of intense and continuous holding, the youngster's behavioral distress disappears.

But the injury to bodily functions from inadequate touching remains. Susceptibility to disease and a broad range of serious weaknesses in the body were induced by the separation. The infant has experienced changes in body temperature and heart rate. EEG and brain-wave patterns have also been disrupted, as have sleep patterns. Much evidence suggests that there are disturbances in immune function as well. Denial of maternal contact leads to physiological dysfunction, with possible serious consequences for the infant's long-term resistance to disease.

The lack of a mother's touch causes psychological problems as well, which scientists believe are similar in humans deprived of mothering. Two groups of monkeys were experimentally deprived of tactile contact from their mothers, one for intermittent periods of fourteen days throughout childhood, the other for longer periods. Infants in both the long-term and short-term groups grew up with serious behavioral problems. The degree and type of their problems depended on the duration of their deprivation.

In the short-term case, monkeys had a great deal of difficulty in becoming confident and independent individuals later on in adulthood. The opposite is the case of those who

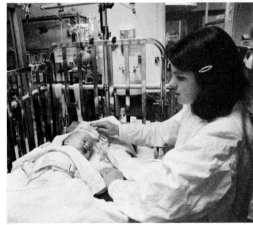

An orangutan cuddles her baby. Experiments have demonstrated that infant primates deprived of their mothers' touch become depressed and susceptible to disease.

Scientists are discovering what many people have known instinctively all along—that touch is a healing process. A gentle hand can have a calming and healing effect on an infant confined to a hospital bed.

experienced long-term separation. As adults, the animals tend to avoid one another, and when they do come into contact physically, the interactions tend to be uncoordinated and exceedingly aggressive. These monkeys grow up with great difficulty learning to cooperate with others, and their usual contact veers from violence to isolation.

Researchers believe that, with respect to touch, severe deprivation is almost certain to have comparable severe consequences for humans.

Hunt for the Legion Killer

A mysterious killer made headlines when it claimed victims of an American Legion convention that had gathered in Philadelphia to celebrate the Bicentennial in July of 1976. Within two weeks of the convention, "Legionnaires' Disease," as it was dubbed, had claimed twenty-one lives.

I just felt . . . unusually tired. The next morning, they used a thermometer. They tell me it peaked at 108.
— an American Legionnaire

July 21, 1976, commenced the Bicentennial convention of the Pennsylvania Department of the American Legion. It was held at the Bellevue Stratford, Philadelphia's most venerable hotel. Twenty-five hundred veterans exchanged decades of separate histories over three days of lingering meals. The troops parted on July 24. By August 2, twelve among them were dead.

The Center for Disease Control (CDC) is a federal agency based in Atlanta, Georgia. Within its maximum-containment laboratories, the most lethal microbes known to humans are raised and studied. In response to a call from the Pennsylvania Health Department, a team of microbial investigators was dispensed throughout Pennsylvania, collecting descriptions of every case of legionnaires' disease.

Not all the victims were vets; yet each, at some point during the convention, had passed through and lingered in the Bellevue Stratford. The disease they contracted seemed much like pneumonia; it was an inflammation of the lungs, accompanied by fevers that rose to 104 and 105 degrees Fahrenheit. Yet the severe diarrhea, the

kidney and liver soreness, and the sense of confusion that also attended victims were bafflingly unfamiliar.

By August 4, twenty-one people had died.

Lung samples from the victims were rushed back to CDC. Virologists and bacteriologists conducted more than 400 separate tests without isolating a single virus or abnormal bacterium. Pathologists examined hundreds of dye-stained slides; no unusual organisms were found. Using computer as well as lab techniques, toxicologists tested for thirty-five metallic elements and 35,000 organic compounds deadly to humans—without success. In a corner lab, Dr. Joseph McDade noted an occasional rodlike organism on his slides. No doubt a bacterium, or some contaminant unworthy of pursuit. He dismissed it.

By August 6, the death toll had reached twenty-four.

The public hunt was less scientific but just as intensive as the professional one. Mail poured into the CDC. Every even letter seemed to have the cause: plastic dentures; onions in the salad; Philadelphia toilet paper. Every odd letter seemed to have the cure: acupuncture; or dilute vinegar; or massive doses of the allegedly infallible vitamin C. Political rightists suggested that the nation was witnessing a Communist attack on veterans. Political leftists were convinced the army and the CIA were exercising lethal biological powers.

By Christmas, there were 221 recorded cases of legionnaires' disease and thirty-four deaths in all.

Four days before the New Year, Dr. McDade returned to his lab to look, yet again, through the slides made from slices of diseased lungs. The "occasional rodlike organism" he had noted and discarded four months earlier had multiplied into a colony. When he mixed a sample with blood from infected legionnaires, human antibodies—one of the body's natural defenses—began to attack. The bacterium was named *Legionella pneumophila,* and scientists knew they had found, in a new species, a new killer.

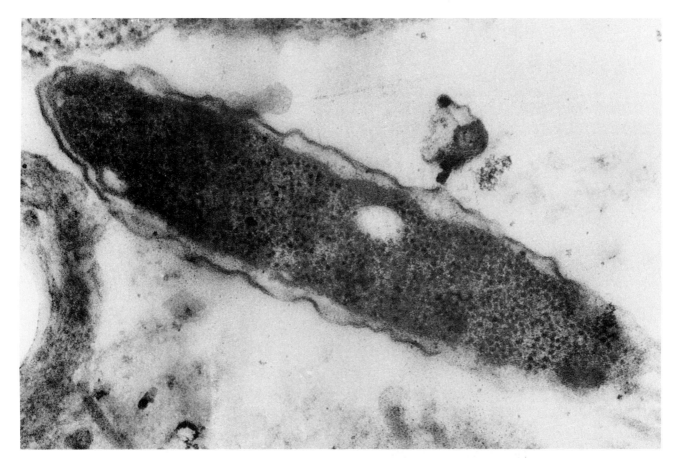

Epidemics are spread in a limited number of ways: from person to person, by food or water, through insect bites, in the air. People on the sidewalk only several feet from the Bellevue Stratford remained uninfected. Those whose social visiting was conducted in the hotel lobby stood at higher risk. But air is air is air, isn't it? Why would a bacterium favor inside over outside?

The hotel's cooling system, like most, combined hot gas with cold water to form vapor; this vapor could be sucked into the ducts that fed rooms with cool air. Perhaps, in the heat of the Bicentennial summer, water-living bacteria were being transported to their victims in the most luxurious style possible: through an air-conditioned spray.

Preventive measures proved the theory correct: minute amounts of chlorine kept cold water free of *Legionella*. Chlorinating systems were installed both in air conditioning and aerosol shower units; without disinfecting, each gives forth a fine, deadly spray capable of carrying bacteria through the air. It seemed, with the discovery of chlorination—

and the use of erythromycin, an antibiotic effective against the bacteria—that the hunt for the legion killers had reached a successful end. . . .

Yet, since 1976, four additional *Legionella*-like organisms have been isolated. All are adapted to water. All induce fever and pneumonia. And unchecked, all cause death.

This electron micrograph shows a highly magnified legionnaires' bacterium, *Legionella pneumophila*, growing in a laboratory yolk sac at the Center for Disease Control in Atlanta, Georgia.

After two years, the legionnaires returned to Philadelphia for another convention. But this time they took precautions. Shown here is Pennsylvania American Legionnaire Charles Rowlands exhaling into a machine to check his pulmonary system.

A Plague on Our Children

Eve DeRock, a teacher who lives in the heart of Oregon's commercial forest region, remembers vividly the day in March 1977 when an International Paper Company helicopter flew overhead and sprayed a herbicide known as 2,4,5,T throughout the valley. The company wanted to kill grass and underbrush and clear the land for a planting of trees.

"The grass died on my part of the valley as well as International Paper's," she said. "A week later, my cows began to drop their babies; they aborted. The fish died in the creek, and we found quail hens dead on their clutches and grouse that died with their young. And our dogs began to bring in slicks [aborted fawns] from the woods. I felt that I was sick from summer flu. . . . I didn't have a whole lot of strength. It got worse, until long into fall. My body became sore, and the inside of my body was sore, and finally, I came down with chills, fever, and convulsions."

Shortly before dawn on March 17, 1978, the clatter of helicopters filled the sky over the fishing village of Portsall, on the northwestern tip of the Brittany coast in France. Curious townspeople who headed outside were greeted with a pungent stench that left them short of breath. By the time they got down to the harbor, a two- or three-minute walk for most of them, their eyes were watering, their faces stinging, their heads aching. The air was filled with the smell of oil.

In the early morning light, fishermen could see their boats moored in the harbor, tossing about on waves of black muck. Turning seaward to see where it was coming from, they looked out on a peculiar sight: although it was the time of spring high tides when incoming surf was especially heavy, and although winds were blowing at near-gale force, the fishermen could see no whitecaps. The sea was black, covered with a blanket of oil that was gushing from a stricken supertanker, the *Amoco Cadiz.*

One spring morning, a New Hampshire fish and wildlife worker was stocking streams and ponds with trout from the state's fish hatcheries. At one small lake, he was startled when a batch of young trout he had just released turned desperately back toward shore. They squirmed and writhed and seemed to gasp for air. Within minutes, they were dead, the apparent victims of acid rain.

This is testimony from the forefront of the environmental crisis. Each incident, by coincidence, occurred in a setting far away from most of us—a rural forest, a foreign coast, a mountain lake—but the problems each portends are really no farther than the air we breathe, the water we drink, the food we eat. These and similar incidents have varying causes and effects, but they all have one common denominator: the oil (and other fossil fuels) manufactured by natural Earth processes over millions of years and now, through the magic of technology and chemistry, recovered and altered to serve the needs and convenience of humans.

Oil and its by-products provide us with fuel for our cars and power plants, fertilizers and pesticides for our farms, solvents for in-

Polluted water empties into a river in the Northwest from an outfall.

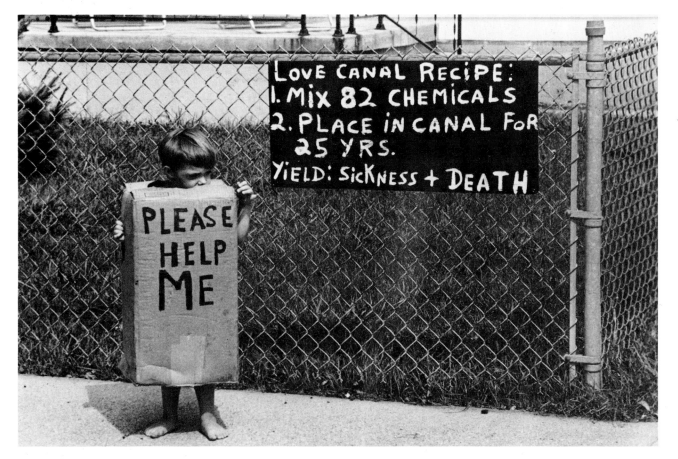

dustry, fibers for clothing, and plastics for just about everything. Some of these products, however, are also among the most toxic substances ever made, and they are turning up in our air, food, and water, in the tissues of fish, birds, and other wildlife, and, indeed, in our own bodies. When they accumulate in high enough concentrations—in some cases, a few parts per billion or even a few parts per *trillion* may be enough—they can cause rashes, nervous disorders, miscarriages, and cancer. The incidence of birth defects in exposed laboratory animals and humans suggest that these chemicals are so lethal they can invade our DNA structure and alter our genetic makeup—the traits we pass on to our children.

Thus, after two decades and billions of dollars' worth of projects and programs designed, often with success, to improve the quality of the environment, we are faced with an environmental threat more insidious, more persistent, more complex, and more costly than ever before.

Science is inextricably wrapped up in this dilemma. After all, it is science that has given us the tools and the technological wherewithal to manufacture the exotic chemicals that sometimes turn into poison; to extract the crude oil from deep within the Earth and construct the supertankers that transport it; and to build the huge power plants that burn oil and other fuels to supply our electricity. Science also plays a major role in assessing the impact of our careless use of this technology, taking biological body counts in the wake of oil spills, measuring the particulates in the emissions of a power plant's tall smokestacks, determining the extent of chemical contamination of our water supplies. And it is to science that we will look for solutions to this crisis, for ways to recycle or safely dispose of chemicals, to reclaim affected lakes and salt marshes, to save wildlife, and to heal and cleanse our bodies.

The problem of chemical pollution has been dramatically illustrated in such areas as the Oregon forests, where the forest industry sprayed tons of dioxin-laced pesticides over an entire region, and the Love Canal in Niagara Falls, New York, where pollution

Love Canal neighborhood residents angrily and graphically demonstrate their plight as victims of a massive chemical disaster in 1978. Their homes, the neighborhood school, and the playground were built around and on top of a giant chemical dump. Miscarriages, birth defects, cancer, and other ailments became so prevalent that the neighborhood was evacuated.

Rancher George Neary of Chico, California, stands among the remains of his stricken herd of cattle. In December 1978, officials of the U.S. Department of Agriculture and the State of California sprayed his herd, against his will, with the pesticide toxaphene. Of Neary's 800 cows, 100 died, 600 aborted the calves they carried, and all surviving animals lost an average of 300 pounds, failed to reconceive, and had to be sold at a massive loss. In addition, the water supply of the area was contaminated, and the government eventually spent $1.1 million to clean up the waste. Neary is suing for damages, and while the case is still pending, he has already received a vindication of sorts—the federal government recently initiated steps to ban the lethal pesticide.

from a long-standing Hooker Chemicals and Plastics Company dump forced the evacuation of a whole neighborhood. The Love Canal, which was capped in the early 1950s and sold to the city school board for a school and playground, contained dioxin, PCBs, and dozens of other chemicals.

These materials are composed of hydrocarbons, simple but flexible compounds of hydrogen and carbon that are the basic structure of most petrochemical products. Because the carbon atom can link by four different bonds to other carbons or hydrogen, or to both, hydrocarbons can easily bind together to form rings or long, complex chains of molecules. To get the maximum yield of gasoline and other products from crude oil, the petrochemical industry has devised techniques that chemically change the hydrocarbon molecules during the refining process. The addition of heat and catalysts cracks the structures into smaller molecules or reforms them into new ones. Each new structure is, in effect, a different chemical. Some are copies of substances already present in nature; some are new, synthetic materials foreign to nature.

Through this and similar processes, the industry produces thousands of synthetic chemicals—eighty billion pounds of them annually. They have spawned a vast new industry of pesticides, herbicides, fertilizers, fibers, and plastics. Some save lives; some threaten life. One chemical by-product of

these processes is benzene, a molecular ring that is the basic structure of both dioxin and PCB, two of the most dangerous and widespread contaminants on the planet.

Dioxin is formed in a side reaction when benzene is chemically changed to produce the herbicide 2,4,5,T. PCB (polychlorinated biphenyl) is produced when two benzene rings are heated and bound together and then chlorinated. It is heat- and fire-resistant and has been used in the manufacture of power transformers as well as fluorescent light bulbs and televisions. Dioxin shows some tendency to biodegrade in sunlight, but evidence at the Love Canal and high levels of dioxin found in fish, for example, suggest that some of its components share one major harmful characteristic with many synthetic chemicals: they don't break down in the environment. PCBs, in fact, are so persistent that they have been found all over the planet, from the Arctic to the Antarctic, from Japan to the Hudson River. They have also been found in human semen and in the milk of nursing mothers.

2,4,5,T was developed as a plant killer by the military during World War II, but its first wartime use was in Vietnam, where, mixed with another chemical, it became known as the defoliant Agent Orange. Forty-four million pounds of the herbicide were sprayed on South Vietnam to destroy vegetation that provided a natural hideout for North Vietnamese soldiers. One effect: claims of widespread illness and birth defects among residents of the affected areas and among American troops who served in those areas. Millions of pounds more have been sprayed on American forests, farms, and ranchlands to kill unwanted vegetation, resulting in similar complaints. "They have babies other places, move here, and have miscarriages," said Debby Marano of Alsea, Oregon, who had four miscarriages herself in four years. "They have miscarriages here, move away, and have babies. It's just quite a coincidence."

Meanwhile, the petrochemical industry ignored indications as early as the late 1940s that the substance was unusually toxic. And government studies in the mid-1960s clearly linked 2,4,5,T to birth defects in rats and mice. The results of that study were kept secret. Finally, in 1970, in response to public pressure, the Defense Department halted use of the herbicide in Vietnam, and the government banned its use in home gardens and near bodies of water. 2,4,5,T was termed a potential hazard to women of childbearing age.

But the government ban placed virtually no restrictions on the industrial use of 2,4,5,T. That was good news to the forest products industry, which placed the long-term economic benefit of the herbicide in the hundreds of millions of dollars. In the late 1970s, when citizens protested the heavy use of the chemical in the forests of the Pacific Northwest, the industry refused to listen.

"We're not aware of any substantiated reports of such human health effects," spokesman Logan Norris said in 1979. "When we compare . . . the levels of 2,4,5,T that are in the forests with the levels that would be required to produce human health effects or effects on other kinds of animals, we don't find those levels present."

In truth, scientists had reached no consensus on such levels, and independent studies on the question differed markedly from those conducted by the industry.

Another spokesman, Cleve Goring, charged that activists challenging the safety of the spraying were guilty of "chemical McCarthyism." He explained: "This is basically to attack chemicals by innuendoes and outright lies simply because chemicals are being used as a means of achieving environmental goals. The attack is not scientific. It is purely emotional. The public does not understand,

Agent Orange

Between 1963 and 1971, 10.65 million gallons of the herbicide Agent Orange were sprayed over Vietnam. The defoliant, whose major constituent is dioxin, was used to clear the dense Asian jungles that provided cover for Viet Cong soldiers. Some 2.4 million American soldiers were exposed to Agent Orange during the defoliation operation, and several thousand of them were subjected to significant amounts of the chemical. The 1,200 men who conducted the spraying for "Operation Ranch Hand" received the most direct exposure, but the General Accounting Office, in response to a growing debate over the herbicide's deleterious health effects, concluded in 1980 that at least 5,900 Marines

were within a third of a mile of sprayed areas during the Air Force's missions, and that an additional 16,100 Marines were within a third of a mile of those areas within four weeks after the spraying.

Since the Agent Orange story was first publicized in 1978, exposed veterans have poured into Veterans Administration facilities complaining of symptoms ranging from skin rashes and numbness to liver dysfunction, decreased sex drive, and severe mood swings. Many vets suffering from cancer believe that their exposure to Agent Orange is responsible for the disease, and others blame the herbicide for birth defects in their children.

The Veterans Administration acknowledges that dioxin, at high levels of concentration, is

a very dangerous carcinogen that also causes birth defects, but the VA argues that the soldiers' exposure to Agent Orange was not significant enough to induce such reactions. The debate will not be resolved until researchers complete controlled studies of the long-term cancer rates of soldiers who were exposed and the birth defect rates of their offspring—a process that will take many years. Until then, the vets are left with a grim sense of foreboding and no VA compensation.

The human costs of dioxin contamination are immense, but the environment also pays quite a price. Land polluted with dioxin must lie fallow indefinitely, because microorganisms that can break down or utilize such manmade toxic substances have not yet

evolved. All this may change if research being conducted by Dr. Ananda Chakrabarty of the Chicago Medical Center ends in success. Dr. Chakrabarty took a soil sample from a dioxin-polluted area of the Love Canal and introduced microorganisms from this sample into a flask. He introduced other microorganisms that contained plasmids known to help break down chemicals similar to dioxin. Over time, he added more and more dioxin until only those microorganisms that could live off of the chemical survived. These in turn may be eventually reintroduced into polluted areas to help consume the toxic waste. This offers the possibility of restoring land like the Love Canal to its original condition, something that has never been done before. Much work must still

be done, however, before microorganisms of this type can be produced in sufficient quantities to attempt such a recovery.

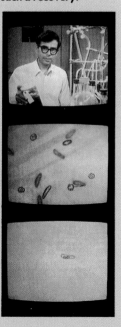

and we unfortunately haven't successfully communicated to the public as yet. Take 2,4,5,T, for example. 2,4,5,T is about as toxic as aspirin. The safety factor in the use of 2,4,5,T is far greater than the safety factor in the use of aspirin. Now, does the public understand that?"

Bonnie Hill, a resident of Alsea, Oregon, never did understand that. When she and other women in her neighborhood realized that eight among them had suffered eleven miscarriages, they put together a chart. It showed that in 1976 and 1977, when the spraying was heaviest, the miscarriage rate was 40 to 45 percent, two to four times the national average. They wrote a letter to the Environmental Protection Agency, which responded by launching a study of its own. On February 28, 1979, the EPA announced an emergency ban on 2,4,5,T and another herbicide called Silvex on forests and rights of

way all over the country. While scientists had known for a decade that dioxin caused birth defects in animals, the EPA did not act until effects on humans were found.

Dave Trotter, a Vietnam veteran, doesn't understand Cleve Goring's argument either. "I was discharged with a nervous disorder, and since then, I had two stillborn children, one of which was severely deformed. I've come up with rashes. . . . I have respiratory problems, shakes. . . . Weight fluctuations, they just continue. I went for medical help, and nothing I've done so far, especially for the rashes, has helped."

Scientists are not certain why the effects of dioxin might last so long, although it is possible that the chemical concentrates in fatty tissue remain there until the victim loses weight. Then the fat breaks down, releasing the dioxin to begin its work on other parts of the body. Dioxin's connection to birth

Environmental Disasters: A Recap

January 1981–U.S. government declares that acid rain has intensified and spread over the Northeast during the last thirty years; rainfall in that area is now ten to one thousand times more acidic than normal.
June 1979–An exploratory well off the Yucatán Peninsula blows, releasing 30,000 to 45,000 barrels of oil a day. Creates a 400-mile slick that eventually pollutes Texas beaches and fouls the Gulf of Mexico with 3.1 million barrels of crude–the worst oil disaster in history.
March 1979–An accident at the Three Mile Island nuclear power plant results in a partial meltdown of the core and release of radiation

into the surrounding area. The first major nuclear accident in the United States.
April 1977–Blowout at a North Sea oil rig causes a 180-foot-high fountain of oil, which creates a 45-by-30-mile slick containing 8.2 million gallons of oil.
December 1976–The *Argo Merchant* runs

aground and splits in half off Nantucket Island, spilling 7.7 million gallons of oil into the rich Georges Bank fishing area.
July 1976–Blast in a chemical plant in Seveso, Italy, spews the deadly chemical dioxin over the city and forces a mass evacuation. Hundreds are injured.

July 1971–The "red tide," an invasion of masses of microscopic organisms that discharge a toxic red substance, hits Tampa Bay, where tens of thousands of dead fish float to the surface. In six days, 450 tons of fish are cleared.
1970–1974–Allied Chemical Corporation dumps the deadly insecticide kepone into the James River, Virginia, and eventually shuts down the multimillion-dollar fishing industry. Allied was fined the maximum amount allowed by law, $13,375,000, for its four years of pollution.
June 1969–Endosulfan, a potent insecticide, pollutes the Rhine River in West Germany and the Netherlands, killing millions of fish.
January 1969–An oil platform in the Santa Barbara Channel fouls

twenty miles of the California coast with an eight-by-twenty-mile slick.
March 1967–The U.S. tanker *Torrey Canyon* runs aground off Great Britain and creates an oil slick that pollutes 120 miles of English coastline.
1960–1964–Pesticide runoff into the lower Mississippi River kills millions of fish and decimates wildlife in the area.
December 1952–Five days of smog blanket the city of London, killing 4,000 people. The worst air pollution disaster in history.
October 1948–The city of Donora, Pennsylvania, is engulfed in air pollution that kills eighteen people in a day. Six thousand cases of illness are reported in addition to the twenty deaths from the disaster.

defects might be explained by its molecular resemblance to other molecules scientists know to cause mutations.

In 1980, efforts continued to determine the validity of claims that dioxin exposure was responsible for the health problems of Vietnam veterans like Dave Trotter and to measure dioxin's effects on chemical workers. Meanwhile, 2,4,5,T and Silvex were still cleared for use on American rangeland and rice crops, raising the specter of still further disclosures about dioxin's deadly effects. A University of Wisconsin study reported that tumors developed in rats that were fed dioxin at levels as low as five parts per trillion. Thus scientists are concerned about reports of the chemical's accumulation in the fat of beef cattle grazing on land sprayed with the herbicide. If the rat data can be extrapolated to people, Harvard biologist Matthew Meselson speculated, the amount of dioxin in the American diet could induce enough tumors to cause the deaths of thousands of Americans every year.

In the case of PCBs, the government's action has been almost as slow, but ultimately more firm than that taken on dioxin. After it was discovered that General Electric had discharged a half-million pounds of PCBs into the Hudson River, Congress passed the Toxic Substances Control Act (TOSCA) of 1976. And in 1979, the sale and manufacture of PCBs were banned.

That still leaves another 50,000 chemicals that the industry has manufactured and put on the market. Very few have been tested to determine whether they are carcinogens or mutagens. Under the provisions of TOSCA, the EPA has only ninety days to rule on the safety of a chemical once a company has announced its intention to sell it. And three years after the law had passed, a testing program on new and backlogged chemicals hadn't even started. There are few agreed-upon standards anyway. What, after all, constitutes a safe chemical? One that is proven not to cause cancer? One that might be carcinogenic at one hundred parts per million, but is apparently safe at ten parts per million? Where do we draw the line, and based on whose evidence?

"One of the things we found that is very difficult for us as an agency, because we are on the frontier of knowledge with regard to many chemicals, is [getting] data at what levels the human organism begins to take effect," said the EPA's Chris Beck. "We have to do studies. We have to gather information to make those determinations, because those determinations are not only for the purposes of protecting people's safety. . . . When this agency picks a number, ... behind it are tremendous costs: there are economic costs; there are social costs; there are behavioral costs. And we've come to learn that it takes time to do this."

This picture of an industrial plant was taken in the 1940s, when smoke spewing from stacks was a sign of a healthy economy, a symbol of industrial might and progress. Now we are paying the price of an attitude that neglected the concerns of our environment and our personal health.

A victim of the *Amoco Cadiz* oil spill off the coast of Brittany in March 1978.

When the *Amoco Cadiz* spilled its cargo of 68 million gallons of oil into the sea, the environmental impact was immediate and devastating. Clean, white beaches were coated with thick, black goo. Sensitive salt marshes, estuaries, and oyster beds were saturated. The estimated toll on marine and wildlife was staggering: 20,000 birds, dead; 10,000 finfish, dead; millions of sea urchins and razor clams, lying in windrow after windrow along the beach, dead; thousands of clams, lobsters, crabs, periwinkles, limpets, and sea worms, dead; acres of seaweed beds and marsh grass, dead.

Such body counts, however, are only part of the story. Assessing the long-term impact and the nonfatal injury to marine life is where the science gets tricky. Hundreds of millions of gallons of oil have spilled into the sea from tankers, barges, oil rigs, pipelines, and other sources since the *Torrey Canyon* broke up off Cornwall, England, in March 1967, sounding the first international alarm to the perils of oil. Scientists who flock to spill sites to study the effects are confronted with a variety of obstacles they don't face in the lab. In contrast to its effect on marine life, a spill's most direct impact on humans is economic; that can color the observations of fishermen and resort owners, for example, who have much at stake when the damages are added up. Second, scientists collecting samples of water, sediments, and marine life at the scene of a spill almost always lack one crucial piece of evidence: a control sample taken from the same area *before* the spill occurred. Thus they have no base from which to measure the damage and, later, de-

termine the extent of recovery. Moreover, oil changes rapidly in the water; it evaporates, emulsifies, mixes, disperses, and compacts. At each stage, it reacts differently with the sea and its organisms, posing constantly changing conditions for the scientist.

To make matters more complicated, investigators cannot always depend on past experience to guide them, for no two spills are alike. Their effects on marine life depend on what kind of oil is spilled, where it is spilled, what season it is spilled in, what weather conditions are present, what kinds of organisms live in the area, and what methods are used to clean the spill. When the *Argo Merchant* ran aground off Nantucket, Massachusetts, in December 1976 and spilled 7.6 million gallons of heavy industrial oil, high winds that foiled cleanup efforts pushed much of the oil out to sea; the environmental damage was minimal. When the barge *Florida* went aground outside West Falmouth, Massachusetts, in September 1969, it spilled about 200,000 gallons of No. 2 fuel oil, much of which found its way into the sediments of the town's rich scallop beds. More than a decade later, some polluted beds were still closed to shellfishermen.

Scientists studying the *Amoco Cadiz* spill determined that 70,000 tons—about one-third of its cargo—evaporated. Another 50,000 tons were sucked into vacuum trucks during the months-long cleanup. That left 100,000 tons of oil still unaccounted for even after the beaches were relatively clean. Scientists figured that about half of it seeped into the life-supporting sediments along the 100 miles of beaches, inlets, and marshes. The rest, it was discovered, mixed with the sea and sand in heavy tides and formed a fine, oily silt that settled on the ocean floor. Just how long the oil buried in sediments and lingering on the bottom would remain and what its effects would be were not known months after the spill, but leading French scientists predicted it would be a decade before the sea and its regenerative powers could cleanse the coastline and replenish it with sea life.

Such an estimate assumes that Brittany will be spared the damage of another spill. Despite the immediate massive impact of spills like the *Amoco Cadiz,* scientists consider the slow, chronic dripping of oil from sewer outfalls, harbor spills, and storage tank leaks more damaging to the environment because the affected areas never have time to recover from one dose of oil before the next one comes along. Large tanker spills are not considered chronic to a specific area because that kind of lightning has rarely struck twice in the same place. The people of Brittany, however, have good reason to dispute that. The *Amoco Cadiz* marked the fourth time in eleven years that *la marée noire,* the black tide, had battered the Brittany shore. In 1976, two other tankers spilled oil off the coast. And in 1967, the *Torrey Canyon's* cargo also reached Brittany,

The bow of the stricken supertanker *Amoco Cadiz* looms over the French fishing village of Portsall on the northwestern tip of Brittany. The ocean swarms over the tanker's deck *(inset above)* while an army of citizens with only primitive equipment *(inset below)* works to clean up some of the 68 million gallons of crude oil that spilled from the hull.

causing widespread damage. For eleven years, until the *Amoco Cadiz* incident, the *Torrey Canyon* spill of 33 million gallons stood as the largest on record. Thus Brittany now lays claim to a unique and sad distinction: her shores have had to absorb the two largest tanker spills in history.

Like oil spills, acid rain is happening around the world. Unlike oil spills, its damage is indicated in part by an eerie beauty: acidified lakes turn a deep, clear blue and seem almost pure. Actually, they are almost sterile, nearly devoid of fish and other aquatic life.

Acid precipitation is caused by sulfur and nitric oxide emissions generated by the combustion of fossil fuels. The predominant source of these emissions is the giant smokestacks from gas-, coal-, and oil-fired power plants, particularly those in the industrial Midwest. The pollutants spewed into the air ride the prevailing winds to the Northeast and combine with atmospheric moisture to form sulfuric and nitric acid, which falls as acid rain or snow when the clouds collide with mountain peaks in New York, northern New England, and Canada. In the Adirondack Mountains of upstate New York, the most severely affected area of the United States, acid rain has already claimed more than 200 lakes. Other states in the northeastern quadrant of the country are feeling similar effects.

The lakes and streams of the Northeast are particularly vulnerable to acidity because the region's geology is granite-based and has no buffer like limestone to neutralize the effect of acid. Thus, as the rain and snow fall, the pH—the measure of acidity—in the lakes and streams gradually decreases. Adirondack lakes are currently more than forty times more acidic than "pure" rain.

The acid itself is enough to discourage some fish, but it also leaches aluminum and other metals from the sediments, poisoning the fish. Studies also show that high acid levels have especially harsh effects on the reproductive cycles of mature fish and on fish eggs and fry.

Much of the concern about acid precipitation has focused on its impact on lakes and wildlife, but recent developments suggest that its harm may reach far beyond rural lakes and woodlands to farmland and vegetation, to water supplies and human health. Indeed, some research indicates that acid precipitation leaches nutrients from leaves and thins their outer covering, making them vulnerable to insects and disease. Massachusetts officials are concerned about the acidification of the Quabbin Reservoir, the water supply for more than two million people in the Boston area. The economic costs of keeping the water potable are likely to be severe. And just as the acid leaches metals into rural lakes, so it will also contaminate municipal water supplies.

Some scientists are trying to mitigate the effects of acidity by developing stocks of fish more tolerant of acid, and by treating the lakes themselves with lime and other neutralizers. But it is likely the problem will abate only if sulfuric and nitric oxide emissions are substantially reduced by burning low-sulfur fuels and by installing scrubbers on all power plants. It is a costly solution, but many feel it is a necessary one if we are to alleviate the effects of an environmental problem that some scientists rate at least as significant as chemical pollution.

Chemical waste is another significant contributor to environmental hazards. Even if

Brook trout, confined within a wire cage, are asphyxiated in an Adirondack stream that has been polluted by acid rain.

we learn to use petrochemicals safely and avoid life-threatening substances like PCBs and dioxin, the environmental crisis will not be diminished unless we find a way to get rid of chemical waste. PCBs may be banned, but that hardly ensures their safe removal from the environment.

Indeed, 90 percent of the nation's chemical waste is dumped illegally or on unsafe sites. The result: hundreds of communities with contaminated water supplies; dozens of "cancer towns," neighborhoods where, some experts believe, chemical pollution has caused unusually high incidence of debilitating disease; and an environmental disaster like the Love Canal. Two hundred thirty-seven families were evacuated from the perimeter of a chemical dump they had lived with for years, unaware that its poisons were inflicting them with birth defects and disease. In July of 1982, amid renewed controversy, approximately two-thirds of the evacuated area was deemed

fit for rehabitation, while one-third remained dangerously contaminated.

Scientists and engineers are developing the technology to construct "safe" chemical waste-processing facilities, but companies that want to operate them are finding stiff local opposition. The horror stories of chemical pollution are now commonplace across the country, and few people are willing, no matter how earnest the guarantees of safety, to risk a chemical dump in their own backyard. Yet safe disposal sites will be necessary if we are to avoid more Love Canals. Otherwise, we may face a generation of children like the young boy who in February 1979 stood up in the midst of an emotion-packed hearing on the Love Canal and said: "I want to know if I can go home and live with my mother and father again. I had to go to my grandmother's house. I want to know if I'm going to grow up to be a normal man."

An infrared aerial view of the Love Canal shows the disruption of vegetation (red) affected by the poisons that leached from the dump site and spread throughout the neighborhood along underground streams.

A Conversation with Helen Caldicott

Dr. Helen Caldicott resigned her post as a pediatrician at Boston's Children's Hospital in 1980, but she insists she is still practicing medicine. Her work as an antinuclear activist, she says, is in fact "the ultimate medical practice."

Caldicott now devotes herself full time to educating people about the medical hazards of the nuclear industries, such as the link between radiation and cancer. She says of her present occupation: "It is not just an extension of my work as a physician: it *is* my work as a physician, because, as Howard Hiatt, dean of the Harvard School of Public Health, says, nuclear war will create the final medical epidemic.

"The medical dictum states that if you have an incurable disease, the only resource is prevention," she says. "So, we *must* prevent nuclear war from happening."

In that sentiment she is far from alone in her profession. Caldicott is president of Physicians for Social Responsibility, a national antinuclear organization whose membership has billowed from a few hundred to many thousand in just three years. The scientific expertise of doctors gives them credibility with their audiences when they speak on the "medical effects" of nuclear war and energy production. Furthermore, Caldicott says, the Hippocratic oath to which physicians swear gives them a moral obligation to speak out.

Caldicott has become a leading figure in the growing international movement for disarmament. She insists that scientists as well as doctors must accept roles as stewards of our planet. She knows that the tradition of emotional detachment prevails in science, but says the world can no longer afford such an attitude.

"If you're a good doctor, you remove your emotions from the practice of medicine, because otherwise they will bias the way you practice medicine," she says. "Similarly, scientists don't allow their emotions to impinge upon their scientific work, nor do they allow themselves to consider what their work will lead to. But it is *not* appropriate, as we approach the annihilation of the

species by the works of man and the scientists, to be unemotional about it."

Albert Einstein, whose theories contributed to the development of the atomic bomb, vocally opposed the post–World War II arms race and urged scientists to speak out on the hazards of nuclear weapons. But many scientists today maintain that science should be "value free," that they should concern themselves only with their research, not with its impact on society.

Caldicott disagrees. "Scientists have developed the theories that have allowed the nuclear weapons and their delivery systems to happen," she says, slapping the arm of her chair for emphasis, "and thus they apply their science in ways which are medically contraindicated and which I consider totally amoral. They are creating a global gas oven in an unconscious way."

The resolve with which she approaches this issue was a trait she showed even as a child in Melbourne, Australia. In 1950, at the age of eleven, she had a natural curiosity and a love of biology, and she knew "absolutely" that she would someday become a doctor. She first became aware of the nuclear threat when as a teenager she read *On the Beach,* Nevil Shute's fictional account of the annihilation of humankind in a nuclear war. The novel, set in Caldicott's hometown of Melbourne, terrified the young woman, whose atomic angst increased as she later read thoroughly factual newspaper accounts of the world arms race.

After medical school she married William Caldicott and had three children. Then in 1971 she learned that the French had been conducting atmospheric tests of atomic bombs for five years in the South Pacific. Inspired by Bertrand Russell's autobiography, which told of his "ban the bomb" campaign, Caldicott protested the French tests in an angry letter to a local newspaper. She knew that children and fetuses were ten to twenty times more susceptible than adults to the effects of radiation and was angered to learn that Australian drinking water showed high levels of radioactivity from the fallout. As a pediatrician and a mother, she spoke out in the media on the health hazards of radiation, pressured public officials, and persuaded 75 percent of Australians to oppose the tests. After two years of protest, the French agreed to conduct their tests underground.

The Caldicotts moved to the United States in 1977, when Helen Caldicott and her husband, a radiologist, accepted positions at Children's Hospital and Harvard Medical School, respectively. But Helen Caldicott quit her job three years later because of what she calls a conflict of interests. "I treated children who had cystic fibrosis, the commonest fatal genetic disease of childhood," she says, her blue eyes intent. "And I kept thinking, 'Why am I keeping them alive, when they could be killed by a nuclear war in the extended lifespan I've provided them?' It was bad medicine."

Scientists, too, should examine the ethics of their trade, says Caldicott. But they often do not consider the long-term consequences of their work, because of the tremendous specialization in the sciences today. Researchers concentrate on one highly defined area, and no single scientist or engineer has an overview of the whole industry.

Half of America's scientists and research engineers work for the military or industries related to it, according to Caldicott, and not enough of them have publicly opposed the proliferation of nuclear arms. On the contrary, she says, "It's absolutely well known that Los Alamos and Lawrence-Livermore are the ones who are leading the arms race. They don't want a complete test ban treaty. They want their grants." Technological developments in weaponry often predate the political rationale for their use, she says.

Science and technology can just as well be turned toward the benefit of humanity as toward its destruction, says Caldicott. "Two-thirds of the world's children are malnourished and starving. The world's in a terrible mess. We have the technology and the expertise to save ourselves and make life better for everyone."

Although she misses her practice as a working pediatrician, she says she won't resume it until the world is rid of the bomb. That is a monumental goal, and the constant crusading sometimes tires her. To renew her energy she spends time with her family. "That's where I get my strength," she says, smiling. "The media attention sometimes drives me crazy, but I know that it's necessary, and if it's achieving the end goal, then it's okay."

In her lectures throughout the country, Caldicott catalogues the hard, cold facts of nuclear weapons and then makes an emotional appeal to the compassion of her audience. She wants to "crack open the souls" of her listeners, including the scientists among them. "Scientists are all basically good people," she says. "They have kids that they love, wives that they love, and they care about their own lives. We doctors treat them when they're dying or when their kids are dying. We see what they're like: soft, frightened, loving human beings.

"I just want them to open up their souls and look at this thing from the perspective of their humanity," she says, gazing off into the distance, "and stop being scientists for a while, and understand what we're doing to the planet."

Of Pests and Poisons

The earliest humans were no doubt annoyed by the stings of wasps and the bites of mosquitoes and lice. They fell victim to insect-borne diseases and suffered the loss of stored foods to insect pests (although Paleolithic food gatherers were probably more tolerant of insect-damaged food than are today's fastidious shoppers). But it wasn't until the development of agriculture that people and bugs came into serious conflict. Annual losses to insects now exceed $5 billion in the United States alone; damage worldwide must be many times more costly. The control of insect pests has become a big business.

The notorious desert locust, a kind of grasshopper, is one of the worst offenders. In the belt of arid country that stretches across northern Africa to India and Central Asia, occasional periods of heavy rain can trigger an extraordinary change in the locusts. The light-colored, solitary, sedentary insects will give rise to a generation of darker, gregarious, migratory young. These wet-phase individuals can reproduce themselves at an astonishing rate, and the huge swarms that result spill out of their territory and invade neighboring lands. Wherever they alight, the locusts strip the countryside bare. The descent of a locust plague can spell disaster and famine for local people.

Locusts can't be exterminated, but they can be controlled. The United Nations Food

Population explosions of locusts may occur when the weather is uncharacteristically rainy in the arid lands of North Africa and western and central Asia. Enormous hordes of locusts then form and take flight. Where they land, they devour almost every living plant in their path. Plagues of locusts threaten local residents with famine and incalculable economic loss, and outbreaks of the pests are monitored diligently by the U.N. Food and Agriculture Organization in Rome.

and Agriculture Organization (FAO) keeps a careful watch on satellite-recorded weather conditions that favor the irruption of locust populations. In its headquarters in Rome, the FAO collects reports of local outbreaks of swarms and tries to predict where they will travel. It coordinates eradication programs, usually involving the aerial spraying of insecticides, and funds research on new, more effective control measures. In the end, though, the only real check on locust numbers is the inevitable return of drought conditions.

In Africa, south of the arid locust country, another insect places severe constraints on agriculture: the tsetse fly. Unlike the locust, which poses episodic problems, the fly creates chronic difficulties. The tsetse is a carrier of parasites called trypanosomes ("tryps"). When it bites a cow or goat or sheep, the fly injects the tryps, which cause the mammal to become anemic and, finally, to die. Because of the fly's wide range in sub-Saharan Africa, much of the continent is unsuitable for keeping livestock.

Traditional control has been expensive and labor-intensive. Most of Africa lacks the infrastructure that would permit the efficient distribution and administration of available anti-

In many parts of Africa, the tsetse fly makes cattle herding a perilous undertaking. Trypanosomiasis, a deadly disease of livestock, is transmitted by the fly when it bites. Efforts to control both the tsetse and the disease it spreads have so far met with only limited success.

tryp drugs. In relatively affluent Zimbabwe, the prosperous cattle industry is protected by a combination of DDT application and the maintenance of a cumbersome network of barrier fences that exclude game animals, which are living reservoirs for the parasites, although themselves immune to trypanosomiasis. But few other countries could afford such costly measures, and cheap and easy techniques are avidly sought.

Some researchers are pursuing the development of an effective vaccine. Others are experimenting with the captive propagation of male tsetses, which are sterilized and released; it is hoped that unfruitful unions between wild females and lab-sterilized males will cause the fly population to plummet. The discovery that the flies are drawn to the smell of cattle, like moths to a flame, has prodded inventors to develop a "tsetse trap," a device baited with attractant chemicals isolated from the breath of cows and equipped to electrocute or sterilize the insects it traps.

Not everyone welcomes the eventual control of the tsetse. This "guardian of Africa" has helped preserve great tracts of land from conversion to livestock pasturage and the overgrazing and erosion that usually accompany such conversion. Wild game animals don't destroy the range, they are immune to

fly-borne tryp disease, and they are more efficient at processing the plants they graze and browse into meat. For these reasons, some ecologists and wildlife managers favor the adoption of "game ranching," which might feasibly yield greater amounts of animal protein for hungry people and simultaneously prevent the destruction of the remaining savannas and preserve examples of the diverse fauna of the African plains.

The use of DDT against the tsetse fly has been rather limited, primarily because of the difficulty of distributing the chemicals and keeping up the machinery needed for spraying. In marked contrast, the widespread use of insecticides has been a cornerstone of the intensive agriculture practiced in the United States. The post–World War II availability of DDT, then hailed as a miraculous boon, ushered in a revolution in insect control. That substance, and others even more deadly, soon came to be applied indiscriminately by the nation's farmers, who were exultant over the prospect of freedom from economic losses to inimical insects. Their euphoria was short-lived. The insecticides were nonspecific—they killed not only pests but also beneficial species, such as predatory insects (which act as natural controls on populations of harmful bugs) and bees and other pollinators, whose services to agriculture

Thousands of readers were introduced to the intricate interconnections between life and its environment through the beautifully written books of biologist Rachel Carson, who directly inspired the development of the ecology movement. Her prize-winning *Silent Spring,* first published in 1962, alerted many to the problems posed by the unrestricted use of pesticides.

more than balance the damage wrought by the pests. And the new insecticides were long-lasting. They persisted in the ecosystem and accumulated, often to lethal levels, in the bodies of animals near the top of the food pyramid. The populations of many birds declined so drastically that Rachel Carson was led to prophesy the coming of the "silent spring." Alarm mounted over the progressive toxification of the environment, and the use of some pesticides was somewhat curtailed. Certain poisons, including DDT, were banned for most domestic applications (although the government generously allowed U.S. firms to continue to manufacture them for use overseas, from where they return to us in foodstuffs imported for American consumption—the "pesticide boomerang").

Perhaps the most disturbing outcome of the prodigal use of insecticides was the exacerbation of the very problem the chemicals were employed to solve. After an initial decline, populations of pest species started to explode. The sprays themselves had favored the differential survival of resistant individuals, which fostered new hordes of poison-immune pests. And the destruction of beneficial insects removed an important check on pest numbers, making necessary still more dosing with killer chemicals.

Farmers have turned to science to help them get off the pesticide treadmill. A return to practices of sound crop management by itself can reduce losses significantly. Take the case of cotton, the crop on which half of all the insecticides used each year is lavished. The planting of varieties that mature early, before the pest populations can build to destructive levels, and the plowing under of the stalks that remain after harvesting, thereby denying pests the shelter they need to get through the winter, are techniques that were followed before the advent of chemical sprays and to which many modern farmers are now turning. The study of the life cycle of a pest species also often reveals a brief period of greatest susceptibility, when a single spraying can preclude a plague later in the season.

Enlightened farmers and entomologists (scientists who study insects) have further discovered that even apparently heavy infestations of pests may still lie below the threshold at which economic loss becomes intolerable, and that restraint from spraying allows the beneficial bugs, which can do a better job of keeping the pests under control, to survive. This realization that certain bugs – up to a limit – can help the situation has led to the artificial propagation of insect predators and parasitoids. Many California citrus groves, for example, are kept free of injurious scale by tiny wasps that are raised in captivity and then released. The females lay their eggs in the bodies of the scale insects, which are consumed from the inside by the wasp larvae that hatch.

Especially exciting are experiments involving the identification of chemicals that could be used to disrupt the normal development and behavior of insect pests, while sparing other species. Researchers are investigating antijuvenile hormones, which would induce pest larvae, like many caterpillars, to metamorphose prematurely, with fatal results. Other workers are attempting to isolate and synthesize pheromones, species-specific sex attractants that bring the males and females together. Pheromones can be

Bye Bye Blackbird

Poultry farmers and sheep ranchers have an (unfounded) antipathy for birds of prey. City dwellers have a more legitimate dislike of pigeons, starlings, and house sparrows. But the hatred of blackbirds (a family that includes several different species of blackbirds proper, as well as grackles and cowbirds) among agriculturists in the southeastern United States borders on the hysterical. No matter that the birds' huge populations were made possible in the first place by the conversion of forests, swamps, and marshes to extensive monocultures of corn and rice. No matter that the annual loss per acre to the birds is still several times less than the waste that normally results from mechanical harvesting. And no matter that the birds' value as destroyers of insect pests throughout the year outstrips by far the damage they do during the growing season. The birds are per-

ceived as pests, and eradication measures are undertaken at great expense and with little efficacy, an all too characteristic response of a civilization alienated from nature.

The sight of crop-dusting planes is becoming rarer over our nation's agricultural lands. As recognition of the dangers and counterproductive results of profligate pesticide use mounts, farmers are turning to more controlled schemes of pest control.

used to entice the males to a sterilizing trap, or can be spotted throughout a field to produce an aphrodisiac cloud. The confused, overstimulated males are unable to distinguish between the actual females and the artificial lures, and the mating system then collapses.

More than three-quarters of all known animal species are insects. Ubiquitous and abundant, they are key members of just about every terrestrial ecosystem. The benefits that accrue to humankind from insects are incalculable and outweigh the damages. That some species compete seriously with people for limited resources is nevertheless inarguable. Some control will always be necessary, although we must be wary of killing cures.

Caution: Poison

Most of the poisons used to snuff bugs fall into two great classes of chemical compounds. DDT is probably the best known of the chlorinated hydrocarbons, which also include dieldrin, toxophene, chlordane, and lindane. These toxins, absorbed readily through the chitinous cuticle that encases an insect's body, attack the central nervous system, producing convulsions, paralysis, and death. Because the chlorinated hydrocarbons are poisonous not only to insects but to a great variety of other species, because they break down so slowly, because they are carried easily in water and in airborne dust, and because they tend to concentrate in the fatty tissues of organisms, ecologists fear the effects, long of term and wide of range, that these chemicals will have on living systems throughout the world. Less dangerous are organophosphates like malathion, parathion, and phosdrin. All toxins of this class render inactive the enzyme that breaks down acetylcholine, a substance important in the transmission of nerve impulses. An insect so poisoned twitches to death. Because the structure of organophosphates is unstable, they break down quickly and pose less of a hazard to ecosystems. Still, like the chlorinated hydrocarbons, they often aggravate problems by eliminating the predators of injurious insects and promoting the natural selection of poison-resistant strains of

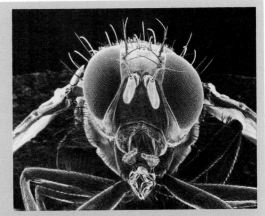

precisely those pest species they were intended to control.

Death Watch for the Whale

The nineteenth-century Yankee whalers sailed out of New Bedford and other New England ports. Ignorant of such modern concepts of resource management as rational exploitation and sustained yields, they took a terrible toll of the marine mammals they hunted and drove many populations to near or total extinction.

The bowhead is one of the most critically endangered of the great whales. It is estimated that not more than 3,000 remain in the shallow, plankton-rich waters of the Arctic. Twice each year, the whales pass through leads in the ice of the Bering Straits, between their summer feeding grounds in the Beaufort and Chukchi seas to their winter range in the Bering Sea to the south. And for well over a thousand years the people of northwestern Alaska have hunted these migrating bowheads.

Until quite recently, entire Eskimo villages depended on whale meat through the long Arctic winters. The whale's blubber yielded precious fuel. Every part of the carcass was used, down to the skeleton and the baleen (the flexible "whalebone" with which the bowhead strains its food from the sea). Today the bowhead remains central to Eskimo culture, and an elaborate system of beliefs and rituals has grown up around the annual hunt for this vital mainstay of the econ-

omy. Taking perhaps forty whales a year, too few to make an appreciable dent in the population, the Eskimos coexisted with their great quarry for centuries.

Enter the Yankee whalers. The first ship to penetrate the Straits arrived in 1848. The fat, sluggish bowheads proved easy marks, and the whalers filled 1,800 barrels—more than 60,000 gallons—with oil in just one short month. The word quickly spread, and more than 20,000 bowheads fell to the whalers' harpoon guns over the next sixty years. The high prices then being fetched for whale oil and baleen (used to stiffen ladies' corsets) lured even some of the Eskimos into the commerce.

By 1910, the whale stocks were so depleted that the industry collapsed, and the Eskimos returned to their subsistence hunting of the remnant herds. But the forces of "civilization" were again to intrude, this time in the 1960s, when the construction of the trans-Alaska pipeline and other projects on

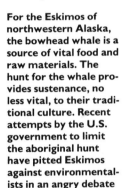

For the Eskimos of northwestern Alaska, the bowhead whale is a source of vital food and raw materials. The hunt for the whale provides sustenance, no less vital, to their traditional culture. Recent attempts by the U.S. government to limit the aboriginal hunt have pitted Eskimos against environmentalists in an angry debate over issues of cultural and biological survival.

the North Slope provided a new source of jobs and money to the local Eskimos. The whale hunt had historically been limited because only a few men could afford to equip and maintain a crew. The sudden infusion of new wealth enabled more Eskimos to become captains and mount whaling expeditions, and the annual catch of bowheads started to climb.

The International Whaling Commission (IWC) had forbidden almost all hunting of bowhead whales since 1931. The Eskimos' subsistence hunt was the lone exemption. In the decades prior to 1970, the Eskimo whalers had been taking only about fifteen animals each year. But in 1976, forty-eight whales were killed, and twenty-eight more in 1977. Add to this the number of whales struck but lost: forty-three in 1976 and seventy-seven the following year. Most of the struck and lost animals are assumed to have perished later. This escalation of hunting pressure on the already endangered bowhead population alarmed conservationists, who responded by pressing more strenuously for an end to the hunt.

Fearful that growing and uncontrolled Eskimo whaling might drive the bowhead to extinction, the IWC in 1977 extended its ban to the aboriginal hunt. The Eskimos were infuriated by the imposition of the new zero quota. They charged officials with interference in their traditional way of life, abrogation of hunting rights promised by treaty,

The Business of Extinction

I see you got your brand-new leopardskin pillbox hat . . . and your penknife with the rhinoceros-horn handle, and your pet whooping crane. The destruction of precious and rare wild animals for their conversion into apparel, curios, trinkets, and trophies is unconscionable. The removal of wild animals from natural breeding populations so that they may serve as exotic pets for people bored with cocker spaniels and guppies is no less deplorable. The collection of animals for these frivolous purposes, in combination with the even more dangerous practices of habitat destruction, has reduced most populations of large mammals, birds, and reptiles to levels where their numbers could not flesh out an average American small town. Perhaps all this will be checked by the recently enacted Convention on International Trade in

Endangered Species of Wild Fauna and Flora (CITES) and by the burgeoning concern among citizens of the industrialized nations, on whose behalf the worst excesses of environmental abuse are committed.

The Plight of the Whales

As whales were slaughtered at a rate of one whale every twelve minutes—more than 100 whales a day—many whale species neared extinction by the 1970s. The International Whaling Commission (IWC), formed in 1946 to regulate whaling on a global scale, currently bans the hunting of blue, gray, right, humpback, and bowhead whales, except by Eskimos and other native peoples. The United States put an end to all commercial exploitation of whales in U.S. waters in 1972 with the passage of the Marine Mammal Protection Act. Meanwhile, Russian and Japanese whalers continued to kill almost 40,000 whales a year, nearly 80 percent of the world catch. In a 1979 meeting the IWC established a 40-million-square-mile whale sanctuary in the Indian Ocean and a moratorium on all factory-ship whaling. Russia and Japan voted against the moratorium, but Russia announced plans to terminate whaling within five years. While the United States and several other nations work for a total ban on commercial whaling, the permissible commercial take set by the IWC dwindled to 13,851 whales in 1981, down from an annual quota of 46,000 in 1973. In 1982 the IWC voted to ban commercial whaling beginning in 1986. However, Norway and Japan continue to oppose the moratorium and threaten to withdraw from the commission—a move that could weaken it considerably.

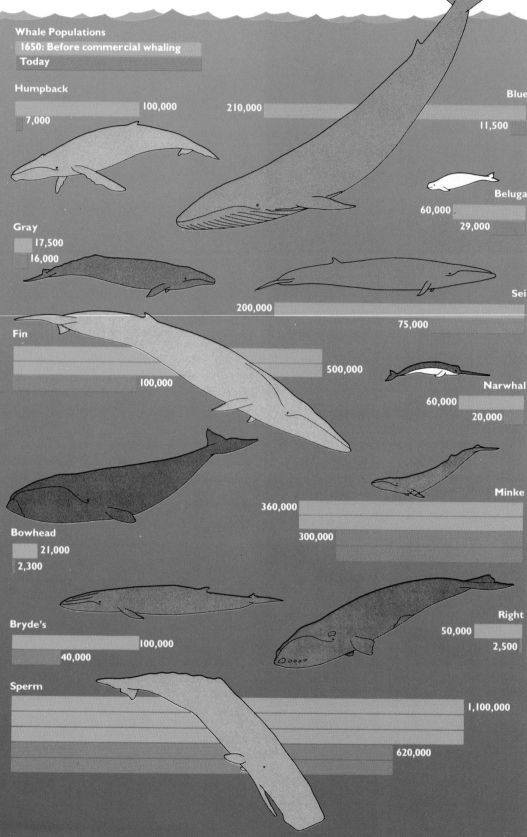

Whale Populations

1650: Before commercial whaling
Today

Humpback
100,000
7,000

Blue
210,000
11,500

Beluga
60,000
29,000

Gray
17,500
16,000

Sei
200,000
75,000

Fin
500,000
100,000

Narwhal
60,000
20,000

Minke
360,000
300,000

Bowhead
21,000
2,300

Right
50,000
2,500

Bryde's
100,000
40,000

Sperm
1,100,000
620,000

cultural imperialism, even racism. In the acrimonious debate that followed, some conservationists antagonistically dismissed the Eskimos' arguments, claiming their traditional culture was already debased, that Eskimos who watched color TV, drove snowmobiles, and speared whales with harpoons fitted with explosive heads had forfeited any legitimate right to the subsistence hunt. Others in the conservation community, more sympathetic to the Eskimos' position but nevertheless convinced that continued hunting *at this time* would depress the whale population to the point of no return, also supported the zero quota.

The Carter administration found itself between a rock and a hard place. It had gone on record as being committed both to the conservation of whales and other threatened species and to the vigorous support of the rights of human minorities. The U.S. commissioner to the IWC persuaded his colleagues to restore a limited quota of twelve whales landed or eighteen struck (later raised to eighteen and twenty-seven for the 1979 hunt), a solution that mollified neither the Eskimos nor the conservationists. The resentful Eskimos declared the quota so low as to deny their subsistence needs and to rend further the fabric of their society. Some captains threatened to defy the quota. The conservationists maintained that *any* exploitation of an endangered stock was impermissible, and that the government's equivocating stance had compromised, perhaps fatally, the credibility of U.S. leadership in the movement to impose a moratorium on whaling worldwide.

Population studies of the bowheads have been undertaken, but the data garnered so far are inconclusive. The Eskimos continue to insist that the herd is large enough to withstand greater hunting pressure. The conservationist position was best expressed in the statement of the International Union for the Conservation of Nature / World Wildlife Fund to the 1980 IWC meeting: " . . . although there are few biological absolutes, extinction is one of them; and

extinction at the hands of a subsistence hunter is no less final than at the hands of an industrial harpoonist." A final irony is that oil and gas development in the Arctic is accelerating even as the bitter debate over the Eskimo hunt continues. A major oil spill could destroy the swarms of plankton the bowheads feed on, clog the delicate filter hairs on the whales' baleens, and obviate any future discussion of the need to guarantee the survival of both the species and the cultural integrity of a people.

Old-fashioned whaling was fraught with danger for the crew. Modern techniques have reduced the hazard, although whalers continue to work under conditions of great difficulty. Concern over the fate of the gentle and intelligent whales has sparked a growing international movement to stop the slaughter.

The Wolf Equation

Now virtually extirpated from its former wide range in Mexico, the contiguous United States, and much of Canada, the North American wolf hangs on in Alaska, where for thousands of years it has made its living hunting the peculiar deer called caribou. Enter people, with their settlements, their quest for oil and natural gas, their ingrained prejudice against the wolf as vermin and competitor, and their own fondness for caribou meat, hides, and trophy racks. The numbers of caribou plummeted, and people of course blamed the wolf. In fact, most evidence suggests that caribou herds have traditionally passed through cycles of boom and bust, and that repopulation of depleted herds has been accomplished through the immigration of deer from neighboring areas. Humans today take a greater toll of caribou than do

wolves, and exacerbate the situation by impeding migration between herds. Rather than curb their own deleterious activities, people have rushed instead to redress the imbalance by slaughtering "excess" wolves.

Preserving the Balance

Living things don't exist as isolated individuals. Every organism, be it a bacterium, an alga, or a protozoan, a fungus, a plant, or an animal, passes its life in various associations with others of its own kind and of different kinds. Neither do living things exist in a vacuum, but in the context of an environment with particular physical and chemical characteristics. Complexes of organisms interacting with one another and reacting to and modifying their environment are called ecosystems, "the basic units of nature on the face of the Earth," according to Alfred Tansley, who coined the term in 1935.

Implicit in this concept of a system of dynamically interacting components is the notion that a change in one will effect changes in others. One of the ways that scientists learn about ecosystems is by observing the results of changes, both deliberately induced and naturally occurring. Human beings, by far the most numerous large animals on Earth, are agents of enormous and pervasive change.

We induce changes not only directly in ecosystems where we reside, but also indirectly in ecosystems where no person has ever set foot.

We humans thus play an active rather than a passive role in our environment. And as the main characters, we must take the responsibility not only for those resources and environments necessary to our existence, but also for the environments and resources of the entire global population.

Ecology, a relatively new science, is often called the study of the economics and sociology of nature. More precisely, it is that branch of science concerned with the interrelationships of organisms to their environments, with the total pattern—and ultimately the health—of an ecosystem. When we examine a sampling of four such ecosystems—unlike except that all have been affected by the activities and work of people—we see the precarious balance that ecologists strive to preserve.

One "spaceship" as seen from another. This stunning view of the Earth was photographed from *Apollo 17* in December 1972. From the Sahel, the vast belt of arid steppe south of the Sahara, visible in the upper left quadrant, to Antarctica, the frozen southernmost continent, no place on our planet has escaped being affected by the hand of *Homo sapiens.*

The River The Colorado rises in the Rocky Mountains of the state that shares its name. Joined by its tributaries, it makes its way west and south toward the Gulf of California, where it once emptied. Now, this once-great river no longer reaches its delta, but instead sinks, a stagnant trickle, into the sands of Baja California.

The Colorado once delivered an average of 14 million acre-feet of water each year to the gulf (an acre-foot is the amount of water required to flood an acre of land to a depth of twelve inches). Where did the Colorado go? Into aqueducts and diversion tunnels, which yearly spirit 5.5 million acre-feet away from the river's natural basin, to the west to the southern California megalopolis, to the east across the mountains and into the Platte River, and thence to the Missouri, the Mississippi, and the Gulf of Mexico. Where? Into irrigation ditches (5.6 million acre-feet), to wet the fields and farms of the seven states that claim water rights (and a few farms in Mexico, which must depend on water from California's All-American Canal because, beyond the Imperial Dam near Yuma, the Colorado doesn't flow anymore). Where? Into underground aquifers (0.5 million acre-feet), by percolation through the porous sandstone banks and bottom of Lake Powell, the gargantuan artificial lake backed up behind the Glen Canyon Dam and burying beneath its waters a landscape once considered among the most glorious in the world. Where? Into the roots and out through the leaves of phreatophytes (another 0.5 million acre-feet), thirsty plants that suck up ground water (as opposed to soil moisture) and that have become permanently established along banks and beaches since technological shackles have put an end to the seasonal floods of the free-flowing river. Where? Into the air (1.6 million acre-feet), lost as vapor to the desert sun and wind from the surface of lakes Powell, Mead, Havasu, and other impounded reservoirs. The scant 0.3 million acre-feet remaining from the original 14 million is stored in these several reservoirs.

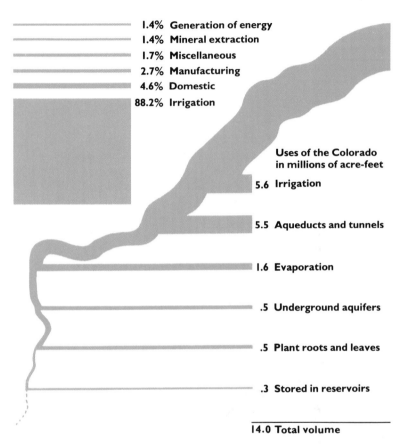

1.4%	**Generation of energy**
1.4%	**Mineral extraction**
1.7%	**Miscellaneous**
2.7%	**Manufacturing**
4.6%	**Domestic**
88.2%	**Irrigation**

Uses of the Colorado in millions of acre-feet

5.6 **Irrigation**

5.5 **Aqueducts and tunnels**

1.6 **Evaporation**

.5 **Underground aquifers**

.5 **Plant roots and leaves**

.3 **Stored in reservoirs**

14.0 **Total volume**

Flowing through the naturally arid American Southwest, the Colorado River is one of the most extensively tapped of the world's major streams. So great are the withdrawals of water for irrigation, power plants, and domestic, industrial, and other uses that the Colorado no longer reaches its former delta at the Gulf of California.

Does a river really need to reach its mouth, however? Are not the benefits derived—generation of hydroelectric power, flood control, provision of water for agricultural, industrial, and consumer needs and incidental creation of playgrounds in the desert—justification for the manipulation of a river's natural flow? Alas, the sparkling triumphs of engineering seem a trifle tarnished on closer look. As dams and ditches and tunnels were constructed in rapid order, problems arose. Water release from Hoover Dam, for instance, was adjusted to Los Angeles's demand for electricity. This frequently resulted in insufficient water for Mexican farmers downstream. So another big dam, Davis Dam, was built to control and store the flow from Hoover and ensure water for Mexico. As reservoirs behind dams rapidly filled with silt (with which the Colorado is especially

The intensive agriculture in southern California's Imperial Valley is made possible by the diversion of Colorado River water through the All-American Canal. The border with Mexico is clearly seen in this color-enhanced LANDSAT photo. The difference in color reflects differences in season planting practices, in intensity of land use, and, probably most important, in availability of water, most of which is withdrawn upstream for U.S. use. The dark area is the Salton Sea, a shallow depression that was filled with water in 1906 by the flooding Colorado, its change in course facilitated by newly constructed sluicegates.

heavy and for whose reddish color the river is named), their capacity for holding water diminished, which required the construction of additional dams, diversions, and treatment works downstream. The progressive alteration of the Colorado has been called a monumental series of technological fixes, a spiral of new engineering projects needed to remedy problems with the old.

The massive intrusion of technology along the Colorado has completely altered the ecosystem of the river and valley. Fish and other aquatic organisms are killed when cold water is released suddenly through sluices, and rapid siltation buries their eggs and fry. Wholesale changes in the vegetation accompany the stabilization of a river formerly given to seasonal floods. The water grows ever saltier, as salts leached from irrigated fields wash back into the river, rendering it toxic to river life and to crops downstream. (One suggested fix, the damming and desalination of the naturally salty Little Colorado, would drown parts of the Grand Canyon.) Aquatic life in the Gulf of California is robbed of the nutrients it once received from the river.

As usable water dwindles, various schemes for augmenting the supply have been proposed, ranging from the introduction of water diverted from Canadian rivers to cloud seeding over the southern Rockies (which would result in increased snowfall, more avalanches, greater erosion, and profound disturbance of montane animals and plants). Still, demands for water grow daily more shrill, for drinking, for power, for irrigation, for shale oil extraction. At some point we, the dominant characters in this threatened ecosystem, must ask how much longer we can continue to apply extraordinary measures to sustain large cities and intensive agriculture in the naturally water-poor American Southwest.

Where the Sea Meets the Land

In many parts of the temperate world, the interface between land and sea is occupied by the salt marsh community. Especially well developed along tidal inlets and behind the dunes of the barrier islands that fringe our Atlantic and Gulf coasts, salt marshes are nursery and kitchen to the rich life not only of the immediate inshore area but of the entire continental shelf. These marshes are deceptively simple in appearance, usually covered by near-uniform stands of cordgrass. But this grass grows so rapidly and supports such a complex web of other life that the marshes rank as some of the most productive of the Earth's biological communities. Owing to our draining, dredging, and filling, to outfalls of

heated water and chemical pollutants, the precious salt marshes are also among the most threatened.

The Polar Desert and Ocean

Unusual international harmony reigns in Antarctica, that frozen continent where stands the South Pole at 90 degrees below the equator. Scientists from a dozen nations cooperate under the auspices of the Antarctic Treaty of 1961, investigating the meteorology, biology, paleontology, and geology of this forbidding polar desert.

A polar desert? For all its ice, Antarctica receives very little precipitation each year. The extremely cold air cannot hold much moisture. What little snow falls is retained, however, since the same cold air prevents melting and evaporation. The snowfalls of millennia past are thus preserved, and scientists, by drilling cores, can obtain a frozen record of global climate and atmospheric conditions over the last thousand centuries. Gases and dust in the atmosphere circle the globe; some drift to Earth, others are carried down in rain and snow. Antarctic ice cores reveal the accumulated evidence of events like volcanic eruptions. They also bear ominous witness to the results of twentieth-century human activity: the snow in even so remote and pristine a place as Antarctica brings with it radioactive fallout and chemical pollutants from elsewhere, offering striking support for the metaphor of our planet as Spaceship Earth, a single, great, interconnected system.

Antarctica and the Southern Ocean are an ideal laboratory for ecosystems analysis. Except for some hardy microbes, life is confined to the ice shelf and the surrounding sea. It is abundant. The populations of the few species that can withstand the rigorous conditions are typically huge. Food chains are relatively simple. Swarms of plankton are food for hordes of shrimplike crustaceans called krill. The krill in turn feed schools of squids and fishes, flocks of penguins and other birds, and herds of seals and whales. Killer whales and leopard seals are the top predators. The droppings of all these animals nourish the plankton. The food chain is actually a great circle. Because of the simplicity of the Antarctic ecosystem, scientists can study it easily and extrapolate their findings to other, more intricate natural systems.

Yet one broken link in so perfect a chain can render the others vulnerable. The great whales in the Antarctic have already been hunted to near-extinction, and the large-scale krill-harvesting schemes now being contemplated by some of the developed nations could have dire and far-reaching consequences for the balance of life in the Southern Ocean. Environmentalists are urging caution; they push for the monitoring of Antarctic populations so that changes might be detected before they prove irreversible.

Although the frigid Southern Ocean teems with life, the ice-covered land is near barren. Some mosses and algae and only two species of flowering plants make up the entire flora. The animal life is also poor: some mites, wingless flies, called springtails, and a few other invertebrates, almost all associated with the seasonal penguin rookeries, maintain a precarious existence.

Antarctic Food Web

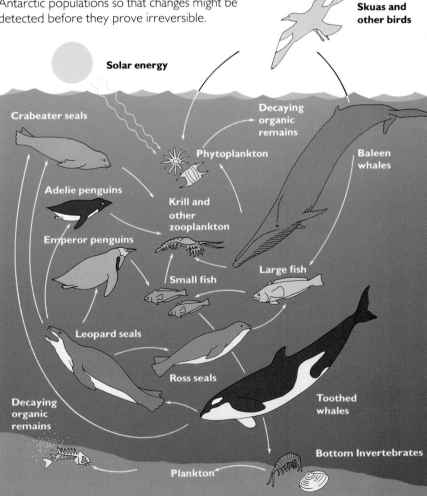

Skuas and other birds

Solar energy

Crabeater seals

Decaying organic remains

Phytoplankton

Baleen whales

Adelie penguins

Krill and other zooplankton

Emperor penguins

Large fish

Small fish

Leopard seals

Ross seals

Toothed whales

Decaying organic remains

Bottom Invertebrates

Plankton

Arid lands are only marginally useful for human habitation. The availability of water is **the principal limiting factor. Not only is rain in short supply, but it tends to fall unpredicta-** **bly. For dwellers in the world's dry regions, the search for water is an incessant activity.**

The Subtropical Desert Bordering the Sahara Desert on the south is a great belt of dry, sparsely wooded steppe, the Sahel. Not a true desert, this arid country at the desert's edge has for millennia supported a small but vigorous human population of no-madic pastoralists and dryland farmers. As a rule, the lower the rainfall, the more unpre-dictable it is. Traditional Sahelian cultures were attuned to the inevitable periods of drought. When the rains failed, the nomads moved with their camels, cattle, goats, and sheep to greener pastures. Where rainfall was slightly higher and more dependable, dryland farmers raised drought-resistant mil-let, sorghum, and sesame as subsistence crops and gum arabic trees as a cash crop. Pe-riodic fallowing of the fields prevented de-pletion of the soil.

Then, in the 1970s, the Sahel suffered a terrible five-year drought. Vegetation with-ered. Soils turned to dust and blew away. Livestock starved to death. A quarter-million people also died of starvation, and thousands of others became refugees. But the Sahel has always endured droughts. At least twenty-two droughts have been recorded since the sixteenth century, of which five lasted as long as or longer than the drought of the 1970s. Never before had there been such utter devastation. Why?

Some of the blame must be placed on the well-meaning but short-sighted policies of contemporary governments, which pro-vided improved medical and veterinary services and encouraged "modernized" agri-culture. As a result, human populations and their herds of livestock increased, and clear-and-plow farming was extended into areas of marginal suitability. When the inevitable de-mand for more water arose, the govern-ments drilled deep wells. Farmers settled near the wells. More farmers on less land led to smaller plots and shorter fallowing pe-riods. Nomads congregated near the wells with their herds. The concentration of live-stock led to soil compaction and range deteri-oration. As rising oil prices drove up the cost of kerosene fuel, people turned more and

more to local stands of trees and shrubs for firewood and charcoal. By the time the drought finally hit, pastures were trampled and overgrazed, farm soils were exhausted, and the landscape was denuded of its protective cover of woody plants. The subsequent erosion transformed formerly productive arid areas to utterly barren desert; people, domestic stock, and wildlife suffered mightily.

Desertification is a worldwide problem. More than 23,000 square miles of land are converted to unproductive desert every year. A full quarter of the Earth's land surface is at risk, on every continent. By the end of this century, the loss of cultivable land, combined with the inexorable growth of world population, will leave less than half the amount of cultivated land per capita than was available in 1975 (a year not notable for universal adequate nutrition).

All is not hopeless, though. New techniques of improved irrigation and rediscovered ancient techniques for capturing and storing water in arid lands point to the possibility for sustained human use. The United Nations Conference on Desertification has reckoned a cost of $2 billion a year to halt and reverse the process. A lot of money? Global expenditures for military research and development now exceed $550 billion yearly.

Their life adjusted to periods of drought, Sahelian nomads once moved their herds whenever a prolonged lack of rain denied them adequate pasturage.

Desert Dwellers

The deserts of the world are home to a surprising diversity of life. Since they are rooted in place, plants can't burrow to seek shade or travel to find water, as do many of the animals. They must cope in other ways with conditions of great heat and prolonged drought. Some drop their leaves in the dry season; others grow small, waxy leaves resistant to water loss. Spines and hairy buds protect against water loss in many desert plants. Cacti and other succulents store water in their leaves, stems, or roots. Like the giant saguaro cactus, many succulents have shallow roots that soak up water quickly after a rain. Other desert dwellers, like mesquite, grow long roots that tap water sources as deep as eighty feet.

And the beautiful ephemerals, which carpet the desert floor after the rains, sprout, flower, set seed, and die in the brief period before drought returns.

In a controversial experiment of mammoth scale, Daniel Ludwig acquired Jari, an enormous parcel of land in northern Brazil, and stripped it of much of its original forest cover. In place of the hundreds of tree species that grew in the rain forest, Ludwig installed plantations of just two kinds—gmelina and Caribbean pine.

The Tropical Forest Humid equatorial forests in general, and the great Amazonian forest in particular, are ecosystems of unimaginable complexity. Probably half of all 5 to 10 million species of life on Earth (the fuzziness of "5 to 10 million" is an indication of just how profound our ignorance of our world still is) live in tropical rain forests. Unfortunately, forests throughout the world are being felled at a frightening rate. Whereas forests covered one-quarter of the Earth's land surface in 1955, this has dropped to one-fifth today and will diminish further to one-seventh by the year 2020. Tropical forests are now disappearing at a rate variously estimated as between 25 and 100 acres a minute!

Temperate forests, with their rich soil and relatively simple biotic community, can usually regenerate quickly after being cleared. Tropical forests return slowly, if at all. Why should this be? Tropical forests grow through the year in a seasonless climate of constant rain and warmth. Fallen leaves, twigs, and fruits, and the droppings and corpses of animals, decompose very rapidly on the forest floor (decomposition is much slower in cool temperate forest soils), and the nutrients released are recycled immediately, absorbed by the roots of the rain forest plants. Although the forest itself is wildly luxuriant, the underlying soil is thin and poor in nutrients. Almost all the organic matter in the ecosystem is locked up in the bodies of living trees and other plants. When the forest is cut and removed, the nutrients are lost. The exposed soil is drenched by heavy rains, which wash away whatever organic matter remains. The heat of the sun then bakes the sterilized soil, and an impermeable crust forms. For these reasons, almost every scheme to convert large stands of tropical forest to pasture or farmland ends in failure.

Unlike polar ecosystems, with their huge populations of very few, wide-ranging species, tropical forests harbor small populations of a dizzying variety of species, many with rather restricted distribution. When large tracts of rain forest are destroyed, many local populations may be driven to extinction. Even when a secondary forest grows up on a cut-over site, it is usually poorer in species variety than the original.

Undeterred by the sorry record of agricultural projects on tropical forest soils, the richest man in America purchased a huge tract of land in northern Brazil and set about creating a tropical dreamland. The tract, Jari, was bought by Daniel Ludwig in 1967 for $3 million—that's less than $1 an acre for a piece of property larger than Connecticut. Much of this vast holding was cleared of its magnificent forest, so species-diverse that more than 400 different kinds of trees grew there. It was replaced with a managed plantation of just two kinds of exotic trees: gmelina, a fast-growing species from Southeast Asia, and Caribbean pine. A complete prefabricated pulp mill cum effluent treatment works and chemical recycling plant was floated from Japan to Brazil and up the Rio Jari to be installed on Ludwig's land. By 1980, the oldest trees were ready for harvesting and a daily output of 750 tons of bleached kraft pulp for a paper-hungry world was predicted.

Ludwig's boosters praised his bold scheme. To be sure, if his experiment in tropical plantation forestry were to succeed, it could have taken pressure off remaining tracts of virgin Amazon forest, and plantation forestry is generally acknowledged to be less damaging than many other possible uses of rain forest land. (The systematic recycling of wood and paper products and wholesale changes in wasteful practices of consumption would also help to check the rapacious destruction of the world's forests.)

Ludwig's detractors were less sanguine. They doubted that the plantation soil could survive several successive harvests without massive infusions of fertilizer. They questioned the long-term economic feasibility of his methods, which were still extremely intensive of labor and energy. And they wondered about the wisdom of replacing a rich and complex forest with a simplified monoculture of genetically uniform individuals. Such monocultures are notoriously susceptible to pests and epidemics, and the critics fear the inevitable application of remedial biocidal poisons on the Jari plantations would adversely affect the neighboring forests.

If successful, tropical tree farming could remove logging pressure from those parts of the Amazonian rain forest that remain undisturbed. Still, critics fear the consequences of practices of intensive arboriculture on the fragile soil and the life of the nearby forests.

Still Waters

Few wild mammals have as profound an effect on the country they inhabit as do beavers. The houses built by these large rodents dam streams, control flooding and erosion, and create countless freshwater ponds. The ponds become breeding and feeding grounds for a rich variety of insects and other invertebrates, fishes, amphibians, reptiles, birds, and mammals. Ephemeral features of the landscape, the ponds silt up in time and become meadows, which are then invaded by shrubs and trees from the nearby woods. But unless exterminated by humans (who covet their lustrous pelts and sometimes regard them as pests to agriculture and forestry), the beavers continue to build and dam, creating the ponds that serve as magnets for other life in the neighborhood.

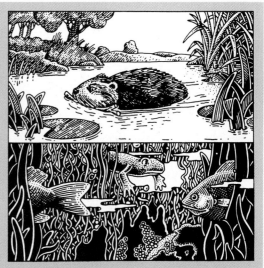

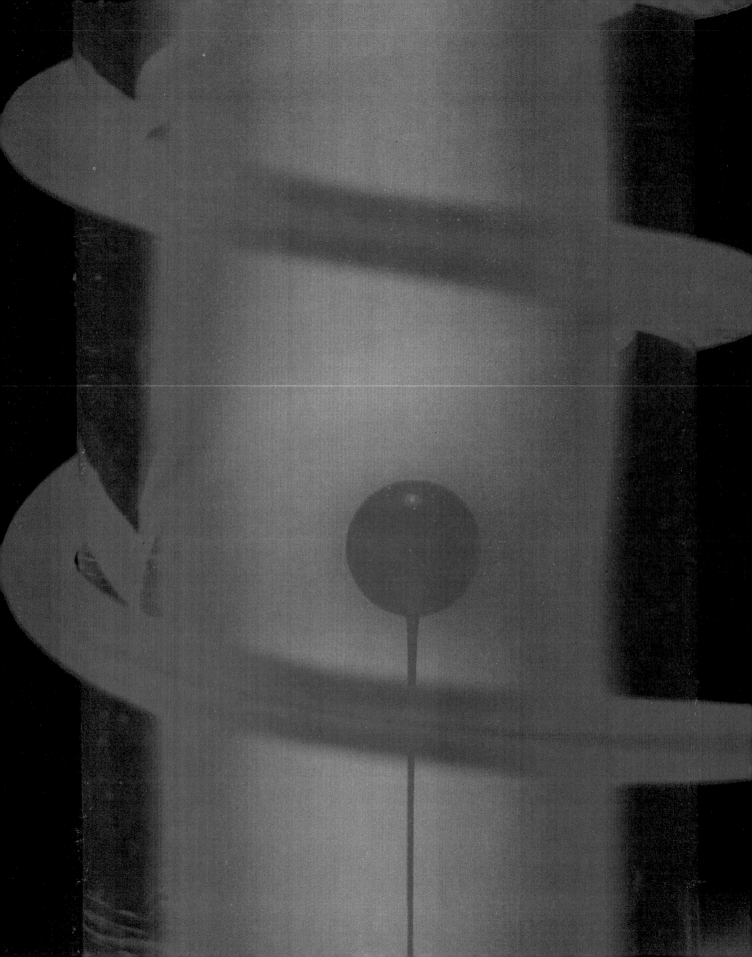

Part IV

T E C H N O L O G Y

Ultimately, science means progress, and the major progress of science in this century is technological. From advancements in winemaking and energy alternatives to artificial intelligence and genetic engineering, technology explodes around us, amazing us daily with dazzling new capabilities. Science fiction, it seems, has become reality.

We are poised on the brink of one of the most exciting times in scientific history. Technology is making our lives easier, more comfortable, more convenient. At the same time it is leading us into uncharted territory. Life as we know it is about to be redefined, for science has pushed its own frontiers beyond anyone's imagination. The ancient question of "Where did we come from?" has been transformed into "How far can we go?" In extending life's possibilities, science must create new rules, new ethics, a new order of things. Machines and humans have existed side by side for almost one hundred years now; what happens in the next ten years could very well determine the future of life on Earth.

A glass fusion pellet of deuterium and tritium is the target of high-energy laser pulses in this experiment at KMS Fusion in Ann Arbor, Michigan.

Children of Icarus

(Above) Otto Lilienthal designed this antecedent of the modern hang glider. In building his *Gossamer Condor (right),* Paul Mac-Cready pushed glider technology to its limits and added a new dimension: pedal power.

In 1959 a British industrialist named Henry Kremer announced that he would give a substantial sum (eventually to reach $86,000) to whoever first managed to fly a man-powered aircraft in a figure-eight pattern around two pylons one-half mile apart. Over the next two decades there was considerable speculation about what his motives might have been. He could not have hoped that the "age-old dream of man-powered flight" would be achieved thereby, because that had been done before, and often. Both British and German aeronautical engineers, among others, had built and flown man-powered craft in the 1930s. And besides, whatever age-old dreams might still be unachieved certainly had nothing to do with the exhausting struggle to keep some huge ungainly semikite lumbering through the air for barely more than a mile. Some, recalling that Kremer was a physical-fitness enthusiast, suggested that the British Aeronautical Society—which wanted to see some research and development going into light aircraft—may have fascinated him with the vision of thousands of British youths pedaling their way to manly vigor across the English skies.

Whatever his reasons, the prize was large enough to attract the efforts of a number of groups in Japan, Europe, and North America. The problems these teams had to attack all rose from a single unavoidable reality: humans, compared with other means of powering aircraft, develop very little power for their weight. This is not necessarily a fatal difficulty; according to aerodynamic theory, a small amount of power can pull a heavy weight up into the air if it has the assistance of a large wing. The problem with that solution was that ordinarily one would expect such a wing, like any other sizable structure, to weigh a good deal itself. That could not be allowed here; the theoretical margins for man-powered flight are narrow in any case, and the Kremer conditions had reduced them to a fingernail's width. The wing had to weigh practically nothing.

So the core of the challenge was not just to build a new kind of airplane. What was required was to invent a new engineering

science: the design, assembly, and testing of large, superlight structures. This was not appreciated immediately. The groups that decided to try for the prize were either aeronautical engineering students at different schools—the Massachusetts Institute of Technology, the Royal Air Force College, Nihon University in Tokyo—or teams of professional engineers attached to large aircraft design companies. Both groups tended to follow the traditional route in aircraft design: they analyzed the specifications, laid out a design that promised, at least theoretically, to meet them, built a test model, and flew it. If and when difficulties appeared, they went, as the famous saying has it, "back to the drawing boards." The problem attracted exceptionally intelligent and imaginative students, and, while it is hard to generalize about all the innovative ideas these groups proposed, as a rule observers were consistently impressed with the high standard of their analytical ability and design intelligence.

Unfortunately, none of these remarkable airplanes was able to meet the Kremer standards. Steady progress was certainly

made; each year's versions flew a little farther. But none had come close to navigating the full figure eight.

In July 1976, a national gliding champion named Paul MacCready was relaxing on a family vacation, thinking idly about an article he was writing on the soaring abilities of such birds as hawks. MacCready was intimately familiar with the theory and performance of hang gliders and began comparing them with birds. "I had just done some calculations on how much horsepower a powered hang glider uses as it flies," he said later. "A good one is down to 1.2 hp, or even less. . . . All of a sudden I put a few things together. . . ."

What MacCready had suddenly seen was that if one simply tripled the dimensions of a conventional hang glider, the power required to maintain a level glide would be around one-third of a horsepower, easily within the capabilities of a human being in reasonable physical condition. MacCready decided to try for the Kremer prize. He set out the basic equations governing the design of such a plane and solved them. The *Gossamer Condor*, as the aircraft came to be

Icarus's undoing came from flying too close to the sun. His wings of wax melted and he plummeted into the ocean. Paul Mac-Cready's craft the *Gossamer Albatross* keeps to a sober distance from both sun and ground, eventually safely navigating the English Channel.

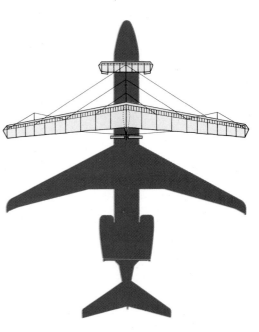

(Top) **MacCready, right, and an associate stretch sheets of ultralight plastic over the spars of the *Condor*'s wings. *(Above)* The *Condor* lifts off under pedal power above Shafter, California. *(Right)* Many attempts ended ignominiously, but the easy-to-repair *Condor* was always ready in a few days for another try.**

called, would have to have a ninety-six-foot wingspan—wider than that of a DC-9—and yet could not weigh much more, in total, than about seventy pounds!

When MacCready started to build the *Condor,* he made the most radical departure from his competitors' designs: he basically slapped it together out of cheap, available materials. Using cardboard—the kind found in candy boxes—balsa wood, Styrofoam, sheets of thin plastic (Mylar) that could be attached together with adhesive tape, piano wire, and aluminum tubing, he had a version ready to fly in only five months. "In essence, the *Condor* structure was six sticks and seventy-two piano wires connecting everything to everything else," he said.

MacCready took this approach, as he has said, because he wanted to win the prize, and it only made sense to do so in as cheap—in his words, as "quick and dirty"—a way as possible. It is probably not irrelevant that, unlike his competitors, MacCready is a small-business owner, the president of his own aeronautical design company. As such, he might have placed a higher value on both time and money than students or hobbyists. In any case, his cheap, simple structure turned out

(Left) **MacCready rigs the wiring used to hold the ungainly craft together.**

(Right) **One of the** *Condor'***s forerunners, a pedal-powered craft tested here by Mac-Cready's son in the Mojave Desert.**

to be exactly what was needed. It meant that the plane could be repaired and modified quickly after each flight. In the other planes a crash was a serious setback; with the *Condor,* practically nothing could happen to it that couldn't be fixed in few days at the very most. And it was through these crashes that MacCready gradually learned what was necessary where, and how strong or weak it needed to be. "If something didn't break, we made it thinner," he said. By the time he finally won the prize in August 1977, there had been nearly 500 test flights involving twelve significantly different versions of the plane.

After his success he received a tremendous amount of attention in the media, which reported that he had shattered the barriers of man-powered flight. This not only was wrong—probably hundreds of people had by then flown under their own muscle power—but also submerged his real achievement. What Paul MacCready had done was find a way to gain experience instead of merely applying the logical consequences of theory. The chronic human tendency to do it "right," letting professional standards rather than firsthand experience dictate problem

solving, is a far more crippling barrier to accomplishment than the inability to fly with unaided muscle tissue. And it was that barrier that MacCready broke when he won the prize that Henry Kremer announced eighteen years before.

Bird Brains

The fascination with flying led long ago to intense observations of birds, and we owe to them much of our elementary knowledge in aeronautical design. Yet birds are a continuing source of amazement to those who study them. In addition to the astonishing distances covered by birds in migration, their ability to navigate is uncanny. In the last few decades, ornithologists have begun to understand the variety of environmental cues to which birds respond. Most migratory species come equipped with an innate sense of compass direction, which assists first-time migrants in getting their general bearings. With the help of their extraordinarily keen sight, birds apparently scan the land and sea below and "memorize" the features for reference in subsequent flights. Many birds seem to orient themselves by the position of the sun and stars and the direction of the winds. And the accurate homing ability of those species that can navigate in the absence of visual signals—during overcast weather or when experimentally "blindfolded"—leads researchers to consider

the possibility that they can perceive slight differences in the Earth's magnetic field. To identify the avian sensory receptors of magnetism is one of the aims of ongoing research into the marvels of migration.

(Above) The *Albatross* skims above the English Channel while escort boats monitor the flight.

(Right) The *Penguin*, MacCready's first solar-powered plane, soars above the desert floor.

Henry Kremer sent his congratulations to MacCready for "a brilliant design" and promptly announced a new, £100,000 prize for the crossing of the English Channel in a human-powered aircraft. Again MacCready rose to the challenge, but this time with a different approach. No longer was "quick and dirty" the Gossamer program's motto; MacCready wanted to "build ultimate" with the most sophisticated materials available. But he needed financial backing to pursue such a project, and DuPont provided it, along with all the ultralight plastics used in the new plane's construction. Gone were the aluminum tubes and cardboard put together by trial and error. Instead, the Gossamer team carefully sculpted the new craft, *Albatross,* from carbon fiber—reinforced spars, Kevlar cloth, and polystyrene foam.

The very different requirements of the new Kremer Prize necessitated this radical departure from the *Condor's* construction procedure. The *Condor* needed to fly a mere 1.15 miles in the calm air of California's Central Valley; the *Albatross* had to fly 22.5 miles between England and France through the more turbulent air space found over the Channel. Yet the underlying theme of the program remained unchanged: it was, as MacCready states, to "approach a problem with a knowledge of the fundamentals but without the deadening influence of prior detailed expertise and prejudice."

As pilot Brian Allen pedaled the *Albatross* across the Channel in the early morning hours of June 12, 1979, the Gossamer team again succeeded where others had failed. The flight was not without difficulties—at times the plane plunged to within one foot of the sea swells—and pilot Allen tested the bounds of his physical capabilities as he pumped away at 80 to 100 r.p.m. for two hours and forty-nine minutes, just one minute shy of his calculated absolute endurance limit. Yet succeed they did, and MacCready was ready for his next aeronautical coup.

The Gossamer team had assembled two planes for the Channel crossing—the *Albatross* and a second, smaller plane that was more airworthy in turbulence and wind. This plane, dubbed the *Gossamer Penguin* but referred to by team members as "the sports car," was shipped back to California after *Albatross's* successful flight. The *Penguin* soon became the object of MacCready's next flight of fancy. Intrigued with the notion of solar-powered aircraft, he constructed a 45-square-foot panel of photovoltaic cells and placed it above the wing of the plane. He added a small electrical motor to the contraption, and on May 18, 1980, MacCready's eighty-pound son Marshall took off in the *Penguin* and flew 100 yards. The plane eventually made a flight of more than two miles in August of that year, but by then MacCready had turned his attention to the construction of a specialized solar-powered aircraft.

The newly designed plane, the *Solar Challenger,* looked far more conventional than the earlier human-powered products of MacCready's imagination, but it was no ordinary piece of engineering. NASA loaned the project 16,128 Sectrolab silicon cells, which were spread over the wings and horizontal stabilizer. These surfaces departed radically from traditional airfoil design; they were nearly flat, to maximize the solar cells' exposure to the sun. Again DuPont underwrote MacCready (this time for approximately $725,000), and again DuPont provided all the synthetic plastics used in the *Challenger*'s construction.

The result was an aircraft with a 47-foot wingspan whose solar cells produced approximately 3,000 watts of electricity, enough to generate about four horsepower. This was sufficient to lift the *Challenger* and its pilot 11,000 feet above the French countryside one sunny July morning in 1981 and send it off toward England at 43 m.p.h. The plane stayed aloft five hours and twenty-three minutes on its 230-mile flight from a village outside Paris to a Royal Air Force Base near Canterbury and encountered difficulty only when unauthorized aircraft carrying hordes of photographers flew too close and buffeted the feather-light craft.

MacCready was once again the aeronautical trailblazer. For the first time, a sunlight-propelled aircraft had been engineered that flew unaided by batteries or any other form of stored energy. Although he commented that "flying an airplane with solar cells is just about the most ridiculous use for solar energy that I can think of," the project was not without purpose. "We wanted to point out just how much solar power can do," states

MacCready, and the flight of the *Solar Challenger* certainly did just that. But perhaps a larger achievement of the entire Gossamer experience lies in its testament to the power of unbridled human creativity. No one would have believed in 1976 that within three short years a human-powered plane would cross the English Channel, to be followed two years later by a craft propelled by the sun. The guiding principles of the Gossamer project—single-mindedness, persistence, flexibility, and confidence—provided the guidance necessary for the team to realize goals left unattained by thousands of years of human aspiration.

The most famous of those aspirants, a figure from Greek mythology named Icarus, fell to Earth for having flown too close to the sun. The heat melted his magnificent wings, and he was hurtled out of the sky. Millennia later, MacCready and his associates turned foe into friend and proved that human ingenuity can overcome nearly any adversary—even one of astronomical proportions.

The *Solar Challenger* flies high above the Atlantic in vindication of Icarus and his fall. *(Inset)* Workers cover the *Challenger*'s wings with hundreds of thousands of dollars' worth of solar cells on loan from NASA.

The Great Violin Mystery

Antonio Stradivari (1644–1737). Few details are known of the life of the most famous craftsman of the violin. The trademark that distinguishes his instruments is shown above.

(*Facing page*) A violin crafted by Guarneri del Gesu of Cremona in 1731.

Here is an instrument that grieves over one note, dogs a second, and still has resources left for shameless flirtation with the third. A master of musical clarity, it leaves no questions unexplained or unanswered. Yet the violin itself is encased in mystery. The source of its design is as uncertain as its originator; the reasons for its sound as uncomprehended as the reasons modern instruments lack a brilliance characteristic of the work of the old masters.

Cremona is a small town in northern Italy. Along a single alleyway, in the seventeenth century, the workshops of Italy's greatest violin makers clustered together in artistic union: the Amati family, Guarneri del Gesu, and Antonio Stradivari, who, over the course of three-quarters of a century, refined the violin into its finest form.

Each Stradivarius violin began its musical career as a small wooden box coated multiple times with varnish. An architectural hourglass defined by its sides, it had, in addition, a neck and a gently arched back; a sound post slightly off center, wedged like a cone between the top and back; F-shaped holes in the top, reinforced by a stiff beam, the bass bar; a bridge, tailpiece, fingerboard, and four strings. All the violins of the masters shared this fundamental structure. Yet each produced, from a single form, its own superior sound, its own cut and edge. With their deaths, the Italian masters left their tools—but not the secrets—of their craft. With their deaths ended the golden era of violin making.

Over the last century, technology has joined art in a hunt to reclaim the great violins. What subtle features held in common explain their uncommon tones? Some said that it was the wood. Legend held that the masters toured the forests of nearby Lombardy, rapping on trees to find the finest spruce for their violins' tops, the finest maple for the backs. Perhaps, under their fingers, the wood brought its own collaborative magic to their work.

But wood, whatever its other powers, cannot conjure uniform tones, for it is not a uniform substance. Within one plank, variations in texture and composition cause the wood to vibrate at markedly different frequencies: magical, perhaps, but unstable as well.

Other hunters trailed a second scent: the way the wood was treated and the components of its varnish. They claimed there was hocus-pocus in the pure Venetian turpentine; razzle in the mixture of resins; or else, in the most appropriate style of magicianship, some secret ingredient. When infrared spectroscopy became a viable technique for isolating and identifying the components of chemical mixtures, scrapings of the old varnishes were analyzed. But the graphs showed no tricksters in the grain; instead, they revealed a composition much like any good varnish used today.

A third set of investigators turned to the techniques of construction. A violin's sound begins with a single taut string, a simple object moving in a simple way. Its delicate motion is too subtle, though, to radiate sound effectively without amplification: the string must be attached to a structure that can magnify its sound without distorting its movement. This is the acoustical function of the violin's box. When strings are plucked or bowed, the strings, the box, and the air within the box all move in a complex vibrational web, setting up threads of frequencies that run parallel and perpendicular to each other.

Devices attached to an instrument can measure the thickness of wood in the top and back of the box. Minute graduations of width affect the violin's movement and its sound, often dramatically. Using this discovery, one investigator in the early twentieth century prepared a set of maps. Averaging measurements of many separate tops and backs, the investigator, Simoni Sacconni, noticed a pattern in the great violins—they all shared a strict symmetry in design. The backs in particular showed a regularity in thickness, tapering uniformly off from a broad center to thin edges. Sacconni reasoned that, as the masters had conceived them, these graduations forced air into intentional routes and produced intentional sounds. His maps became considered the musical key to a musical problem, and the problem itself seemed one only of composition. Confident violin makers began to follow the maps; cartography became the foundation for construction. The resultant instruments were indeed improved. But none echoed, even faintly, the sought-for tones of its predecessors.

Jack Fry is a physicist at the University of Wisconsin. He works with subatomic particles. He also has a mean way with a fiddle, and for twenty years, as an amateur musician and violin maker, he has studied the sound of violins with the patience of a scientist and the passion of an artist.

A dual relationship to the instrument leaves Fry with an unusual perspective: he feels respectful of, but not romantic about, its workings. His violin is an object of wood and glue—no magic dust, no blessed strings—that moves mechanically under the influence of stresses and forces; its artistry is the product of its physics. Fry chose to analyze his box of wood and glue as a system whose aesthetic properties are determined by its physical

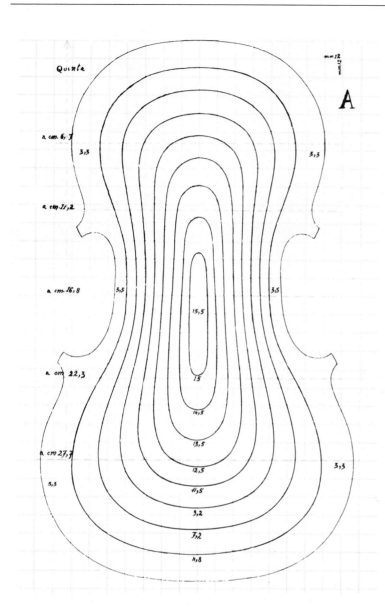

A topographic map outlining the symmetrical bulge in the back of a violin. This drawing, taken from Sacconni's *The Secrets of Stradivari*, characterizes his emphasis on the symmetry of violin construction.

Physicist Jack Fry used a trampoline as a conceptual model to understand the motion of the violin's back. The athlete represents the violin's sound post, transmitting the motions of the strings and driving the back. Fry found that the off-center placement of the sound post (the off-center circle on the trampoline) required an asymmetrical graduation in the thickness of the violin back in order to amplify sound efficiently. On the trampoline, this was modeled by removing selected springs on one side, as shown in the last frame.

ones: stiffness, distribution of mass, and other factors cause oscillation, and oscillation in turn radiates sound outward. No room is left, in this analysis, for rabbits in the instrument's hat or aces up the performer's sleeve.

Fry began his research with a fundamental question: how is sound amplified by the movement of air within the violin's box? In less than a year after posing it, he had run several hundred experiments and ruined scores of instruments. Concentrating on the varied thicknesses of different tops and backs, he constructed his own careful maps. Contrary to Sacconni's work, though, Fry's findings suggested that asymmetrical graduations allowed a back to move most evenly and efficiently and regulated the loudness of the sound. For corroboration, he turned—with considerably more care—from the ruins in his laboratory to some of the old Italian instruments still in use. Using instrumentation as delicate as only a subatomic physicist can devise, he measured their internal thicknesses and found a similar asymmetry present in all. The Fry formula for asymmetric amplification had been confirmed—from the inside out.

In the scientific order of things, data precede abstraction. Having verified a hunch, Fry next developed a simple model to explain the violin's complex motion. Generalizing its range of tones to three categories—low, high, and middle—he devised three corresponding theories to justify their differences. At the lowest frequencies, as a vibrating string swings right and left, the box moves in internal opposition with itself: its back rises while its top depresses; then the top rises while the back depresses. Air is driven out of and into the box through the F holes

carved into its top; this bellowslike motion creates waves that amplify the lowest bass sounds. Fry called this the "breathing" mode.

At high frequencies, motion is confined to a single area: only the small patch of wood between the F holes is light enough to respond to rapid vibrations of the strings. Exercising the right of all discoverers, Fry named this mode as well: the "supertweeter" controls a violin's brightness. Middle frequencies are amplified in yet a third way, a "rocking" mode in which the top is driven by a seesaw-like motion.

Not only the box, but each component attached to it (bass bar, bridge) or removed from it (F holes), can be analyzed by laws of physics relevant to its separate function. Each piece plays its own subtle role in movement, and, once identified, each can be manipulated with this role in mind. Tiny adjustments control the timbre, volume, and brightness

that together produce a final impression, a final sound. Each piece carries its own part; but physics is the baton raised over them all.

Science can duplicate results, but not always intentions. It is difficult to know whether the craftsmanship of the Amatis or Stradivari rested on technical understandings as specific as those Fry has quantified. The challenge, as he conceives it, is a synergistic one: to combine scientific parameters with artistic ones, structural means with harmonious ends. Whether the old masters were spokesmen for the heavens or progeny of Newtonian physics may never become clear. If their instruments became replicable, though, their private secrets may take second place to a more public feat: the return of a golden musical era long believed to have been buried with its creators. It would be the first time, would it not, that the old masters found themselves in the role of second fiddle?

Leonard Sorkin *(left)* of the Milwaukee Fine Arts Quartet watches as physicist Jack Fry uses a modified postage scale to measure the asymmetry in the back of Sorkin's Guarneri del Gesu violin.

Mind Machines

Human beings have long prided themselves on their superior intellect. While no one knows when our prehistoric ancestors first became conscious of their unique brain power, many people are beginning to realize that we may be the last humans to take this uniqueness for granted. Our intellectual supremacy is being challenged—not by a more highly evolved biological being, but by our own mechanical brainchild: the computer.

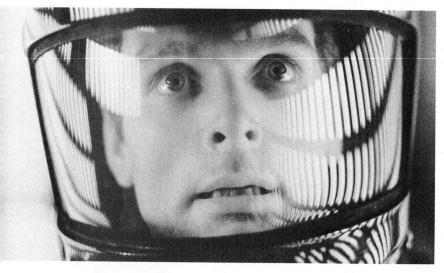

But is the computer just a calculator? Or is it capable of judgment, intuition, common sense, and learning? Will computers ever think like people? Could they match or even surpass the human mind? What effect will increasingly smart machines have on life as we know it? Computer scientists and philosophers alike debate whether artificial intelligence represents our downfall or our destiny.

In the movie *2001: A Space Odyssey,* the machines outpaced their human creators. The HAL 9000 computer had a mind of its own, a ruthless mind. It stopped at nothing and eventually had to be destroyed. HAL was a product of writer Arthur C. Clarke's imagination, but someday an artificial intelligence, a machine beyond the human mind, may become fact.

"I think what we are doing now is in a sense creating our own successors," says

Clarke. "We have seen the first crude beginnings of artificial intelligence. It doesn't really exist yet at any level because our most complex computers are still morons — high-speed morons, but still morons. Nevertheless, some of them are capable of learning, and one day we will be able to design systems that can go on improving themselves. At that stage, we have the possibility of machines which can outpace their creators and therefore become more intelligent than we."

While some scientists say that on the basis of present knowledge and achievement such predictions are utterly ridiculous, others say it is only a matter of time before they are reality. Indeed current doubts about artificial intelligence are viewed as analogous to the skepticism once evoked by talk of space travel and gene synthesis. "From the beginning of science," says Donald Michie of the University of Edinburgh, "there have been people telling you that this or that boundary line is a sacred one and can never be crossed. . . . This vitalist attitude of uncrossable frontiers is behind the question of whether there are human mental abilities which could never be simulated. I personally see no more reason now to be discouraged by this vaguely expressed feeling than scientists have been in the past."

In an anatomical sense, a computer is simply a giant adding machine. But like humans, the computer has a memory where its knowledge is stored. Its tiny circuits hold billions of pieces of information. The circuits understand only the presence or absence of electrical impulses, so all information must be put in those terms. The memory or knowledge base can be thought of as a maze of lights, some off and some on. Like the dots and dashes of Morse code, the lights can represent anything, even words.

Everything is represented by numbers within a computer. Words are encoded according to a standard put forth by the American National Standards Institute. The word *NOVA,* for example, would be encoded as 78798665, with each successive two digits

(Top) An astronaut confronts HAL, the rebellious computer in the 1968 movie *2001: A Space Odyssey,* based on a novel by Arthur C. Clarke. "We are . . . in a sense creating our own successors," says Clarke *(bottom)* of the increasing sophistication of thinking machines.

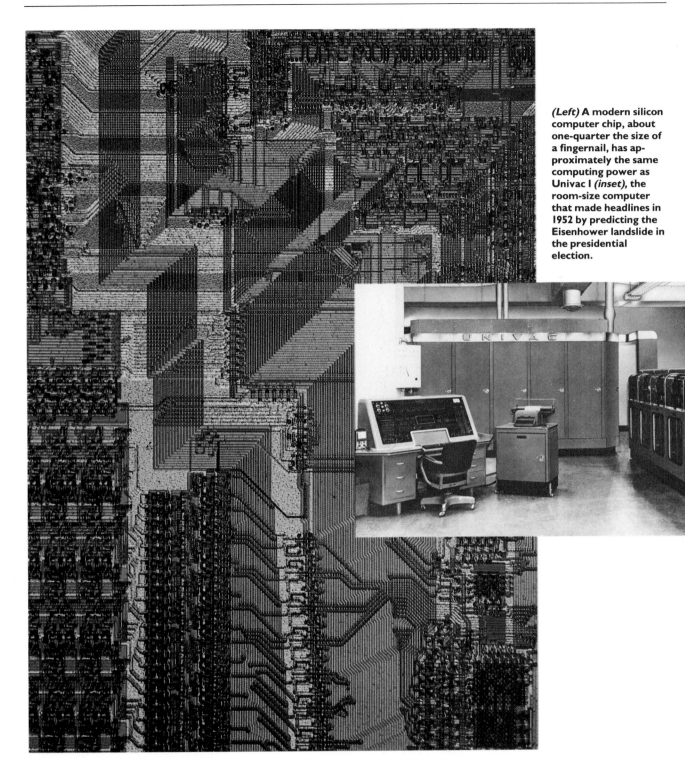

(Left) A modern silicon computer chip, about one-quarter the size of a fingernail, has approximately the same computing power as Univac I *(inset),* the room-size computer that made headlines in 1952 by predicting the Eisenhower landslide in the presidential election.

representing a letter. Within a computer these numbers are manipulated by instructions that are also represented by numbers. The numbers are reconverted to words or signals at the computer's output device — usually a printer.

During the past thirty years computers have become smaller and cheaper. In 1951, Univac I occupied a space ten cubic feet in size and cost $701,000. Today the same amount of computing power can be stored in a tiny silicon chip that costs just $19.

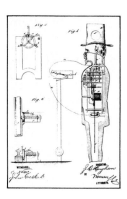

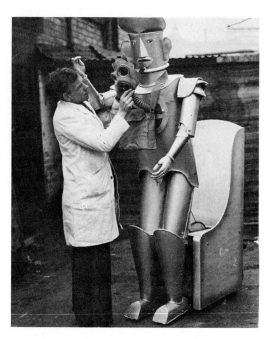

In the early days of their development, intelligent machines tended to be built on anthropomorphic principles. The robots shown on this page could imitate various human functions, from walking and talking to distinguishing colors and smoking.

The quest to build intelligent machines began long before the computer age. Traditionally they have been cast in the image of humanity. The Green Lady was built by eighteenth-century craftsmen to entertain the royal courts of Europe. An artist, she could draw but one picture. Her complicated and graceful moves are rigidly controlled by a hidden mechanism. Early computerized robots were equally inflexible; they could perform only a single kind of task. In 1968, however, researchers at the Stanford Research Institute built Shakey, a more sophisticated robot. Dismantled now, the versatile Shakey could understand English commands and devise a way to carry them out, even in an unfamiliar environment. Using its power of vision, Shakey could, for example, find and retrieve objects.

More recently, computers have been built that can imitate and even replace such human functions as speech and eye-hand coordination. Researchers at NASA's Jet Propulsion Laboratory are testing a computerized wheelchair with a mounted arm that can understand and respond to human speech. Another NASA project is the Rover. As the representative of Earthbound explorers, the four-wheeled vehicle with a mechanical arm and grasping hand is designed to make its way over planetary landscapes collecting rock and soil samples. With an on-board computer, the Rover can coordinate mobility and vision using a laser rangefinder and two television cameras.

Computer intelligence is undeniably mechanical, and for Joseph Weizenbaum of the Massachusetts Institute of Technology, it is nothing more. "Computers are prodigious calculators," he says. "They can solve huge systems of differential equations, invert very large matrices, and do other such mathematical

things." In his view, however, "there's an enormous difference between judgment and calculation that computers can't cross."

Proponents of artificial intelligence argue, though, that the fundamental processes of human brain cells are just as mechanical and that, as with the computer, it is their combination—the program—that counts. Artificial intelligence is based on the faith that there are rules underlying every aspect of human life, rules that can be uncovered and turned into programs and that can thus give machines intelligence for more than number crunching. Critics of artificial intelligence, however, doubt that human life is reducible to rules. Herbert Dreyfus of the University of California says that "it's not a bunch of facts any more than somebody who knows how to swim knows the rules for swimming. We don't know the rules for being a human being or the rules for how to move or stand up. We just embody those rules. Yeats said something very relevant to this. He said that we can embody the truth but we cannot know it. We'd have to know it to be able to tell a computer what it is to be a human being."

But even for those whose confidence in artificial intelligence is steadfast, teaching computers to do more than calculate and remember facts remains a challenge. The first few experiments in artificial intelligence clearly illustrated that computers could play a good game of chess and solve calculus prob-

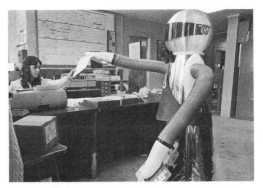

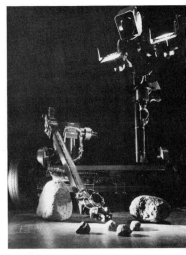

lems. "But," says Marvin Minsky of the Massachusetts Institute of Technology, "it was much harder to get the programs to answer simple questions in ordinary language, the kinds of things that any child can do to solve simple, everyday common sense problems."

To have common sense intelligence, a computer would need many skills, including language. Stanford University's Terry Winograd has written a computer program to experiment with language understanding by a computer. He designed a world full of three-dimensional blocks of different shapes and colors, with a kind of hand that could move them around. The computer had no intrinsic knowledge of the block world, so Winograd filled its knowledge base with information about the blocks and their relationships to one another. It is by correctly carrying out commands to pick up or move blocks that the computer proves that it understands English. But even in this limited world, the process of understanding a command is no simple matter. When Winograd types in a request such as "Pick up a big red block," the computer has to go through several different phases of analysis.

First the computer locates the words in the dictionary that is part of its program and deciphers the structure of the sentence—the subject, the verb, and the object. Then it analyzes the meaning of the sentence, which involves converting the specific words to a set of concepts that it has about the blocks world—what the objects are, what can be done with them, what their colors are—so that it can construct a program for carrying out the action. Finally, the computer "reasons" about the action in order to know what has to be done to carry it out.

In addition to knowing the rules of the block world, the program has to master the rules of language, which are not always clear-cut. When it is given a more complicated command such as "Find the block which is taller than the one you are holding and put it into the box," the computer must figure out what is meant by words like *one* and *it*. According to Winograd, we use these words in normal, everyday language in a way that has to be interpreted within the context in which they appear. To comprehend the command, the computer has to consult a set of rules of thumb written by Winograd that describes how people use these and other similar words. Winograd learned, in trying to program a computer to use language, that he had to examine exactly what people do when they use language. He had to explain words that we would consider too obvious to need explaining.

Shown left to right on this page: the 1968 robot "Shakey"; a voice-activated manipulator arm for quadraplegics; a housekeeping, dog-walking robot; a roving spacecraft designed to explore the surface of Mars. These are all "intelligent" computers that not only imitate human functions but have some of our built-in flexibility as well.

Les Smart Cards

In the town of Saint Malo on the coast of Brittany, the French government has begun a daring experiment to launch every French household into the computer age. The townspeople are receiving, free of charge, computer terminals provided by the state-run telephone company. The French call the idea *télématique*, and it is putting computer power in the hands of people in their homes. The initial reason for distributing the electronic terminals is to replace the telephone book, and so the terminals are, for the moment, used just for locating telephone numbers. But in the future they can be linked to electronic mail, computerized shopping, electronic banking, and a host of other computerized information networks. Meanwhile, French bankers will soon be testing the "Smart Card," a wallet-sized card containing a silicon-chip computer with a memory developed by CII-Honeywell Bull. The card, to be used during shopping and business transactions, can be loaded with credit at a bank and then debited at the point of sale by shopkeepers. The card is protected from misuse

by a secret code known only to the card holder; when someone attempts to use the wrong code, the chip electronically closes down. French government experts are predicting that mass computerization will become as indispensable to society as electricity, and so France is plunging into the future with a policy committed to socializing the computer.

Winograd's program is skillful in handling the imprecision and ambiguity of English, but only if the topic of conversation is blocks. Like us, a computer cannot learn or understand language in a vacuum; it must know what it is talking about. "One of the things that we've learned from writing programs like this one," says Winograd, "is the complexity of the way people understand language, the kinds of connections there are between using your knowledge about what's being talked about and your knowledge of language. . . . As we build more and more complex programs, we look at different aspects of the ways in which people communicate, the ways in which their understanding aids in that communication and takes a part in that communication."

If language and knowledge of the world cannot be separated, how do children acquire language? One theory is that even before they learn to talk, children accumulate a detailed knowledge of routine experiences, called a script. Later they draw on their scripts as a basis for language and conversation. Roger Schank of Yale University believes that computers can learn to communicate in much the same manner. He writes computer programs that can understand stories like this one: "John went to a restaurant. He ordered lobster. He paid the check and left." It is a simple story, even for a child, but surprisingly difficult for a computer.

The main problem with communicating with computers, according to Schank, is that although computers can do certain manipulations, they do not know the facts of human experience. "If we want to tell a story about what goes on in a restaurant, the computer better know about restaurants—what they're for and what goes on in them—so it can fill in the blanks of what I didn't say. If people had to say every piece of information that ever happened, that little three-line story . . . would take hundreds and hundreds of lines because there are assumptions that we share. As human beings having been in restaurants, we know what goes on in them."

Schank finds that the programmer can compensate for the computer's lack of experience by spelling out exactly what goes on in a given situation, such as in a restaurant, by providing the computer with a script. It is by fitting the story into the script that the program conveys the story's meaning. The program matches each sentence of the story to the script and must make assumptions about what happened between story lines. For example, in the story, the main event, eating, is never stated. The program was told only that John ordered lobster; but when Schank asks it what John ate, the program replies, "Lobster." The program has no more trouble

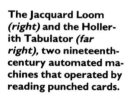

The Jacquard Loom *(right)* and the Hollerith Tabulator *(far right)*, two nineteenth-century automated machines that operated by reading punched cards.

with the question than a person would, because "it has in fact understood the story," says Schank.

Not necessarily so, says Weizenbaum. He believes that language understanding involves much more than the mere comprehension of a string of words. Silences, for instance, are important to human communication. "Even the most ordinary linguistic intercourse among people involves shared experiences. And the fundamental difficulty with computer understanding of language is that there are human experiences which the computer by its very nature, by virtue of its structure and the differences between its structure and the biological structure and needs and so on of human beings, can simply not share. Communication involves sharing."

Schank disagrees. "I think whatever the shared experience is, you have some rule for assessing it. If you have a rule that says, 'Well, I remember feeling that when I was in love . . . ,' I can write that rule into a computer program. Whenever you see something about love, you can assume that the person talking might feel this way and might do this. It's just a question of understanding what people think they know and think they are understanding in a situation."

Patrick Winston of the Massachusetts Institute of Technology is trying a different tack in the effort to give computers human-like intelligence. He has written a program designed to study learning by a computer. By *learning* he does not mean rote learning in which the computer is told the facts it needs to know. Rather, Winston wants the computer to be more involved in the learning process—to do analysis, make descriptions, compare descriptions, and use them to develop models.

To teach a computer how to recognize an arch, Winston began by giving it a model drawing of an arch. The program labeled and counted the parts and the important features and stored this information in its knowledge base. The computer was told only that the drawing was of an arch; it had to figure out for itself what an arch is. Next Winston gave it another drawing, but in this one the two supports are touching. The computer asks Winston whether this drawing is of an arch, and Winston types back that the drawing is not an arch. From the response, the computer can draw an important conclusion: that the supports of an arch must not touch. From

Printout of a computer program.

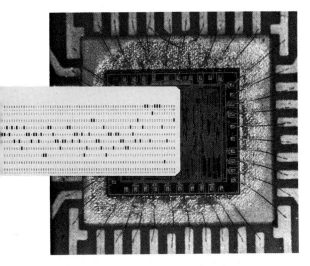

The punched card *(far left)*, an invention of the 1800s, continued to be used until recently when miniaturization of circuitry *(center)* gave computers the speed to process instructions almost instantaneously. The microelectronic revolution has put computers within the means of almost anyone *(left)*.

examples like this, the computer accumulates knowledge about arches. But can it apply its knowledge?

To test what the computer has learned, Winston types in a picture that looks like an arch except that at the top is a wedge-shaped object instead of a flat one as before. He asks the computer whether the picture is an arch. The program analyzes the drawing to see whether it fulfills the minimum requirements of an arch without violating any of the conditions and then responds affirmatively and correctly.

Will this ability to learn go beyond arches? Will computers learn not just by example, but from experience as well? Winston thinks this ability to learn might start a chain reaction of intelligence. "That is," says Winston, "the smart computer might be able to learn to make itself smarter. And that in fact would lead to a kind of intelligence that is very difficult for us to fathom."

Smart machines may be so difficult to imagine in part because people are frightened by their perception of computers as sinister futuristic robots. Arthur Clarke thinks

that even if people understand that artificial intelligence may be a reality, they will try to pretend to themselves that it isn't. Such denial, says Clarke, "is a kind of whistling in the graveyard." After all, our cities are already cognitive cities with many of their functions computerized. An estimated five trillion clerical workers would be needed to do the same work now done by computers worldwide. Artificial intelligence will further this trend by giving computers the ability to program themselves—and maybe explain to us what they're doing. Perhaps we will run things better in partnership with smart machines, or perhaps, as some fear, we will no longer run things at all.

In the German film *Metropolis,* made in 1926, a mad scientist schemes to replace human workers with thinking machines. It is from fantasies like this that our image of the intelligent machine has come. And Hollywood has maintained this Frankenstein motif in an endless series of horror movies starring robots and malevolent computers. Recently, however, there have been exceptions, such as the sensitive R2-D2 of *Star Wars.* What will

Is our increasing experience with computers making us more friendly toward them? *(Right)* The sinister robot in the 1926 film *Metropolis* is from an era when such machines were a total fantasy and yet feared. The more personable C-3PO and R2-D2 from *Star Wars (far right)* reflect a changing attitude that these machines might just as well be our friends.

intelligent machines of the future be like? What function will they serve? How will they influence the workplace, the home, and personal privacy? Is there anything to fear?

At the Stanford Research Institute, artificial intelligence is being applied to industrial problems. Researchers are testing a robot that can be employed on an assembly line. It is thought that intelligent computers will join the work force gradually. They will be assigned first to dangerous and monotonous jobs that are unpopular with human workers, but it is inconceivable that the trend will stop there. Charles Rosen of the Institute believes that the human work force engaged in manufacturing goods will decline from 30–40 percent to 5–10 percent by the year 2000. Arthur Clarke predicts that increased numbers of computers in the labor force will restructure society completely and raise tremendous social and philosophical issues. With much of the routine work, which has taken up so much human time, being done by computers, "what will the people who were only capable of low-grade computer-type work—what will they do in the future? The much more profound question is: what is the purpose of life? What do we want to live for? And that is the question which the intelligent computer will force us to pay attention to."

The decision-making capability of intelligent computers makes them as appropriate to the professions as to the workplace. A computer is currently being programmed by Dr. James Myers and artificial intelligence expert Harry Pople to make medical diagnoses, a skill once thought to require a human touch. Such a program is intuition turned into numbers, judgments converted into calculation. The program is intended to help physicians, not replace them. Yet it is possible that humans may begin placing more faith in the computer's decisions than in their own— or, even worse, that people will begin to abdicate responsibility for decisions to computers. This concern becomes increasingly important as decision-making programs are implemented in medicine and other professions as well. In a world of growing

complexity, governments will use computer programs to guide public policy.

Arthur Clarke even believes that it is possible we may become pets of the computers, leading pampered existences like lap dogs. Regardless, Clarke hopes that we will always retain the ability to pull the plug. "Because if we don't, if we hand over everything to the computers, it will just prove that the computers are designed to be our successors. And that perhaps when they come along, it's our function to become obsolete as our predecessors have become obsolete and been replaced by us. And I feel that if that happens, it will serve us right."

A robot spot-welding automobiles in Japan with not a human in sight.

A Conversation with Patrick Winston

Over the last thirty years, electronic computers have been getting faster, more compact, and cheaper. They are performing individual calculations at hundreds of thousands, even millions, of times less cost than the first computers of the late 1940s. Inevitably, under such intense pressure from headlong technological development, we have come to use computers for more and more tasks. At first there were the elephantine jobs, such as processing the vast accumulations of numbers from the United States Census or issuing millions of Social Security checks each month. But today's nimbler computers switch telephone calls, process words into type, handle the accounts of innumerable small enterprises, manage warehouses and factories, play games, and give children increasing autonomy in their schoolwork. Computer-guided robots are even beginning to replace people in repetitive or dangerous factory tasks.

This unending revolution of the computer has had a constant intellectual accompaniment. This is the work of a small band of irreverently bright researchers, most clustered in universities and "think tanks," who seek to use computers and computer programming to understand the basis of human intelligence while improving the capabilities of the computer. These researchers have called their field artificial intelligence.

The iconoclasm and intellectual playfulness of the leaders of artificial intelligence research may have helped underscore a notion of the field as elite, zany, almost Martian. And yet, the science-fictional overtones of artificial intelligence lately have been dissolving like mist on a summer morning.

One scientist who has watched the revival of the notion that artificial intelligence might have a constructive connection to the workaday world is Patrick Henry Winston. A thirty-nine-year-old native of East Peoria, Illinois, Winston has been the director of the Artificial Intelligence Laboratory at the Massachusetts Institute of Technology for nearly ten years. He took over the position from Marvin Minsky, still a leading member of the

laboratory, who is regarded by many as one of the presiding geniuses of artificial intelligence. The M.I.T. laboratory, along with similar groups at Carnegie-Mellon and Stanford universities and the Rand Corporation and Stanford Research Institute, is one of the Olympuses of the field.

The research in the laboratory ranges from Winston's work on his theory of reasoning by analogy to a major effort to make more efficient use of now-vast computer memories by doing multiple parallel searches through them, instead of "accessing" one memory cell at a time. There are studies of the acquisition of natural-language syntax, computer vision, and other aspects of designing and using robots, and such examples of what is called expert problem solving as advanced programming for automated design of the very large-scale integrated electronic circuits of the years ahead.

Winston says it is slightly mysterious why attitudes toward artificial intelligence work have shifted so greatly from skepticism to cheerleading in just a few years. He says, "It's hard to explain why there has been such a rapid change from lunatic fringe to respectable. Now the field is eagerly sought out instead of being regarded as screwy. It seems to be at the intersection of cognitive science, psychology, and philosophy."

Behind the change, Winston says, there is a "shift from technology-push to need-pull," a shift from computer technologies as a solution looking for a problem to "the only hope" for solving certain problems—chiefly those of the productivity needed to maintain America's suddenly challenged economic preeminence in the world.

The productivity problems concern not only workers in factories, but also the activities of experts, whether they are engineers designing computer systems, doctors seeking more accurate diagnoses of their patients' illnesses, or managers of oil fields seeking more detailed analyses of the production of each well in order to get more oil more cheaply. The search for productivity, Winston thinks, will lead to the replacement of many factory workers by robots and the surrounding of experts with computerized tools for everything from bookkeeping to planning and efficiency suggestions.

While supervising the laboratory's administration, Winston has been studying the learning process by working out, through programming increasingly capable computers, model problems such as teaching a computer to put together a few simple building blocks to make an arch, or to make comparisons between similar situations in Shakespearean tragedies, such as *Macbeth* and *Hamlet,* in each of which the murder of a king is followed by revenge.

Most work on learning, Winston says, concentrates on a single object and its purposes. But Winston sees himself working on parts of objects and the relationships between those parts. These studies reinforce Winston's notion that "you can't learn much unless you almost know it already. You learn by combining what you have already learned." He thinks that rules, being composed of pieces that are put together, are almost known before they are learned.

Winston believes that "we might be on the verge of serious learning in computers." Some time ago the proliferation of electronic pocket calculators made it unnecessary for people to go thorough one of the classic tortures of grade school education: long division. Now, computers bid fair to take over one of the brain-busting problems of late adolescence: the calculus. Winston comments tersely, "Computers do it better."

Already the user of a pair of terminals at his desk in Technology Square, across the street from M.I.T.'s sprawling Cambridge campus, Winston also has just begun using a personal computer at home for confidential budget analyses. All these machines help him with another problem. "I can't spell worth a damn. The computer can find all the mistakes."

The revolutionary implications of increasingly intelligent computers will take getting used to, Winston said. "People have a remarkable ability to adjust over generations,

Patrick Winston (*seated*) working with colleague Joel Moses.

but a poor ability to adjust in a few years." The result of work on artificial intelligence could be "an increasingly vigilant system, able to listen to all phone conversations and read all mail. There is a danger of invasion of privacy."

To Winston, the reasonable way around such potential problems is "to make sure that people know how these things work. We want people not to be mystified by intelligent computers. To be sure, the kiddies with their home computers will feel less shock than their elders, but we hope people, by learning how computers work, will accept them and not be afraid of them, and learn not to trust the computers when they are not trustworthy."

To enhance general understanding of computers, Winston said, "we need to develop the technology of explanation. Computers of the future will have to be able to explain how their conclusions were reached in a humanlike way. If the machine is a sly fox and gives only the conclusion, we've failed."

In high school Winston's main interests were sports, science, and mathematics, "but I always had a fascination with the question of intelligence," he says. So when he arrived at M.I.T., he began taking courses in the anatomy of the nervous system, information theory, and psychology. It was only after some time that study of the M.I.T. catalogue revealed to him that there was something called an Artificial Intelligence Laboratory, "where people used computers to understand intelligence."

Asked about his attitude toward working at one of the world's leading technical universities for the last twenty years, Winston said, "I was scared to death when I stepped in the door. I haven't stopped being scared yet." Winston appears to enjoy running scared.

The Pirates of Privacy

The increasing use of computers by governments and businesses raises a major concern: the threat to personal privacy. In his Privacy Address of 1974, President Nixon spoke of the dangers posed by the misuse or abuse of data technology in both private and public hands. At that time, the United States government alone stored information about its citizens in over 7,000 government computers, and the names of over 150 million Americans were in computer banks scattered across the country. Nixon said, "Advanced technology has created the possibility for new abuses of the individual American citizen. Adequate safeguards must always stand watch so that man remains the master and never the victim of the computer."

Half a century ago, Hitler's Gestapo rounded up more than six million Jews from all over Europe. The lists that identified their religion were stored in shoe boxes. Today, ethnic group, race, religion, income, and political persuasion are all stored in computer files. Never have the governments of the Western world known so much about their citizens. The memory of the Holocaust is still vivid, at least in the minds of Europeans, and in September 1980, the Council of Europe signed a treaty of rules protecting privacy of computerized personal data.

In the United States, however, to be admitted into a hospital for emergency treatment is to sign a blanket release, turning over all information gathered about the patient to his or her insurance company. That information may ultimately end up in the computer files of welfare agencies, credit reporting agencies, and a giant data bank shared by life insurance companies, called the Medical Information Bureau.

The biggest record keeper of them all is still the federal government. At last count, there were more than 5,800 personal data banks holding an average of eighteen files on its citizens. Congressional concern over the increased use of computers in federal record keeping led to the signing of the Privacy Act in 1974. This act provided that each individual has both the right to find out which federal record systems keep files on him or her and the contents of those files, and the

Racks of data tapes stored in a computer room illustrate the sheer volume of information available from computer files.

right to correct the files. The listing of the federal file system is published yearly by the Federal Register. For an American citizen to see his or her files, 5,800 separate letters must be written. Few people have the time, inclination, or the $1,160 in postage required to exercise fully this right to privacy.

In 1978, Congress passed the Foreign Intelligence Surveillance Act. With some exceptions, it requires warrants for all wiretaps conducted within the United States. In the area of international communications, however, the privacy of Americans corresponding with foreigners is not so well protected. One of the nation's largest secret intelligence agencies, the National Security Agency (NSA), was created in 1952 to collect information that moves in the air or across wires: telephone messages, cable traffic, telex correspondence, and radar. According to Morton Halperin of the Center for National Security Studies, the NSA gathers that information and analyzes it in order to learn about the activities of foreign governments. Sometimes the NSA personnel will simply read unclassified and uncoded messages; sometimes they must break diplomatic codes. It is said that the NSA's computer capacities are twenty years ahead of private industry's.

And the government is not the only party keeping files. The buying and selling of mailing lists have become big business and big politics. Millionaire Richard Viguerie heads a direct mail company that raises over $30 million a year through mailings for his conservative clients. As part of the bargain, he gets to keep all the names on his clients' lists. According to Viguerie, his magnetic tapes hold the names and addresses of 25 million people, and of those, maybe about 4.5 million have been identified as being conservative on one or more issues. "When a new conservative cause comes along, we don't have to go out there and invent the wheel all over again. We've got a list of people we can go right to. Direct mail to the conservative movement means a great, great deal. . . . There wouldn't be a conservative movement in this country—there

certainly wouldn't be something called the New Right—without direct mail . . . it's our method of communication.

And communicate they do. Viguerie's company can get out a mailing of 100,000 letters in a day, and while only about 3 percent respond, Viguerie does not feel that he is invading the privacy of the 97 percent who do not. "I feel strongly that it is not an invasion of privacy, no more than advertisements in the magazine you subscribe to . . . or advertisements in your daily newspapers, or advertisements on radio or television."

Direct mail may one day be replaced by Telephone Broadcasting Systems's new marketing technique, the telecomputer. A marriage of the telephone and the computer, the telecomputer plays a canned sales or collection pitch in a conversational context, prompting responses from the phone answerer. Paul Greenberg of Telephone Broadcasting Systems says, "Our experience has been that people really enjoy talking to the computer . . . especially when they hear that first prompt . . . and they know it's a tape. It just does something to them. It makes them answer any question that we ask them."

"There certainly wouldn't be something called the New Right without direct mail," according to Richard Viguerie, head of a direct mail company whose clients are exclusively conservative. His company's computer file includes a list of over 25 million addresses, arranged according to the conservative issue the residents support.

Computer banking could provide a prospective "spy" with a wealth of information critical to the security of a nation, as well as to the privacy of an individual.

At the same time that direct mail and telecomputer calls are invading domestic privacy, electronic threats to data communications privacy and security are increasing as international institutions computerize. Citibank, for example, America's second largest bank, has more than 200 branches in 100 countries hooked together with their private lease-line network. With major switches in Hong Kong, London, Bahrein, and New York, Citibank computers process over $30 billion a day.

Citibank, like many other financial institutions, is committed to electronic information and transaction processing. But electronic transfer systems are vulnerable to wiretapping. Not only can computer thieves tap financial data lines to gain access to and manipulate electronic fund transfers, but computer spies can tap lines to collect confidential information about individuals and institutions as it travels between computers. All it takes is the right equipment and access to the right location.

Wiretaps can occur at three points in data communications systems. In an inside job, an institution's main computer can be tapped directly. Outside, a minicomputer can be attached directly to phone lines. Finally, invaders can break computer access codes and then dial into the system by public telephone. There are several million home computers in the United States today, any one of which could be a fine tool for wiretapping.

Several years ago, Paul Armer of the Charles Babbage Institute was a member of a workshop attended by computer and law enforcement specialists. The group was asked to pretend that they were consultants to the Russian secret police and design a system that would keep track of all Soviet citizens and foreigners within the U.S.S.R. "After considerable study," said Armer, "the workshop concluded that the best system to build for the KGB . . . was an electronic funds transfer system, for the reason that electronic funds transfer systems know not only what you're buying, but where you are in real time at the time you're making your financial transaction."

Some privacy experts acknowledge these threats to personal privacy but consider them beyond most existing computer capacities. The NSA, however, recognizes them and routinely makes codes to maintain communications security and breaks codes to conduct its foreign intelligence missions. Some banks too, such as Miami's Amerifirst

Software Bandits

The Apple II, which debuted in 1976 as the first true home computer, opened up a vast technological frontier to droves of eager explorers. And like any good adventure story, this one has its outlaws—the bandits who illegally reproduce software disks by unraveling the intricate protection mechanisms that encase the commercial programs. The thieves never sell their loot—that would violate their code of honor—but trades are everyday occurrences that involve a nation-wide network of conspirators. An investigative reporter, Lee Gomes, traversed this shady side of the computer boom and came back with a surprising portrait of the criminals: they range from adolescent veterans of the Rubik's Cube to business executives who cut their teeth years ago on Ma Bell's nemesis, that little blue box. At Apple II users' meetings, it's not unusual to find a scraggly twelve-year-old and a middle-aged tycoon huddled together over the details of an illicit exchange; authorities estimate that five illegally copied disks

change hands through such trades for every legitimate program sold. What's worse, most experts agree that there isn't a protection technique around that can't be cracked. So until the $25-million-a-year software industry finds its own Matt Dillon, the bandits are riding high through the land of the Apple II.

A minicomputer and a telephone are used to tap into a bank's data line. Information screened on the terminal can then be recorded and phony transactions entered into the computer.

Federal Savings and Loan, have protected their data communications through "datacryption," a method of scrambling data that are transmitted over phone lines so that they cannot be interpreted as they are being passed back and forth. But not one of the world's eight largest banks encrypts all its data communications. While four of the eight encrypt at strategic places, the other four don't encrypt at all. Encryption can be expensive, and one expert claims that the world's financial institutions will undergo a disaster – a Three Mile Island of data processing where billions will suddenly be lost – before encryption systems are installed everywhere.

In the worst-case scenario, in the minds of many, smart computers will one day lead to the kind of life foretold in George Orwell's *1984*. To Paul Armer, 1984 is more a state of mind than a particular set of government capabilities and regulations. "It's a state of mind in which the individual is so concerned about his actions, where his actions are going to be known to government and a large set of people, that he is inhibited . . . in the sense that he is not willing to stick his neck out. In a society of that sort, we'll soon lose our initiative and innovation. . . . Democracy cannot flourish in a society in which everyone is conforming out of fear."

Computer-Age Encryption

The spread of electronic communications systems has brought with it the problem of keeping eavesdroppers and forgers from tampering with private messages from organizations or individuals. Conventional cryptographic systems—mathematical methods of encrypting, or transforming, information into a digital "alphabet" so as to keep it secret—are inadequate because both sender and receiver share a single key for enciphering and deciphering messages. It is hard to keep such a key secret, since it could be stolen while being transmitted to users. The solution lies in the new double-key system, in which each user holds two keys of his own: an enciphering key, which can safely be made public, and a corresponding deciphering key which remains private. For example, A sends a message to B by enciphering it with B's public key. (A does not have to know B's private key.) Only B can decipher the message, by means of his private key. Because of the keys' mathematical complexity, it is virtually impossible for a third party to derive the deciphering key from the enciphering one. Under this scheme it's also possible to authenticate a message by means of a "digital signature," which prevents third parties from forging coded messages. Here's how it works: A enciphers a message with his own private key, then further enciphers the result, using B's public key. Only B can decipher the doubly enciphered result, by using his private key plus A's public key. B now has proof that A sent the

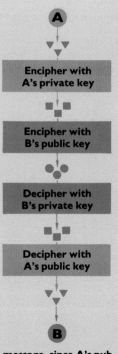

message, since A's public key helped in the deciphering. Since this can be so only if A enciphered the message with his private key, A has, in effect, signed the message.

The Light of the Twenty-first Century

The principles behind certain techno-logical breakthroughs are often understood long before any working models exist with which to demonstrate them. People knew of the power of steam before there were steam engines, and the equations of nuclear physics were scattered across laboratory blackboards long before Oppenheimer went to Los Alamos. But sometimes it all comes nearly at once—a new science, new engineering, new hard-ware. It can happen so fast that society doesn't quite know what to do with it all, leav-ing problems to catch up with their solutions. So it was barely twenty years ago, with the infant science of photonics and a curious de-vice called the laser.

The Ships Used Their Terrible Ray Guns

Future space lasers (top) may be the weap-ons that Buck Rogers dreamed of in 1935 (bottom).

Laser is an acronym that stands for Light Amplification by Stimulated Emission of Radia-tion. Put simply, this is a means of producing coherent light. The light we are most familiar with is incoherent, consisting of varying wave-lengths. (The name *photonics* refers to the tiny subatomic particles called photons, which are believed to compose light waves.) The waves of coherent light are perfectly regu-lar, and their color is pure. This result is achieved by stimulating the emission of uni-form waves of light from the similar atoms of a crystal or, in more advanced applications, a gas. These emissions reinforce one another. The accumulating light radiation is reflected

from high-quality mirrors in a parallel beam and may be precisely focused in a given direction.

In 1960, when the prototype laser followed by only a few years the first theoretical expositions of the principles be-hind its operation, hardly anyone could envi-sion the possibilities that lay ahead. It was soon demonstrated that the color—and hence the wavelength—of laser light could be governed by the substance from which it is radiated. Krypton gas, for instance, pro-duces a red beam. The means of stimulation was also refined. While the earliest lasers depended upon an outside source of light to trigger the stimulation of radiation among the atoms, electricity soon became the activating medium of choice.

Now that the scientists had produced something they could call coherent light, what were engineers to do with it? The first an-swers to this question, though certainly sensi-ble, were nonetheless somewhat tentative and passive. The narrow, parallel, and unwav-ering beam of a laser could be used for sur-veying and leveling in construction. Also, the laser's emission of uniform wavelengths of light made it possible to take a count of those wavelengths in order to measure distances and speeds with incredible accuracy. The dis-tance to the moon could be estimated to within a foot, and the velocity of light itself was calculated precisely at 186,282.397 miles per second. But was this wonderful new dis-covery to make its impact merely as a cosmic measuring tape?

More powerful lasers have pushed the frontiers farther. One of the first pursuits for which laser applications were investigated was that of warfare; the fantasy of the death ray is at least at old as this century. The prac-tice of selective military destruction, as op-posed to all-out nuclear war, has long been limited by the technology of conventional high explosives. The laser appears to offer an alternative that is swift, surgical, and deadly.

A surveyor's laser will not knock down an enemy plane or disable a satellite in space. Problems of intensity and distance

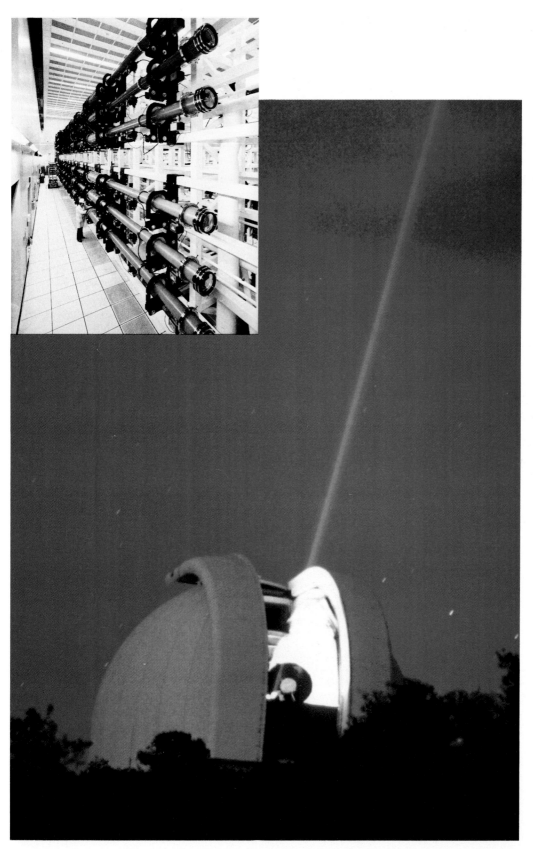

Lawrence-Livermore Laboratory's giant laser, Shiva *(inset),* **is the world's largest. Its twenty amplifier chains, of which six are shown, can produce 40 trillion watts and generate temperatures found at the center of the sun.**

The McDonald Observatory directs a laser toward retroreflectors left on the moon by Apollo astronauts. The beam spreads out a mere two miles on its journey across a quarter-million miles of space and returns to Earth with enough intensity to be easily detected. By measuring the time of the laser's voyage, scientists can calculate to within one foot the distance between the two spheres. Such information, over time, allows scientists to detect the moon's recession from the Earth and movements in the surface of both bodies.

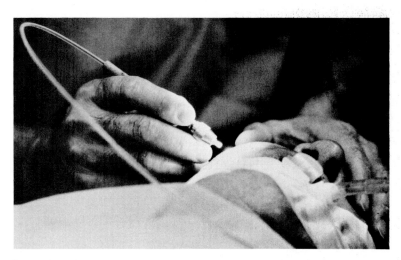

Lasers, here being used to treat a skin malignancy, have revolutionized cancer therapy.

must be overcome. The first of those hurdles was cleared early in the game through the development of high-pressure carbon dioxide lasers, proven to have a piercing effect on oak, firebrick, granite, and even an eight-inch block of plexiglass. That was in 1978, when *Aviation Week* correspondent Phillip Klass estimated that military researchers were working with lasers of several hundred thousand kilowatts of continuous power. More recent developments are largely classified, but it is likely that laser technology has moved even closer to becoming a part of modern warfare. Satellites may already be carrying lasers, although their usefulness, for now, is likely to be limited to engagements with other satellites. Still, it is

unsettling to think that today's nuclear disarmament demonstrators might be protesting the technology of the 1950s while the next generation of "ultimate" weapons takes shape behind closed doors.

Research into the constructive use of lasers has proceeded at least as rapidly. In fact, the notion of the "surgical" capability of military lasers is ironic, for surgery is one of the fields in which lasers have shown the most promise. The laser is a scalpel of incredible sharpness, capable of simultaneously opening and cauterizing an incision under conditions of perfect sterility.

Laser surgery has made particular advances in two difficult areas: eye disorders and the removal of cancerous growths. An ophthalmologist can use a laser beam to spot-weld a detached retina back into place. The hemorrhage of minute optical blood vessels can be prevented by laser cauterization. No steel scalpel, no matter how small or how skillfully wielded, could perform as well as the precision laser in delicate operations such as these.

The laser does not merely excise a malignant tumor; it can obliterate it, attacking a mass of cancerous tissue as a steam jet would a snowball. Nothing remains of a tumor removed in this way except a harmless vapor; this greatly reduces the conventional surgical problem of lingering cancer cells finding their way into the bloodstream or adjacent organs.

Ultimate medical uses of the laser challenge the imagination. Lasers have been used to target individual cells and can even be focused upon cellular components. Since cancer consists of the uncontrolled multiplication of malignant cells, the laser's potential as a tool for research and prevention of this disease may someday eclipse its curative functions.

High-power lasers are not confined to the world of top-secrecy classifications and operating rooms. There are blue-collar uses for lasers as well, cutting cloth, welding and hardening metals, and—in conjunction with robots—increasing precision on assembly lines. Lasers were chosen for the job of cutting the silicon ribbons from which the latest gen-

Laser Light

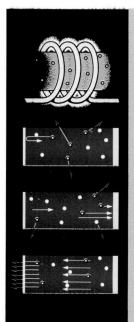

The generation of a laser beam begins with a ruby rod sealed completely at one end and partially at the other with a coating of silver. The rod is embedded with chromium atoms and wrapped in a coiled flash lamp(1). Light from the flash lamp excites the atoms and causes them to emit particles of light, called photons, which either pass out the sides of the tube or reflect back into it off the silver ends(2). Those photons which reflect back into the rod collide with other chromium atoms and trigger the release of more photons, which travel in the same direction as the incoming particles(3). The cumulative effect of these collisions creates a concentrated beam of photons that eventually gathers enough intensity to burst through the partly silver end of the rod(4). The wave motions of these photons are all in phase with one another, concentrating and focusing the laser beam—unlike natural light, which is emitted on numerous wavelengths and is therefore diffuse.

eration of solar electric cells are made. With a material as valuable and hard to produce as single-crystal silicon, it's important that cutting waste be kept to a bare minimum.

It is a somewhat differently educated laser that reads the codes of closely spaced black lines on packages at supermarket checkout counters. The information thus decoded is fed into the cash register, which then records it as a price on an itemized receipt. The same principle underlies the lasers used by some newspapers to translate the compositor's dummy pages into the negatives required by the photo-offset printing process. At the *Los Angeles Times* the original paste-up is "read" by a helium neon laser (that is, light and dark areas are discerned), and the information is converted into electrical impulses traveling along a telephone line. The phone is answered, as it were, by an argon laser, which uses the information to produce a negative of the page, ready for use in the press. With this technology, the composing room no longer has to be in the shadow of the presses, and the advantages of rapid computerized typesetting become easier to exploit.

The synchronization of laser technology and telecommunications has done more than help banish lead type from the printing plant. It also promises to render obsolete the hundreds of thousands of miles of copper wire along which our conversations travel. In conventional telephone technology, sound waves are transmitted as units of electricity along a solid conductor—the familiar insulated copper wire. But light can do the job even more efficiently, since light waves are so short. Low-energy pulsed lasers, emitting millions of bursts of light per second, can vastly increase the message load that can be carried along a single line. Conductive metals such as copper need play no significant part in such a system, because the pulsed laser beams travel along fine, transparent glass fibers. The laser's contributions in speed and efficiency are thus matched by the economic and environmental advantages of eliminating dependence on a finite natural resource.

The awesome force of a laser can slice metal with a margin of error of a few wavelengths of light. It can also pierce granite and cut fifty garments at once without a fray.

Television is also making the laser more prosaic. On a videodisk, the audiovisual information to be transmitted through the receiver is encoded in each groove as a series of tiny hills and valleys. Given the laser's ability to track minute differences in the distances of surfaces at which it is aimed, it was the obvious instrument of choice for "reading" and decoding the disks.

Coherent light, like so many great scientific discoveries, is based upon the most elegantly simple of principles. Its uses already number in the hundreds, even thousands, yet the surface has hardly been scratched. An instrument that can vaporize steel, tally grocery prices, read newspapers, and remove tumors is a protean thing indeed. In some ways, the secret of disciplining light makes atom splitting seem clumsier than the chipping of flint. Of course, arrowheads in their many forms remain one of humanity's favorite products. But the laser is far more than a death ray. It could truly become the light of the twenty-first century.

Many stores now use laser cash registers to tally merchandise marked with these codes.

Scientific Sorcery

Steve Benton invented a holographic plate in 1969 whose image could be projected with ordinary white light rather than a laser. This technological refinement has taken holographic display out of the labs and into the corporate boardrooms—but at a cost of anywhere from $2,000 to $50,000 per image, depending on size and complexity.

Some fifteen years ago, Ray Bradbury wrote a short vignette describing a future in which a family, sitting in the living room after dinner, could conjure up the phantom images of famous persons of the past, with whom they might share an evening's conversation. Bradbury didn't go into what an American family might have to say to an ectoplasmic Aristotle, or he to them: would a computer in the image-generating machinery be programmed with all of his writings, to be scrambled into appropriate responses? But the story did contain hints that the bare rudiments of such magical technology were already in experimental stages.

Today, holograms—three-dimensional images—are a reality. They don't talk, much less talk back, and only recently have they been given color and the illusion of movement. But the miracle is that they exist at all. If there is a point at which, for the uninitiated, science borders on sorcery, it could not be better represented than by the hologram.

Holograms are the work of lasers. The process is somewhat akin to photography, although it does not involve the exposure of a conventional, emulsion-coated film to light. The "print," if that is the word, is made upon glass rather than paper. The creation of a hologram begins with the splitting of a laser beam in two, following which the beams are spread to accommodate the subject. One beam is focused upon the subject, from which it is reflected onto the glass photographic plate. The other beam, called the reference beam, is meanwhile aimed directly at the plate. The exposure lasts only for a second or two. Then the plate is developed in much the same way as an exposed roll of film.

The developed plate looks no different in ordinary light from how it looked before the process was begun. The developing chemicals, however, etch a microscopic granular pattern into the surface of the glass. This pattern corresponds to the wave front projected from the subject by the reflected laser beam. When the reference beam is refocused upon the developed plate, the pattern is deciphered and the picture reappears, with all of the subject's original depth and dimension. (Russian and British experiments point the way toward the viewing of holograms without the reference beam, through the use of ordinary reflected light.)

In a remarkable demonstration of the hologram's properties, a glass plate upon which the three-dimensional image of a model ship had been encoded is deliberately shattered. When each fragment of the plate is viewed in the reference beam, the image revealed is not a corresponding fragment of the whole—not a mast, or a sail, or a portion of the hull—but the entire ship, from differing points of view. If you were to break one of the shards yet again, the pieces would reveal two complete ship images, in three dimensions. The effect resembles the multiplication of the bucket-carrying brooms in the "Sorcerer's Apprentice" sequence of Walt Disney's *Fantasia*, comfortably locked in glass.

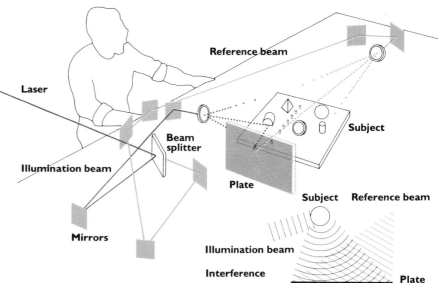

An imprecise but useful analogy might help diminish the eeriness of this phenomenon. When you ride on a bus or subway car at night, you and the other passengers are reflected in the windows. The reflections aren't located at any one place in the glass, nor are they visible from only one angle. Change seats, and all of the reflected images remain, although you see them from a different point of view. If the glass could be broken and the pieces held in the proper places, each would still be capable of the window's full range of reflection. In a hologram, it is as if the lasers were able to fix and retrieve reflections, depth and all.

From here, the development of holograms is a matter of refinement. Improvements in size, clarity, color, and even motion will fall into place as lasers and ancillary technology are better understood. It is still anyone's guess whether, sometime in the next century, Ray Bradbury's talking ghosts will replace video games in the parlor. But perhaps we should rehearse some intelligent questions, just in case.

The production of a hologram. The incoming laser is split into two beams: one is reflected without interference to the photographic plate, while the other is scattered by the subject before reaching the same destination. The scattering of the second beam varies the length of its waves, and the intersection of this altered wave pattern with the original beam creates an interference pattern on the photographic plate. When illuminated with ordinary light *(above)*, the plate projects an indistinct image. But when lasers *(left)* are directed through the plate, three-dimensional images appear; the top of the plate produces views from above, while the lower portion produces views from below.

Artists in the Lab

Perhaps it is a sign of our times—of how tightly high technology is being interwoven into the fabric of our lives. For as our society is becoming more technologically literate, computers, lasers, holograms, and even television are being increasingly exploited as media for artistic expression. These technologies, far from stifling the imagination, are expanding the artist's traditional tools and methods; artists are moving into a new studio, the scientific laboratory, and bringing the visual arts and music into the twentieth century.

Computer programmer David Geshwind stands with a frame buffer. This high-speed memory makes computer graphics possible by enabling the computer to sample a quarter of a million points thirty times a second and display the information on a television screen.

Artists of many media have apprenticed themselves to the computer, and their results are both exciting and fascinating. From computer animation to electronically synthesized music, the new wave of artistry allows its proponents to extend their imaginations. And it remains the human imagination, they contend, that is the focus of their art. Lance Williams of the New York Institute of Technology says that the computer enhances an artist's imagination "in the same way that human calculation was hastened first by writing instruments, then by the slide rule, then by adding machines, and finally by computers."

Other artists concur. "What fascinates me about computer graphics," says David Geshwind, also of the New York Institute of Technology, "is that while I've always been in-terested in the visual arts, with a background in mathematics and engineering, until recently I haven't been able to express those interests. Computer animation gives me the most direct route from conceptualization to visualization and lets me bypass the use of hands and traditional artist's materials."

Ron Hays is a "visual music maker." He takes a piece of music, listens to it, and then transposes it into a visual equivalent. "The technique of computer animation, the technique of electronic animation, the techniques of videography—those are tools just as paintbrushes are to an artist who paints on canvas or chisels with a chisel and hammer," says Hays. "We both sculpt . . . only I do it with cathode ray tubes and photos and electronic imagery."

Artist David Em finds that the computer expands an artist's capabilities. "It gives you the element of control. You don't really have that fine a control in any other medium, but it also gives you tremendous choices. It lets you choose from almost an infinity of colors that you can have on the screen at any given time, and it lets you manipulate those colors in many, many ways. It lets you join together many different disciplines so that . . . you can paint in a single image. . . . You can do very sculptural things that are dimensional, and you can do things that are truly quite architectural."

James F. Blinn, a computer programmer with the Jet Propulsion Laboratory, creates computer animation that simulates what it would be like, for example, to ride along on the *Voyager* as it flies by Jupiter and Saturn. "I go about it in a fairly methodical way, simulating light reflecting off surfaces, and the geometry of objects in space for perspectives and so forth. This is what artists around the Renaissance were concerned with—being able to do more intuitively—but there is a lot of art involved just in the programming itself."

Computers are not the only technology now being used in artistic ways. Holograms are three-dimensional pictures that, until recently, could be viewed only by laser light (see "Scientific Sorcery," pages 224–225).

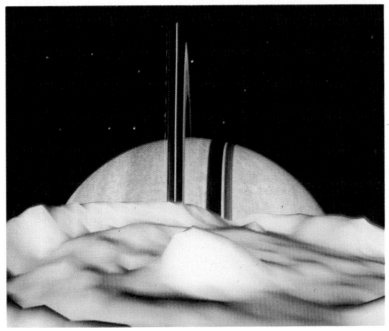

(Above) Artist David Em takes the programs worked out by James Blinn for the Jet Propulsion Laboratory and uses them to create complex compositions such as the work *Nora*.

(Left) James Blinn has made detailed computer movies of the flight plans of the *Voyager* spacecraft through the solar system for NASA's Jet Propulsion Laboratory. Here Saturn is seen from the surface of its innermost large moon, Mimas. The large crater discovered by *Voyager I* is in the foreground.

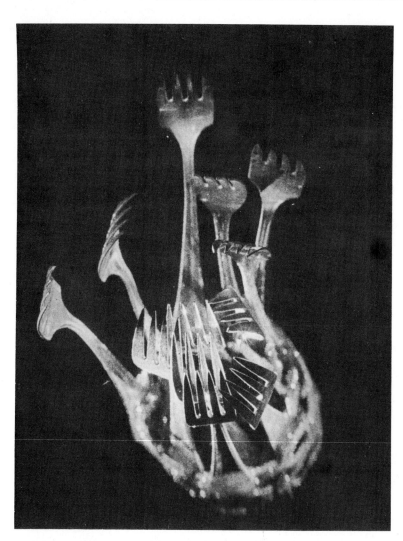

***Equivocal Fork I,* a white-light hologram created by Harriet Casdin-Silver, an artist at M.I.T.'s Center for Advanced Visual Studies.**

When Stephen Benton at Polaroid invented a means of seeing holograms in white light, he opened a floodgate of possibilities in holography.

Harriet Casdin-Silver, an artist at the Center for Advanced Visual Studies at the Massachusetts Institute of Technology, describes her work titled *Equivocal Fork II:* "You can see this hologram—as you can most holography—from side to side. There are particular angles of view, and as you change your angle, you see different versions of the imagery. The forks come in and out for you. Also, you can see this fork plate from way back down the road. As you walk into it, it changes. It will change color as you change your angle of view moving into it—that is, up and down. And as you walk directly into the light, the light will start to swim around your head, so you feel enveloped by the hologram."

Artists and audiences have for the most part stepped gingerly into this frontier of high-technology images and sounds. John Pierce is a pioneer music composer who was first drawn to "this very peculiar, new aspect of sounds generated by computers" in the early 1960s, while researching the processing of speech for Bell Telephone Laboratories. "A lot of people," he believes, "have been intimidated by *everything* that computers do, including the making of music. Indeed, that's what kept good composers away from computers for a long time, and it is perhaps what keeps people away from the music that is produced. Another thing that keeps people away from the music is that people who have done the best work with computers are contemporary, if not avant-garde, musicians. They aren't composing in the tradition of Beethoven or even of Berlioz or later composers; they're composing something that to them is new, using a new means to do new things. And the new things are unfamiliar."

To Lowry Burgess, an artist who teaches at the Massachusetts College of Art, this common tool, the computer, is not really a new instrument but much like an old standard, the cathedral organ. "We see the organ master sitting at his console, playing the keys and pulling the stops," he says, "and what he is really doing is interrupting the flow of air through the chambers into the various pipes to produce a wondrous variety of sound. Just as the organ master sits there, so we see computer artists at their keyboards, essentially interrupting (through the computer) the flow of electrons to produce wonderful images of light and sound—new things for our eyes and ears." Burgess believes that we are entering a new phase in our relationship to technology. Epitomized by the array of mechanical apparel people can wear—the wrist watch, the camera—technology is becoming like jewelry and clothing that is wrapped around us and worn close to our skin.

Yet the new technologies have a magic to them, a staggering magic when one considers what we have accomplished in recent decades. "The new technologies are leading us into more illusionary spaces. Holograms hover in space behind a dim photographic plate. Computers give us music or pictures that are really but numbers stored electronically in the computer's memory," Burgess says. "We can also place our eyeballs on the rings of Saturn; we can put our ears out on the edge of the universe; we can make our hands reach down into the atomic structure." Truly, art and science have finally merged.

As an example of how a concept for a piece of art develops—here a fifteen-second animation—we can trace the steps that led to the new opening for "NOVA." David Geshwind of the New York Institute of Technology relates how the work progressed:

"The production of the new opening for the WGBH science series, 'NOVA,' required the coordination of many computer animation techniques. These covered the range from highly interactive, 'artist-intensive' tasks—such as 'painting' with an electronic palette—to tasks requiring complex mathematics and large amounts of computer time.

"The concept for the opening was first developed as a storyboard. Starting and ending with treatments of the 'NOVA' logo, the animation would be a fifteen-second grand tour of science. It would consist of a succession of images, each representing a major branch of science, and would encompass a change of scale that takes the viewer from the atomic to the cosmic.

(Above) Visual music maker Ron Hays and television director David Atwood work with the tools of a new art form, in the early days of the WGBH New Television Workshop, in Boston. (Left) A single frame from Ron Hays's video piece, *Tristan und Isolde.*

A movie showing a serene day in the life of an island, with rhythmic water waves, changing sky, setting sun, and rising moon, was created entirely on a computer by Nelson Max of the Lawrence Livermore National Laboratory.

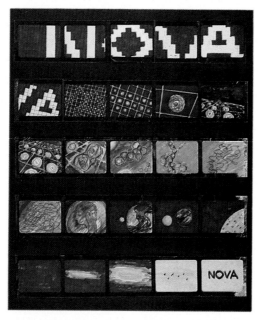

The dazzling new opening animation sequence for "NOVA" began with some graphic notes *(above)* supplied by the show's executive producer, John Mansfield, to its designer, Paul Souza. Translating these sketches and ideas into a logical sequence of coherent images, Paul created a storyboard *(right)*, which he then turned over to the New York Institute of Technology. Throughout the process of producing the opening, both he and I (as technical director of the project) worked closely together to solve the complex technical and artistic problems.

"My first responsibility was to design and 'model' each element in the storyboard— that is, to describe to the computer, using mathematics and programming, the structure and look of each object. Next I had to develop a scheme, using various animation techniques, that would provide smooth motion and a continuous transition between these elements.

"Whereas animation gives the illusion of motion, it is composed of a succession of still frames. Working from programs and data developed in the modeling phase, the computer is instructed to place three-dimensional descriptions of objects in a theoretical 3-D space. By moving these 'objects,' and the theoretical camera position, from frame to frame, and constructing the image resulting from that view, the sequence is built up.

"Each image is constructed in a frame buffer, a device that allows the computer to think that it is accessing standard computer memory, yet which produces a visual result from the information stored in it. For every three numbers stored, representing amounts of red, green, and blue, the frame buffer

The opening effect *(below)* is an example of two-dimensional image processing, in which the move is a flat pan: that is, the perspective doesn't change.

The "NOVA" lettering was created by tracing the logo onto grid paper *(right)*, taking measurements, and then writing a computer program that filled in boxes where

necessary. Six versions of the logo, in the colors of the spectrum, were overlayed and then blurred; a crisp white version was then placed over the spectral trail.

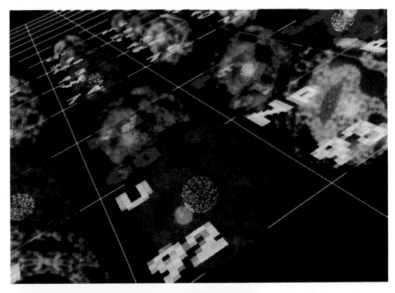

displays one picture element [pixel]. The quarter-million pixels in each frame are lined up in rows and columns to make a video mosaic.

"Since the 'NOVA' opening would be viewed many times, my primary concern was to maintain viewer interest. There is a great deal of information to be absorbed, both in terms of the number of elements and the amount of detail. It is not only *possible*, but it is *necessary* to view it more than once in order to understand all that is happening."

The Earth and moon *(left)*, unlike the other planets (which were flat paintings), were created as three-dimensional objects. Using a technique called texture mapping, paintings of the surface of the moon, the Earth, and a cloud cover were wrapped around spheres. The images of the periodic table *(above)* were composited from several diverse elements. The atoms and grid lines were constructed as 3-D models. The type, showing the symbol and atomic number for each element, was produced using a hybrid technique. The letters and numerals were produced on a video screen using computer typesetting. Small type was enlarged to get a blocky style, consistent with the opening treatment of the "NOVA" logo. This large flat plane of type was then tilted mathematically. The many elements were rendered by the computer separately and then composited to achieve the final image for each frame.

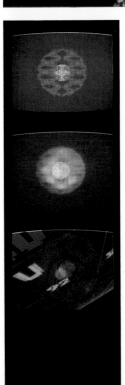

To produce complex three-dimensional movement, the mathematical model of the animated objects must also be 3-D. An atom is made of three parts, each a sphere. A large sphere with a sparse, undulating texture, which is then blurred, represents the electron cloud *(left top)*. A second sphere with re-flected highlights lends "solidity" to the atom. *(left middle)*. The third sphere represents the nucleus of the atom *(left bottom)*. Once the technique for making individual atoms was developed, a structure was developed to cre-ate the **DNA** molecule *(above)*.

The closing sequence *(left)* shows a swirling galaxy enveloped in the Word **NOVA**. As the stars shift in per-spective, a cosmic-sized **NOVA** follows the gal-axy, and we fall back-wards through the **O**. Each 3-D vector frame was treated as a line drawing and filled using a computer program originally developed for automating cartoon an-imation. The line ver-sions from several pre-vious frames were colored red, green, and blue—the video primar-ies—and blurred to give a neon effect before being composited with the solid white **NOVA**.

The Science of Murder

During this century, almost twice as many Americans have been murdered as have died in war. In 1980 there were close to 20,000 homicides in this country. Cleaning up after a murder involves the painstaking work of many people in the police force and in the fields of forensic science and medicine: scientists and doctors who analyze evidence to help solve murders.

Forensic science has a brief, if colorful, history. An early case was the 1849 murder of a Harvard benefactor, Dr. George Parkman, by a chemistry professor at the university. In solving the case, pieces of bone and teeth found in the ashes of the professor's laboratory furnace were used as evidence. In 1892, the first murder case that was solved through fingerprint evidence occurred in Argentina, and in 1910 a Dr. Hawley Crippen, accused of murdering his wife, was found guilty and hanged when a scar on a small piece of skin from remains found in his basement was identified as a known surgery scar on the victim's abdomen.

Today the forensic scientists' evidence ranges from footprints to blood samples, from hair analyses to bite-mark identifications. Their work begins at the scene of the crime, and their first piece of evidence is a body—a dead body.

Before the body is removed to the morgue, the location of every item in the scene is diagrammed, as a prelude to the search for physical evidence that could identify the killer. The killer could have left saliva on a cigarette butt, a good set of fingerprints on a glass, hairs on a hat, blood from a cut. But after identifying the possible sources of evidence, the investigator must protect them. All too easily, evidence is destroyed. If the murderer was smoking a cigarette and threw it in the toilet, say, and if the toilet was subsequently flushed, a piece of evidence is gone. If a police officer picks up the telephone at the scene of the crime, the fingerprint evidence may disappear. Anything that destroys the integrity of that scene can contaminate or destroy potential evidence.

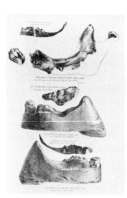

The evidence that convicted the murderer of Dr. George Parkman: his dentist's molds of Dr. Parkman's false teeth. The dentist, Dr. Keeps, was the star witness for the prosecution.

When the medical examiner arrives to take charge of the body, he or she confirms the death and examines the body for injuries. The body is then shipped to the morgue, where it will be examined in detail and dissected at a forensic autopsy. Through the autopsy, the medical examiner seeks to reconstruct the evidence of the death and the circumstances surrounding it. He or she also attempts to recover any evidence, such as a bullet, that is in the body.

At the forensic autopsy, the initial investigation consists of an examination of the exterior of the body for evidence of a fight or a scuffle that might have left marks on the hands or face. If there is a bullet wound, the examiner will inspect it to ascertain what the range of fire was. The medical examiner is also interested in whether the victim took any drugs. This can be determined in a variety of ways. The gall bladder bile, for example, is a rich source for detecting drugs and establishing the narcotics profile of a victim.

Organ samples are sent to the forensic science laboratory, where toxicologists continue the search for drugs and other trace substances. A gas chromatograph—mass spectrometer can find one-millionth of a gram of cocaine in a victim's liver, and from an analysis of this extract the toxicologist can determine exactly how much cocaine was in the body. Technicians can also determine the presence of opiates or barbiturates by testing samples of the victim's urine. Besides testing the corpse for evidence, forensic technicians must also analyze any evidence the police can provide from the scene of the crime. Most often these samples—bloodstains, hairs, footprints, fingerprints, even knife-cut marks—are used to link a suspect or a weapon to the crime.

Using a microscope, for example, a ballistics expert can examine recovered bullets, bullet holes, and gun barrels to match them up. The barrel of a gun is rifled into a spiral of grooves and ridges that leaves unique marks on the bullets it fires. The marks on two bullets fired from the same gun will be as alike as two prints from the same finger. If a gunpowder residue is left around a bullet hole, it can be removed and treated. The pattern that shows up will tell how far away the gun muzzle was—evidence of how the murder was done. Taken further, the powder residue can be magnified with an electron microscope. The character of the particles could lead to a particular batch of ammunition—sold at a particular place, on a particular day.

Likewise, footprints are increasingly conclusive and identifiable. University of North Carolina anthropologist Louise Robbins applies procedures that scientists have used to identify footprints from the past. "There are forty-six points of measurement and one hundred and twenty points to examine for shape," she says. Footprints can also be detected in people's shoes, and scientists like Robbins help police match a shoe to its wearer.

Hair, too, can be matched with increasing sophistication. Different people's hair can differ in color, pigment distribution, texture, thickness, and twenty other characteristics. "The hair is the garbage can of the body," says New York City's forensic serologist, Dr. Robert Shaler. "Everything you eat shows up there." Since hair grows one millimeter a day, Shaler's analysis can tell "if you took aspirin yesterday and drank beer from an aluminum can a week ago."

Perhaps the most exciting field in forensic evidence analysis, however, is the testing of blood samples. Recent developments in serology have produced tests that can pinpoint specific genetic components in blood. Blood experts at a national forensic science research lab in Aldermaston, England, have developed tests sophisticated enough to detect inherited protein components or antibodies left by a suspect's disease history.

Dr. Kenneth Jones of the Aldermaston research establishment says that the tests are so specific as to tell the difference between identical twins with different medical histories. "My twin daughters, who were identical when they were born some sixteen years ago," he says, "now have different blood patterns merely because one of

them had chicken pox and the other one had mumps, and one had influenza when the other one didn't have it. And all the antibodies to those diseases are recorded, contained, and ready and available in each of their bloods."

The life of a forensic scientist is rarely dramatic, but the findings that come out of the painstaking work they do can be.

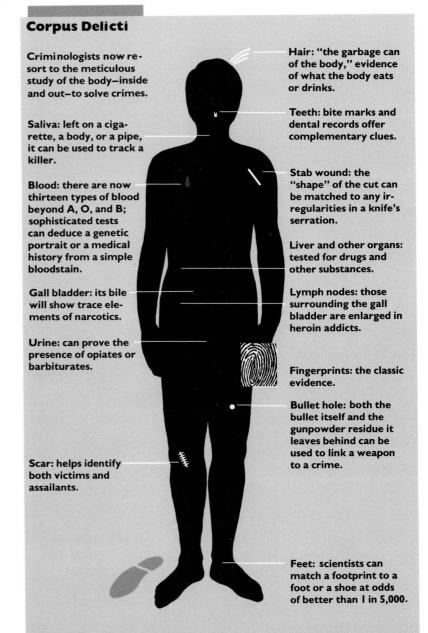

Corpus Delicti

Criminologists now resort to the meticulous study of the body—inside and out—to solve crimes.

Saliva: left on a cigarette, a body, or a pipe, it can be used to track a killer.

Blood: there are now thirteen types of blood beyond A, O, and B; sophisticated tests can deduce a genetic portrait or a medical history from a simple bloodstain.

Gall bladder: its bile will show trace elements of narcotics.

Urine: can prove the presence of opiates or barbiturates.

Scar: helps identify both victims and assailants.

Hair: "the garbage can of the body," evidence of what the body eats or drinks.

Teeth: bite marks and dental records offer complementary clues.

Stab wound: the "shape" of the cut can be matched to any irregularities in a knife's serration.

Liver and other organs: tested for drugs and other substances.

Lymph nodes: those surrounding the gall bladder are enlarged in heroin addicts.

Fingerprints: the classic evidence.

Bullet hole: both the bullet itself and the gunpowder residue it leaves behind can be used to link a weapon to a crime.

Feet: scientists can match a footprint to a foot or a shoe at odds of better than 1 in 5,000.

Ferment in France

O Bacchus, hale father! O Vitus, hearty child!

More grapes are harvested annually, in an international rainbow, than any fruit presently grown. Although the vision of 60 million tons of grapes is a source of delight for juice-stained children, it brings an even greater bloom to the cheeks of a more refined population: wine drinkers.

Winemaking originated more than 12,000 years ago in the Middle East. The greatest quantities are still produced in the Mediterranean, where the Bordeaux region of France has set the world's standard for quality: in these vineyards, it takes over two years to turn a cluster of grapes into a bottle of liquid valued, modestly, at fifty dollars.

The Bordeaux process is painstakingly traditional throughout its long cycle of growing, harvesting, refining, and aging. Both the soil and the climate in which vines are raised are watched meticulously; although they cannot be controlled, subtle variances in the depth of the roots or the strength of the sun can easily turn a grape from conviviality to tartness.

After harvesting, a crusher gently breaks the grapes open, releasing their pulp and seeds; pulp contains color-producing pig-

ments, while seeds are filled with the astringent chemical tannin. This juice, called "must," is pumped into oak vats, where, within a day or two, the wine begins to ferment. Yeast living on the fruit's skin feeds on its natural sugars; the microorganisms multiply rapidly and release carbon dioxide and alcohol in gratitude for the gift of an ongoing meal.

Fermentation continues for three weeks. During that time, the cellar master tests samples regularly, monitoring a critical ratio: as the heavy sugar is converted to a lighter alcohol, the density of must decreases. When the alcohol concentration reaches a peak level, it begins to kill the yeast; they drown, somewhat prophetically, in the waters of their own excess. Any sugar left unconverted in the grapes functions as a sweetener for the finished wine.

The liquid is poured into handmade oak casks and transferred to cool cellars where, for another two years, it is pampered, protected from bacterial contamination, and persuaded, gracefully, to age. Throughout the aging, layers of dead yeast cells collect and sink under their own weight to the cask bottom. A laborious process, "racking," rids the wine of each layer as it forms. In the final days, a mixture of egg whites and salt is poured into the cask; it soaks up any remaining debris, tows it to the bottom, and steadies it for a last round of racking. Once through this rite, the grapes are ready to receive a bottle, a label, and the respect due their advanced status in life.

Most French wine is neither as elegantly produced nor as reverentially consumed as the Bordeaux brands. The greatest quantities are churned out by independent growers, families with less sophisticated equipment and more limited know-how. Their brew is distinguished in one area: consistent inconsistency. It is just as often abundantly acidic as it is deficiently alcoholic. In an effort to narrow the gap between the Bordeaux and down-home cultures, a concerned French government turned to the examples of a latecomer in the fields: the United States.

Attempting to make good wine has traditionally been a gamble, but modern vintners leave as little as possible to chance. This is the quality control laboratory at Buena Vista Winery in Sonoma, California.

If winemaking has a spiritual home, it is the rich vineyards of France. Here, a harvester savors the justification for these endless rows of vines.

In Europe's winemaking regions, the Middle Ages stayed late—the scene above could have been depicted 500 years ago. By the 1960s, the vineyards were ripe for innovation, such as harvesting machines *(below)*, which could take the place of forty workers.

Unburdened by the weights and harnesses of history, the American approach to winemaking is less traditional, and far less tender, than the European. When the country's annual consumption doubled in the 1960s, scientists decided to analyze grape harvesting and wine production as seriously as any technological problem. At the University of California, researchers planted new vines in precise rows, leaving only enough space between them for machines that did the harvesting work of forty men. They pruned the grapes to heights that these same machines could most easily reach. They replaced the natural yeast found in grapes with purebred laboratory strains. They fermented the wine in stainless steel tanks built to hold 200 tons of liquid, and tested the fermentation rate at different temperatures for different lengths of time. They measured the effects of anti-contaminant agents on bacteria and the effects of oak chips on flavor. Racking was done by remote control, and centrifuges replaced egg whites in the final stage of removing solids. Once accustomed to an audience that approved every joke, old Bacchus was humored no longer.

The Europeans responded to American success by introducing, with all respect, their own technological innovations. They pumped crushed grapes into a silo, where an exchanger heated them to 160 degrees. Solids were kept in constant circulation by mechanical stirring. In a single hour, more pigment was extracted by machinery than in three weeks by tradition. Oak tanks were replaced with steel ones, and fermentation speeded from two weeks to three days. Using new machines and new techniques, wines deeper in color, more consistent in quality, and competitive in price were bottled in under a year. Returning to the country that had understood him so well for so many centuries, Bacchus found that even the most loyal of his friends had lost their love for his tricks.

Eighty-eight percent of any bottle of wine is water; take it out. Ten percent of the remaining liquid is alcohol; take it out. Most of what now remains is acid, tannin, and pigment; take them out. That clear, quiet fluid in the bottom of the glass contains the aromatic compounds. Without them, there would be no reason for the pomp of wine masters and the circumstance of wine tasting. They are the magic dust in a process otherwise without mystery; aromatic compounds give wines their individual taste and smell.

Even magic, though, cannot hide beneath the scrutiny of science, and chemists are now able to separate the compounds using the technique of gas chromatography. When an aromatic sample is injected into a coiled tube, its elements rise at different speeds; following independent timetables, each emerges separately at the other end of the coil, where it can be isolated, graphed, and identified. A single aromatic compound can contain up to 200 elements: elusive to the most distinguished of noses, perhaps, but clear to any chromatography chart-out.

How have the revolutions in wine preparation affected those whose interest begins once the process ends: wine drinkers? Long practiced by the few, and little understood by the many, even taste-testing has revised its traditions in deference to the times. Gauche senses are refined, these days, by means of an "aroma box," a sophisticated scratch-'n'-sniff in which chemicals are

Today's fermentation room may look like a refinery, but these vast steel fermentation tanks conceal the same ancient chemistry that has always turned grapes into wine. Below, a more familiar scene: a cellarman draws a sample from an oaken cask.

tagged with precise descriptive terminology. Any rookie can recognize a "fruity" taste. But how many are as confident with "mousy-biscuity" or "catty" scents under their noses?

Since ancient days, when the first dark feet stepped lightly upon the first dark fruit, progress in the wine industry has taken place from the ground up. Appropriate, perhaps, is a closing toast: to the fruits of the vine, the founders of the feast. In spite of the changes that have happened, and those yet to come, may they never lose their grapeness.

Recombinant Revolution

As technologically accomplished as our age is, we cannot deny that the most significant technological advance occurred nine or ten millennia ago, when some imaginative and astute observers learned how to domesticate plants and animals and breed them in desired directions. These discoveries ensured a food supply that was larger and more reliable than any enjoyed by people who survived by hunting and gathering. The presence of this expanded food supply had mixed results: it permitted the development of all the activities we think of as defining civilization, yet it also set in motion events that ended with the complete destruction of the cultures that made the original discoveries.

The "inventions" of domestication and breeding have been called the symbiotic revolution, in that they allowed humans to expand enormously their relationships with other species. Such exchanges of services between species are quite common in nature. Those that the first technological breakthrough allowed humans to form with plants and animals were typical examples. Humans guaranteed to certain species the basic necessities of food, water, shelter, and protection against predators and competitors. In return, oxen and mules came to do man's labor and corn and wheat to bend more willingly to the scythe. An analogy might be

made to the stereotypical relationship between underdeveloped and industrialized nations. Residents of the former provide labor and raw materials and receive in exchange sophisticated, highly finished products (proteins, sugars, oils, fibers) whose synthesis is beyond their technological abilities.

Many observers believe that a second symbiotic revolution is unfolding around us, for better or worse (those vanished hunter-gatherer cultures testify that there can be a worse). If so, it is interesting to note that the new revolution has been triggered by the close study of a molecule whose natural function is precisely to promote and control change. This is the famous DNA, the biological book of life. Perhaps a more exact metaphor would be a cookbook, for each cell of every living thing carries in its DNA the recipes needed for the cell to synthesize those substances that are required either for its own existence or for the subsistence of the entity incorporating it.

Scientists have wished for some time that they could read these recipes, but until very recently they have been stymied, because they had nothing to study. All the imaginable approaches to decoding a gene require a very large number of copies of the gene in question, and getting the number of copies needed seemed impossible. The gene-cookbooks are encyclopedic;

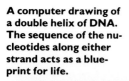

A computer drawing of a double helix of DNA. The sequence of the nucleotides along either strand acts as a blueprint for life.

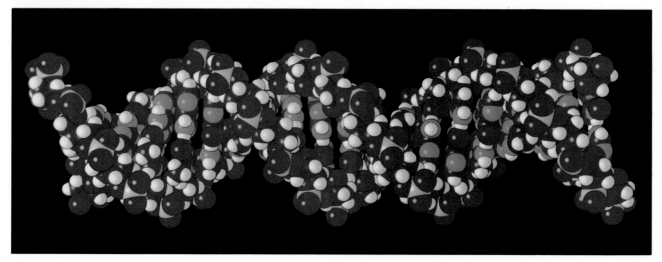

each contains as many recipes as, say, fifty years' back issues of *Gourmet* magazine. There is no index, no table of contents, no recipe titles to help the search. A scientist might be able to get part of one recipe just by taking a gene slice at random, but how could he begin to find the thousands of copies he needed?

After many years of frustration this problem was solved in the early 1970s by research on an apparently quite unrelated problem—the antivirus defenses of bacteria. Bacteria, like people, are plagued by attacks of viruses. However, instead of resigning themselves to fluids and bedrest, bacteria counterattack more directly: they secrete a substance that slices the virus DNA into pieces. In the early 1970s, scientists studying these impressive weapons, which they called restriction enyzmes, discovered two important facts. First, the enzymes from any one strain of bacteria always seemed to slice up DNA in the same way, to make its cuts at the same locations; and second, enzymes from different strains of bacteria make their cuts at different sites. Thus, by working with enzymes extracted from the right bacterial strains, it was possible to get identical stretches of DNA snipped out of a whole group of cells. This discovery was a giant step toward isolating DNA; then came a second discovery that wrapped it up.

In 1973 researchers found a way to insert these snippets into the genes of a normal bacterium, so that when it reproduced, and copied its genes in doing so, it automatically copied the snippet as well. (The name of this process would shortly become notorious: it was called the recombinant DNA procedure.) Bacteria reproduce so quickly that they amount to a biological copy machine. No matter how many gene-copies a scientist may need—a million, a billion—that number of bacteria can be grown quickly. (The application of some restriction enzymes at that point ensures that one doesn't get more than one needs.)

As soon as these techniques were announced, scientists by the thousands, working

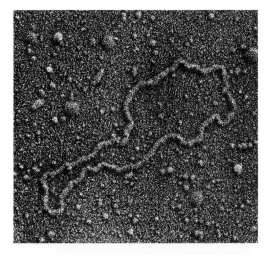

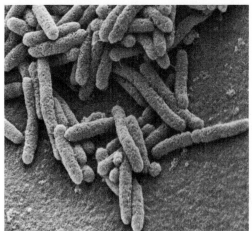

in laboratories all around the world, began copying and translating whatever genetic recipes they found of interest. One of the most important of these wasn't even a normal instruction at all: it was the recipe for cancer. The whole scientific and medical world was very interested in learning exactly what genetic instructions made tumor cells mount their anarchic rebellions against an organism's routine. Plans were proposed to snip genes from tumorous cells and insert them into bacteria for copying.

At this point a number of scientists, including Robert Pollack of Columbia University and Paul Berg of Stanford University, began to worry that these experiments might pose an unprecedented kind of threat to public health. Suppose the bacteria carrying the cancer genes escaped from their culture dish? Was there any chance they might end up spreading cancer? And if so, what was the risk? One in a million? One in ten million?

The recombinant DNA procedure allows DNA material from one organism to be selectively introduced into the DNA of another organism. The desired portion of the original DNA molecule is snipped out by restriction enzymes and inserted into the DNA sequence of a host molecule, in this case a loop-shaped plasmid *(upper left)* from the common bacteria *Escherichia coli (bottom left,* shown magnified **21,000 X**). Returned to its home cell, the recombined DNA molecule rapidly reproduces "cloned" replicas of itself, including the original DNA snippet.

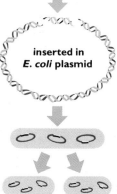

Original DNA (snip)

inserted in E. coli plasmid

reproduced by E. coli bacteria

Paul Berg of the Department of Biochemistry at Stanford University.

Discussion of these and related questions spread through the nation's research community. In early 1974, at the Massachusetts Institute of Technology, Paul Berg and a number of other scientists met to discuss the issues. The meeting ended with the participants asking researchers to postpone performing recombinant experiments that involved copying genes in three categories: cancer genes, genes that made substances toxic to humans, and genes that reduced the effectiveness of antibiotics. Six months later, at a meeting at the Asilomar Conference Center in California, these restrictions were widened and worked out in more detail. For the first time in history an entire subdiscipline of scientists had agreed to refrain from conducting certain experiments in the name of public safety.

The public reaction was intense. Many people believed, rightly or wrongly, that if the researchers had undertaken so uncharacteristic an action, the risks must be far worse than they were revealing. The National Institutes of Health imposed a strict set of regulations as a condition for receiving its funds. Demonstrations were held throughout the country. Several municipalities with local recombinant DNA facilities passed ordinances regulating the experiments. In 1977 sixteen different bills to control the research were introduced in Congress. Scientists began to raise charges of censorship and complain of restrictions on their freedom of inquiry.

During all this controversy, recombinant DNA research in the United States never stopped; it only slowed (and it didn't even slow elsewhere in the world). As experience with the technique grew, and no accidents were reported, the public anxiety abated. Regulations were loosened; research sped ahead. Soon scientists had found ways to make the genes inserted into the bacteria actually function—that is, to produce whatever substance they usually made in the organism from which they had been taken.

With this last discovery the doors to the second symbiotic revolution suddenly stood open. It became possible to make bacteria that produced substances completely different from those made by normal bacteria. One example is insulin. Diabetics are treated by supplementing their body's declining production of insulin with insulin taken from livestock. Animal insulin is expensive and sometimes difficult to procure, and it can cause allergic reactions. But with this new technology, it should be possible to insert the gene for making insulin into a bacterium, "turn it on," and as the bacterium grows into a colony of billions, simply extract a steady flow of all the insulin anyone might need. Better, perhaps the original gene could even be taken from the patient's own pancreas, so that he could receive his own, custom-made, individualized brand.

A second product of this sort is interferon, the biochemical that experiments suggest might be a general-purpose virus killer and thus a cure for diseases ranging from hepatitis to the common cold. (Some evidence even suggests the substance might be useful against some forms of cancer—see

Life: Patent Pending

Recombinant DNA affects medicine, agriculture, and industry, as well as pure scientific research. Large-scale production of scarce or difficult-to-synthesize medical products becomes possible: interferon, the antiviral and anticancer agent; viral vaccines, currently inefficient as well as hazardous; enzymes, like insulin, to remedy fatal enzyme deficiencies; antihemophiliac proteins; pure antibodies against various bacterial diseases; and hormones synthesized only in minute quantities by the body. In agriculture, nonleguminous plants will be modified to "fix" their own atmospheric nitrogen, eliminating the cost and risks of nitrogenous fertilizers; crops like corn will be supplied with missing amino acids necessary to turn them into complete proteins. In industry, we may at last be able to undo some of our own damage; through recombinant DNA, we can clone bacteria that are especially equipped to eat the oil off oil spill slicks. Criticism of recombinant DNA research has been leveled on many fronts. Adversaries accuse scientists of playing God, of endangering public safety by "inventing" possibly toxic, even lethal new organisms, of stanching the free flow of scientific information among colleagues, of sullying the groves of academe with the presence of patent lawyers and crass and grasping agribusinessmen and merchants of pharmaceuticals. The research continues, constrained by regulations that some scientists and other citizens consider inadequate. It's a situation that bears watching.

"The Question of Interferon," pages 140–143.) However, interferon occurs in such minuscule concentrations in the body that world industrial production is far below the level needed to supply an adequate research effort, let alone the amount that would be required if it proved to be an effective drug. An obvious solution is the production of the substance through recombinant bacteria. And there are many other substances in the same category.

The wonders of recombinant technology are by no means restricted to the production of chemical or pharmacological substances. The doctored bacteria can also deliver new services. For instance, it would be convenient to have a "biological janitor" to eat up oil spills. Such an organism could be dropped on the spill, would reproduce up to the limits of the slick, and then would automatically die off when the oil was all gone. There are natural bacteria that eat petroleum, but each specializes in a very particular oil density and composition. When oil is released on the ocean, it tends to pass through

many different states as it weathers and evaporates and fractionates. No one natural strain is sufficiently generalized to be able to deal with all the complexity of a bad oil spill; and if the full set of bacterial strains is dropped on a spill, they seem to interfere with one another in a way that substantially reduces their efficacy. The solution is to extract from each strain the genes that make the enzymes enabling the strain to digest its own particular oil fraction, and then to combine them into a single organism that becomes an "oil spill omnivore." Similar solutions are being pursued for a range of other pollution problems.

A two-day academy forum, held at the National Academy of Sciences in March 1977, sparked controversy as demostrators protested that the issue of recombinant DNA could not be left to the scientists alone.

The Genetic Code

In the body and in the world, proteins exist in formidable quantity: their shapes and functions seem nearly infinite. Yet the building blocks of proteins are twenty amino acids—no more. Instead of many different ingredients, it is the sequence of these few amino acids, strung together, that produces such diversity. Orders are assigned each set of amino acids by master molecules of DNA (deoxyribonucleic acids); orders are conveyed by middlemen molecules of RNA (ribonucleic acid).

"DNA makes RNA makes protein": so runs the central dogma of molecular biology. Four different DNA units called "bases" can be arranged into groups of three to produce sixty-four different triplets. Each triplet represents a specific amino acid; each acid is represented by several triplets; and three additional triplets signal the end of a growing protein chain.
DNA does not synthesize protein directly; it makes a complement copy of its instructions in the form of RNA, a messenger molecule. The order of RNA bases then oversees the stringing of amino

acids. Once bound together, amino acids fold and twist into the complex shapes we recognize as proteins.

Genes are merely linear stretches of DNA. The genetic code—all sixty-four triplets—is inherited and universal:

humbling but true, that human coding mimics that of rabbits and also that of bread mold.

bases

DNA

makes

RNA

triplet

makes

protein   amino acid

Rhizobia are seen colonizing the root hairs of their host legume, clover.

A second example of a new service is the possibility that plants can be bred—using bacterial symbionts—to make their own fertilizer. Some plants, such as many beans and clovers, do this already: they maintain bacterial colonies that transform atmospheric nitrogen into ammonia, which the plants then use to make protein. Agricultural scientists would like to give such important crops as corn this ability to make fertilizer out of air. To do so would require a number of advances in recombinant technology, since what is involved is engineering a new form of symbiosis, a new bacteria-plant relationship, rather than simply giving bacteria alone a new genetic

ability. Nonetheless, many researchers throughout the world are working on the problem.

Meanwhile, the technology of the technology—the skills, techniques, and equipment used to support research activities—has developed a remarkable momentum. A striking example is a device developed at the California Institute of Technology called the gene machine. This is an automated gene synthesizer: all a scientist need do is describe the structure of the gene desired, and the gene machine automatically puts it together in a form all ready for recombinant DNA research.

In 1980 the U.S. Supreme Court decided that genetically modified bacteria were patentable. This removed the last obstacle to the construction of a new industry. One hundred and fifty biotechnology companies had organized themselves by the end of 1981—and the established chemical, medical supply, and drug companies were also setting up their own divisions. The stock of the first such company to make a public offering

The Race for the Double Helix

The basic stuff of genetics is DNA (deoxyribonucleic acid), the universal material that governs heredity in all living things. The form of the giant molecule is a double helix, a sort of spiraling ladder with sides formed of phosphate alternating with sugar, and rungs of paired nitrogenous bases. Variation in the sequence of the bases (of which there are only four kinds, common to all life) yields the myriad genes that specify hereditary information in millions of different species. England in the 1950s was a

hotbed of DNA research. James Watson and Francis Crick were the first to propose (in a stunning 1,000-word essay) a model for

DNA's molecular structure. With their colleague and sometime rival, Maurice Wilkins, they shared the Nobel Prize in 1962 for their

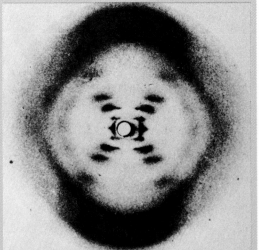

remarkable discovery. (Above) A diffraction pattern produced by X-rays diffracted from the atoms in a DNA crystal. The helical

structure of DNA is suggested by the crosswise pattern of X-ray deflections.

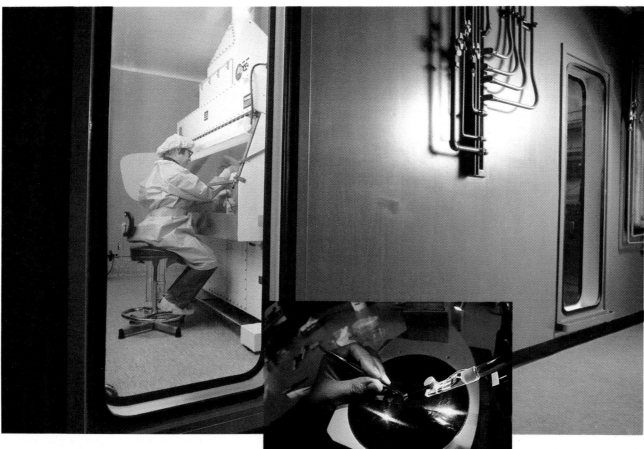

rose higher, faster, than that of any other concern in the preceding twenty years. Investment money—some say in the billions—flowed into this glamorous new field.

The enormous sums involved had a profound effect on the research climate of the nation's universities. Professors who had taken partnership positions in biotech companies suddenly had private fortunes of $10 million or more. Universities such as Harvard and M.I.T. began to strike deals with companies like Monsanto and DuPont, by which the corporations would fund research in return for exclusive licenses on any marketable products. Sometimes universities, drawing on the strength of their in-house expertise, would take equity positions in these new companies. In one way or another, many universities seemed to be edging closer and closer to actually going into the biotech business themselves.

The effects of these commercial tensions on the free flow of scientific information were predictable. "Recently I was at a meeting," an M.I.T. professor said, "where a major corporate patent lawyer announced to the whole group of scientists that the next time they go to a meeting they should have their notebooks certified by a notary public; and they should also consider restricting their conversations with their peers, because they might give away something patentable."

The impact of this high-pressure atmosphere on the rhythms of academic research has become sufficiently a matter of concern that many scientists have been calling for a new Asilomar Conference. The first dealt with the speculative hazards that recombinant DNA research might present. The second would address those hazards that many scientists feel are now a matter of fact—the threat that the commercial exploitation of the research poses to the organization of scientific research. As our experience with the first symbiotic revolution testifies, the social systems responsible for a technological breakthrough have no privileged position against its consequences.

A containment facility to protect the outside world from the living products of genetic engineering. (*Inset*) Recombinant researcher inserts a spliced gene into a frog egg.

A Conversation with Walter Gilbert

Although Walter Gilbert, a physicist turned biochemist turned biotechnological entrepreneur, shared in the 1980 Nobel Prize in chemistry for contributions to working out a research technique —for rapid sequencing of subunits of the chainlike genetic chemical called DNA—he prefers to look back on his quarter-century research career as a succession of attacks on questions.

"I'm question-oriented rather than technique-oriented," he says. He also admits that although he has "never spent very long doing one thing" before moving on to the next problem, there is a unity to the questions he has been exploring: the questions concern the genetic information embodied in the chemical code of DNA and how that information manifests itself in living processes—or is prevented from manifesting itself.

Such problems of genetic control are among the central, and still largely unsolved, mysteries of the world of living things. Even after three decades of explosion of the brand of fundamental biology that sometimes is called molecular biology, the details of genetic control have been worked out only in the simplest of organisms, such as bacteria, and in some of the viruses that prey on bacteria and more complex living cells.

The workings of this system of control have provided intensely compelling work for a rapidly expanding community of biological scientists, many of whom, like Gilbert, have been drawn from other scientific disciplines. In the last few years, such work has begun to show promise of widespread application in industry and agriculture. Genetic recombination techniques—moving pieces of DNA from one species of organism to another—are being used for such purposes as making large quantities of human insulin for therapy of diabetics, making human interferon for tests of its usefulness as a therapy against cancer, or helping to make improved forms of vaccines against animal diseases.

One of the first companies formed to push the application of scientific discoveries about genetic control mechanisms was Bio-gen, N.V., among whose major activities are laboratories in Geneva, Switzerland, and Cambridge, Massachusetts. The company was founded in 1978 and has been supported with capital from such sources as Inco Limited, the Schering-Plough Corporation, Monsanto Company, and Grand Metropolitan Limited. Late in 1981, Gilbert became chairman of the company, and on July 1, 1982, he resigned his American Cancer Society professorship at Harvard University, where he had worked for twenty-five years since receiving his doctorate in physics from Cambridge University in England.

Gilbert calls Biogen "a nontraditional, research-based company." The company is not intended, he said, to develop products that would be turned over to larger organizations, such as drug companies, for mass production and marketing. The company is not designed as a "think tank" or "technology transfer" organization. Instead, according to Gilbert, the attitude of the scientists and venture capitalists who began Biogen was "Let's create an entire industry out of technology." The company was founded, according to Gilbert, only after a founding group of scientists concluded that the advances in molecular biology during the early to mid-1970s were so powerful that the discoveries would have "an appropriate applied phase." Its scientific board (which has thirteen members) has an unusual amount of power over the company's future direction, a conscious decision, Gilbert says, "to create an industry and retain the structures that made the research potent in the first place."

In those advances, and the work that led up to them, Gilbert played a notable part. He was first attracted to biological research by James D. Watson, the co-discoverer of the double-helix structure of DNA, whom he had met in 1955 in Cambridge, England. In the summer of 1960, Watson, then at Harvard, told Gilbert that "there is something exciting going on in the lab."

Gilbert was fascinated. Watson and several colleagues were looking, in competition with others, for solid evidence of the existence of a substance called messenger RNA. This hypothetical copy of the genes in DNA was thought to carry the instructions for making proteins, the substances that do the main work of all living cells, out from the DNA in the cell's nucleus to the site of protein manufacture, globular structures called ribosomes, in the outer part of the cell, the cytoplasm.

In the search for messenger RNA, Gilbert took part first "as a technician and then as an equal participant." He says now that "it was a search for what was then an invisible substance that we couldn't characterize, a hypothetical intermediate that carried the information from DNA to the proteins, whose existence we could suspect and had to convince ourselves that we were seeing." At any given time, a cell would have only a tiny proportion of a particular messenger RNA among its contents.

After a brief return to physics, Gilbert came back to Jim Watson's lab at Harvard to do experiments that showed how ribosomes acted to put proteins together by running along and reading the message embodied in the strip of messenger RNA. This work helped overturn earlier notions of how the amino acid subunits of proteins were strung together.

For his next project Gilbert felt as if he were turning from one "profound question in molecular biology" to the other. The first question, Gilbert says, is "How does the genetic information do anything?" The second is "When does it control anything?" He notes that the second question "is still unsolved in the full sense."

A part of the answer, however, had been suggested in 1961 by the French biologists François Jacob and Jacques Monod, who shared in the 1965 Nobel Prize in medicine. Jacob and Monod said that a gene in bacteria that specified the structure of a protein for digesting a particular sugar could be turned off when the sugar was not available in the environment. This switchoff would be accomplished by a special "repressor" protein binding to a particular small stretch of DNA they called the "operator," thereby preventing

the gene from being copied into messenger RNA and hence stopping synthesis of the sugar-digesting protein.

Gilbert set out to isolate the hypothetical bacterial repressor, in a kind of rivalry with his Harvard colleague, Mark Ptashne, who was looking for a similar repressor in a bacterial virus called lambda. Each found his repressor.

In Gilbert's hands, the bacterial repressor evolved into a tool for looking at the DNA section it bound to. It was possible, very laboriously, to work out the sequence of DNA subunits or bases—adenine, guanine, thymine and cytosine—strung along chains of sugar and phosphate, that form the genetic code of DNA. In 1976, Gilbert and a colleague, Allan Maxam, did an experiment that provided them with a sequencing method that could be extended far beyond their imaginings at that time.

The new rapid sequencing method, Gilbert says, "enables you to work out with reasonable ease sequences of the order of thousands or tens of thousands of bases, so you can know the sequence of any gene." Almost simultaneously, Gilbert had found that it was much easier than had been suspected to use a special DNA-stitching protein called a ligase to link together sequences of DNA. With such techniques, the industrial applications of DNA research techniques seem a lot closer.

Energy Futures

In 1973 the term *energy crisis* became a part of the language, and although public perception of the depth of that crisis has alternately brightened and dimmed during the decade since, few of us will ever again be as complacent as we once were about where the next tankful of gas or kilowatt-hour is coming from. Natural gas and petroleum, we learned, were available either in finite amounts or at exorbitant prices.

Amid all the talk of conventional versus alternative energy systems, and of "hard" and "soft" technologies and scenarios for the future, science has searched steadily for replacement sources of energy. The solutions to this crisis yield all too slowly to reason; they will never be unlocked by the timid or the prejudiced.

When oil first ran short, coal appeared to be our conventional fuel of first resort. But aside from its contributions to acid rain and the atmosphere-heating "greenhouse effect," coal exacts an awful environmental price from the moment it is broken free from the Earth. And unlike any other energy source, it constantly threatens the lives and health of the people who mine it.

More than 100,000 miners have been killed in accidents since deep mining began in the United States. They died in explosions,

like the one that took seventy-eight lives at the Consol Number Nine mine at Farmington, West Virginia, in 1968. They died because of roofs collapsing, or because of the failure of machinery, in what are the most productive but also the most dangerous mines in the world. It wasn't until 1969 that a federal mine safety act was finally passed. The new law empowered government inspectors to monitor mines and mining practices and to shut down unsafe operations. But there are never enough inspectors to get this difficult job done, and budget retrenchments threaten to thin their ranks even further.

Sudden, violent death is only one of the miners' risks. Every day spent in a coal mine increases a person's risk of contracting pneumoconiosis silicosis, black lung disease. Coal dust eats at the very fiber of the lungs; black lung is a progressive disease with no cure. A miner can only leave his work and spend the rest of his days at home, chronically short of breath. Until the 1969 act, those days passed without compensation—only then was black lung recognized as an occupational hazard entitling sufferers to disability pensions. Even now, there is debate over how the disease may be conclusively diagnosed and how to bring down dust levels in the mines.

Not all mining casualties take place in the mines. In 1972, the town of Buffalo Creek, West Virginia, was inundated when a dam built of coal waste collapsed. One hundred and twenty-five people died. The families of miners are hostage to the same dangers that take the lives of their breadwinners.

In the West, where the real future of American coal lies, the miners are safer but the land is not. The western rangelands, beneath which millions of tons of coal lie shallowly buried, are ecologically fragile. No one really knows whether this terrain, once it has been strip-mined, can be successfully restored. Mining companies say yes and point to improved vegetational yields on reclaimed land. Conservationists claim that such "showpiece" tracts are the result of unrealistically heavy fertilization and that the geological, agronomic, and climatological complexity of the

It happened in 1973, and again in 1979—and we'll never look at energy the same way again.

A surface mine in North Dakota. As the vast coal seams of the West are laid open, Americans are questioning the facility with which strip-mined land can be reclaimed.

The major dams along
the Colorado River. (In-
set) Hoover Dam, built
in the 1930s to power
Los Angeles and hold
back the waters of the
new Lake Mead.

high plains precludes any quick and easy rec-
lamation schemes. Disruption of the water table is yet another problem—one acutely felt
by ranchers, whose traditionally secure place is
being challenged by the ascendancy of coal.

If the West is where we must ultimately
decide the viability of using coal as an alternative energy source, it is also where we face
the hardest decisions concerning water policy and the extent of growth that policy supports. Virtually alone among energy sources
currently in widespread use, hydroelectric
power enjoys a reputation for inexhaustibility
and cleanliness. A river flows; it turns a turbine. Nothing is lost, nothing is burned, nothing is released into the atmosphere. But
problems arise with the expansion of a project
the magnitude that fast-growing societies de-
mand, and that is why, as we saw in "Preserving the Balance," (see pages 186–193), the
Colorado River no longer reaches its age-old
delta at the head of the Gulf of California.

To be fair, we must remember that not
all of the impoundments of the Colorado
were created to serve hydroelectric turbines. The prevailing need in the arid Southwest is water, for cities and for agriculture.
But despite differences in their intended
uses, the effects of multiple dams upon a
once-free-flowing river are the same. Both
demands also stem from the same source:
exponential growth.

In 1922, the right to draw water from
the Colorado was divided between the two
groups of states that form its upper and
lower basins. Until that time, only small dams
and sluicegates had been built to serve local
agriculture. The political division having been
accomplished, the real exploitation of the
river could begin. Boulder (later Hoover)
Dam was built just east of Las Vegas. When
this gargantuan project was complete, its turbines provided enough power for all of Los
Angeles—and created Lake Mead.

Hoover Dam was followed by the Parker and Imperial dams, which made possible
Lake Havasu and the All-American Canal.
Next was Davis Dam, between Hoover and
Parker. These were but a prelude to the
colossal Glen Canyon Dam, whose impoundment deepened and spread to become the
largest manmade body of water in the country: Lake Powell. Arguably, this was also the
dam responsible for the greatest depreda-
tion of aesthetic resources. To many, Glen
Canyon was more beautiful in its way than
the more famous Grand Canyon, but now it is
invisible beneath water and muck.

Ironically, it is not only a glorious landscape that is being lost through impoundment.
Lake Powell's porous banks irrevocably absorb millions of acre-feet of water, and the
dry desert atmosphere claims even more
through evaporation. Also, lakes such as
Powell must succumb to inevitable siltation. In
somewhere between 75 and 200 years,
Glen Canyon Dam will simply become a
giant waterfall.

Yet the demands increase—despite evidence that estimates of the Colorado's volume, made in the 1920s, were probably

inflated by a factor of 100 percent. There is simply not enough Colorado water to go around, and desperate remedial proposals have been made. One, rejected by California voters in 1982, centered upon construction of a canal to carry water from the Sacramento River valley to Los Angeles and environs. A real solution, if one exists, would seem to lie with more prudent irrigation, energy conservation, and responsible growth policies. Nevertheless, the results of such a solution would be limited at best.

If there is only so much that conventional energy sources can do, it is not surprising that researchers should turn to new sources, new technologies. Some seem exotic and monolithic; others, along the "softer" energy path, operate on a smaller, local scale. Promising energy alternatives that science is currently exploring range from alcohol and hydrogen fuels to the myriad technologies that make up the solar promise.

Sugar-derived alcohol is one such source, and nowhere has this concept been taken more seriously than in Brazil, the giant among the world's sugar producers and a nation seriously deficient in domestic fossil fuels. Sugar-based alcohol is already powering Brazilian vehicles, most often as part of a gasohol mixture but also on its own, in engines specially built or modifed for its use.

Brazil is discovering that alcohol from sugar cane can help power its economic and agricultural development as well. Cane production is being encouraged on underutilized small farms, and an expanded sugar industry might revitalize the impoverished Brazilian northeast. The accompanying developments in agronomic research could benefit all crops, and the bleak urban job picture would be brightened by new refineries.

Alcohol fuels have further advantages. They burn cleanly and require no lead or sulfur additives because of their inherently higher octane. The air quality in São Paulo, once notoriously poor, has improved significantly since the introduction of alcohol fuels.

Although newer technologies promise greater efficiency, the basic process of making alcohol from sugar is not a difficult one. It involves long-understood methods of fermentation and distilling. And its production could itself be energy-efficient. It is realistic, experts say, to envision plants in which waste products from distilling could be fed to livestock, which would in turn contribute manure to a methane generator adjacent to the plant. Such an installation would, in effect, be partly powered by its own refuse. Louisiana's sugar crop is seen as one potential major contributor to an alcohol fuel program for the northern hemisphere. So are the starch crops of Europe—particularly sugar beets, of which European nations have an annual surplus, and potatoes. Research into similarly renewable sources of liquid fuels, and agriculturally based carbohydrate replacements for petrochemicals, is bound to continue as the century draws to a close.

Another, perhaps even more promising, fuel is not locked in growing plants or in the fossil remains of their distant ancestors. It is hydrogen, and it is all around us in the abundant water of our planet.

Smog darkens São Paulo, Brazil. The use of alcohol fuels in vehicles has brightened the skies over this congested city, but will others follow?

These Brazilian cars run on alcohol. There is sugar cane to burn in Brazil—and it burns cleanly and efficiently.

A hydrogen-powered bus in Provo, Utah. Hydrogen is virtually pollutant-free, but problems of manufacture and storage must be overcome.

Hydrogen is versatile as well as plentiful, and its applications are as diverse as cooking and transportation. The problems with hydrogen lie not in the extent of its supply, or in finding ways to put it to work. Hydrogen's challenge is one of manufacture—how best to isolate it from water and store it conveniently.

We obtain hydrogen by "splitting" molecules of water into their two essential elements, hydrogen and oxygen. This can be accomplished by electrolysis, in which a current is conducted through electrodes immersed in water. The electrodes were formerly made of gold or platinum, but a new solid polymer electrolyte plate has significantly reduced costs. Further problems to be addressed include those of high-volume production and the accessibility of an abundant source of electricity. Even if thousands of electrodes were to be sandwiched into a

commercially viable hydrogen plant, where would the power come from? One answer is off-peak output from nuclear stations; the French appear to be pursuing this method, although in many places the political and economic drawbacks of nuclear energy make its application less certain.

Water molecules can also be split through a combination of intense heat and the application of certain chemicals—an approach that would again involve nuclear power or, to the relief of nuclear opponents, a concentration of thermal energy from the sun. There has also been research into direct hydrogen conversion using sunlight, simple chemicals, and water. At the other end of the technological spectrum is the old gashouse method of burning coal with steam. This produces hydrogen, along with less desirable by-products that would have to be filtered out and kept from polluting the air.

When it comes to actually *using* hydrogen, there are problems of image as well as substance. The image problems stem largely from the 1936 *Hindenburg* explosion, in which the huge German airship, buoyed with hydrogen, crashed and burned at Lakehurst, New Jersey, killing thirty-five people. But scientists are quick to point out that hydrogen, once it accidentally catches fire, is a much more "forgiving" fuel than gasoline or other carbon-based products. Hydrogen burns up and away, and with less radiant heat.

But even if hydrogen's relative safety can be accepted, storage—particularly in vehicles—remains a problem. Hydrogen is a low-volume fuel; in order to carry enough, even in a liquid state, a jumbo jet would have to be lengthened to accommodate new tanks. High-pressure storage of hydrogen gas means heavy tanks, which would be impractical in small vehicles. But lower pressures can be employed if the tank contains a hydrogen-absorbing alloy of iron and titanium, ground to resemble a shiny metal gravel. So there are ways to get around hydrogen's problems, and the costs of these methods, now exorbitant, will seem more reasonable as the price of carbon fuels goes up.

The Road to Happiness

In 1913, Henry Ford applied the techniques of assembly-line production to the manufacture of automobiles. In so doing, he placed possession of a private car within the means of millions of consumers and ushered in an era in which the American landscape was to be transformed almost beyond recognition. The industry spawned by Ford has become the nation's largest—25 percent of the work force is occupied in making, selling, and servicing autos and auto supplies. More than half the land surface in urban areas is given over to streets and expressways, garages, parking lots, and gas stations. And every mile of four-lane highway constructed through the country-

side gobbles up seventeen acres of land, much of it of value to agriculture. One of every nine barrels of oil worldwide goes to fuel American cars, which disgorge 80 million tons of smog-forming pollutants into our air every year. Cars have facilitated the flight to the suburbs, promoted

sprawl, made us a nation of commuters, and signaled the death of many inner-city neighborhoods, businesses, and systems of public transport. Many Americans wonder whether the private car, which once promised to carry us to happiness, might be heading instead down the road to ruin.

The most widely publicized, and the most truly inexhaustible, of the alternative energy sources is the sun. In some ways, solar power is no more exotic or unconventional than oil or natural gas. This is obvious each time a new set of collectors goes up on a suburban roof.

Yet there is much more to solar technology than simply collecting heat. Direct heat is only one of the sun's gifts, only one of the ways in which solar energy serves life on Earth. The sun drives our winds, which in turn move the ocean's waves. It powers the tiny photosynthetic furnaces in the leaves of plants, giving us food to eat and wood to burn. And its rays can act upon manmade converters of energy, creating the most versatile of all our power sources—electricity.

From the earliest coal-fired dynamos to the latest fission reactors, it has been taken for granted that the generation of heat is a necessary step in the generation of electricity. Water boils; steam turns a turbine. This is also the principle behind solar "furnaces," in which mirrors focus sunlight on a boiler. But by far the most promising of all solar electricity technologies is the photovoltaic cell, in which accumulated heat or boiling water plays no part at all. As far back as 1839, it was discovered that an electrical current could be generated by exposing certain light-sensitive materials to solar radiation. But this effect remained a curiosity until, well into this century, silicon was discovered to be the material of choice for photovoltaic cells. Even then, the problems of producing pure silicon, and of handcrafting it into functional solar cells, hampered extensive development.

Now the breakthroughs in cost-effectiveness and efficiency are accumulating, and the number of solar electric homes and public buildings increases yearly. One of the most important accomplishments has been the discovery of a means of extruding a ribbon of silicon, thus avoiding the wasteful and time-consuming cutting and polishing of disks cut from silicon cylinders. Battery storage systems have also been improved.

Localized solar electric systems are working now, but extensive research also focuses on massive, centralized photovoltaic power plants that would orbit the Earth and beam electricity to ground stations via microwave. With the space shuttle in operation, we may be closer to this goal—but we are no closer to deciding whether such vast microwave concentrations would be safe, or if the solar cell satellites would be unacceptably vulnerable in future times that may include space warfare.

(Above) **In areas where sunlight is intense and dependable, large-scale, centralized solar energy installations may be practical. In the Mojave Desert in Barstow, California, an array of solar mirrors concentrates the sun's energy on a boiler, to generate steam for turbines. A computer-screen diagram** *(inset)* **helps technicians focus the array.**

(Below) **A photovoltaic array field, in which modules made up of individual solar cells convert sunlight to electricity.**

(Above) **A solar photo-voltaic house in the Southwest. The roof-top modules provide electricity; passive design reduces energy requirements for heating and cooling.**

(Below) **The winds are another form of solar energy.**

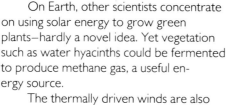

On Earth, other scientists concentrate on using solar energy to grow green plants—hardly a novel idea. Yet vegetation such as water hyacinths could be fermented to produce methane gas, a useful energy source.

The thermally driven winds are also being tapped—not only by updated versions of the familiar windmill, but through the harnessing of the power in ocean waves. This thirdhand use of the sun's energy is being investigated by Scottish scientists, who have discovered that a rocking mechanism, slapped relentlessly back and forth by waves, can be used to drive a generator. The problems are problems of scale: the system envisioned to provide all of Britain's electricity would be as big, and as heavy, as the highway between London and Glasgow. And it would be miles out at sea.

The ocean stores the sun's energy in the form of heat, as well as of motion, and here one of the most ambitious schemes comes into play. It involves the temperature differential between the cold water at lower ocean depths and the warmer water near the surface. Huge floating power stations would work on the cyclical condensation-and-evaporation principle of the air conditioner: warm surface water would boil liquid ammonia, and cold water, pumped from below, would recondense it. During the warming phase of the cycle, the expanding ammonia gas would turn a turbine and pro-

duce electricity. It works, not only in theory but in practice. Large-scale implementation, though, depends upon the manufacture of durable, reliable components and the investment of vast sums of money.

Solar technology is extremely diverse, and different facets of it are in markedly different stages of development. But along the "hard" energy path, researchers are still moving carefully from promise to production in the incredibly challenging and complex realm of nuclear fusion.

Fusion is probably one of the most difficult problems that science has undertaken to solve—and one of the most fundamental. Fusion reactions are the driving force of the universe, the power behind the stars. We are more familiar with nuclear fission, in which energy is released through the splitting of the nuclei atoms in a controlled or uncontrolled reaction. In fusion, the nuclei of deuterium and tritium atoms—two isotopes of hydrogen—join together, giving off an even more impressive amount of energy. In the reactions that fuel the sun, intense solar gravity makes the fusing of these atoms possible. On Earth, a controlled fusion process requires surmounting complicated obstacles, when we try to replicate the conditions of solar temperatures and pressures.

The nuclei of deuterium and tritium atoms are positively charged and so repel each other. The problem becomes one of getting them moving fast enough to overcome their repulsion and get close enough to fuse. High speed means high temperature—as much as 100 million degrees. Then enough of the particles have to be at a sufficient density, for a long enough period of time, for the fusion process to be sustained. These requirements have led to basic research in plasma physics and to the development of some remarkably sophisticated hardware.

Plasma is the state a gas reaches when it becomes so hot that its atoms are stripped of their electrons. This is the raw material of fusion. But what can hold a 100-million-degree plasma together? No physical substance is up

The ELMO Bumpy Torus at Oak Ridge National Laboratory is a working example of tokamak design. The tokamak is one approach to the problem of harnessing nuclear fusion for electrical power generation.

to the task. Scientists have instead turned to confining plasmas within a strong magnetic field, and the most successful of the devices they've constructed for doing so is called a tokamak. The tokamak resembles an enormous doughnut, and it is the magnetic forces generated inside this toroidal device that keep the plasma from dissipating. In effect, the superheated gas "bites its own tail."

Tokamaks are not the only entry in the field of plasma confinement. The tandem mirror holds plasma in a straight cylinder; it could prove to be less complex, but has yet to yield results as impressive as those of the tokamak. Another path of investigation involves what is called inertial, rather than magnetic, confinement. The procedure involves focusing a barrage of laser beams on a tiny pellet of deuterium and tritium within a reaction chamber. The result is a controlled and vastly downscaled version of a hydrogen bomb explosion; a sequence of such miniature explosions would be the driving force in a power-generating plant.

Fission vs. Fusion

Fission is the process in which the nuclei of heavy atoms are bombarded by neutrons and split into two nearly equal parts and several additional neutrons, releasing large amounts of energy. In a nuclear reactor, the energy generates steam that powers an electric generator.

Fusion is the process in which two nuclei of light atoms combine at high temperature to form a heavier nucleus and an extra neutron, releasing an even greater amount of energy. Fusion is known as a thermonuclear reaction, the same process by which the sun and stars radiate tremendous amounts of energy.

In both nuclear reactions, a small amount of mass is converted to huge amounts of energy. (See "$E = mc^2$," p. 265.)

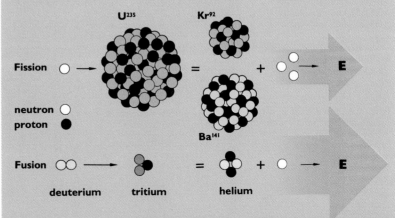

(Above) A model of the Shiva laser system used in fusion experiments at the Lawrence-Livermore Laboratory. This huge device will deliver more than 30 trillion watts of optical power in less than one-billionth of a second. The twenty 150-foot-long laser amplifier chains on the right of the model are focused onto a tiny fusion target located in the center of the four-foot spherical target chamber *(right)*.

The tokamak itself stands to be refined. One improvement will be replacement of the "pulse" system, in which the current that heats the plasma is pulsed into the chamber, with a steady-state current made feasible through the use of radio frequency power. It may well be an improved steady-state tokamak that accomplishes the objectives of the Magnetic Fusion Energy Engineering Act of 1980, which calls for the development of a working (though not necessarily economically feasible) fusion power-generating plant by the end of this century. Commercial applications could follow. Department of Energy and private sector researchers are working during the 1980s to select the optimum fusion concept for this venture and would hope to concentrate in the 1990s on the engineering required to turn fusion energy into electrical power.

Enormous difficulties would be encountered in retrieving the heat energy represented by the frantic neutron orphans of the fusion process and in channeling it to the driving of a turbine. Would helium be the me-

dium? Could tritium be bred as a by-product of the reaction?

Other questions, of a more fundamental and philosophic nature, will have to be asked. Fusion as yet appears safer than fission energy—but advocates of appropriate technology such as Amory Lovins argue that it could be disastrous to have an energy source of that kind, given that our track record in the responsible management of big sources of concentrated energy is not very good. John Holdren of Princeton University has called fusion "a technology that has enormous potential . . . an energy source which can make a substantial contribution." Yet Holdren also ventures that "the process of developing a technology like fusion must draw on all kinds of expertise outside the technological community as well as in it, to be sure that the process leads to a device that society really wants." These words might well apply to all of our energy technologies—coal, hydro, solar, hydrogen, nuclear—as we move toward a time when petroleum will be a memory.

Inside the Shiva vacuum target chamber a needlelike arm holds a tiny amount of deuterium-tritium gas in a target smaller than a grain of sand. For less than a billionth of a second, the laser energy will heat and compress the fusion fuel to temperatures and densities like those found in the sun. This technology is still in its early experimental phase, as are the engineering schemes for harnessing this energy.

Why America Burns

O nce or twice a year, a big fire makes the news. The MGM Grand Hotel's casino in Las Vegas goes up in flames, and eighty-four people die. Downtown Lynn, Massachusetts, burns, and an extensive urban renewal project is lost. These are the spectacular fires, the ones the newsmagazines and the networks cover. But the media and the rest of us tend to forget that during each hour that the MGM Grand or the city of Lynn burned, 300 other fires raged throughout America. Each of those hours of fire cost the country one life, three serious injuries, and $2 million—and each year there are 8,760 such fire-hours, resulting in 12,000 deaths, 24,000 tragic and disfiguring injuries, $20 billion dollars of damage, and unimaginable misery and pain. The young, the old, and the helpless are the most likely victims. Unfortunately, the problem seems to be getting worse.

The United States has more than its share of fires. The fire death rate in Switzerland, with its dense concentration of ancient buildings, is one-tenth as high as America's, and the fire chief of Zurich cannot remember losing an entire residential structure. Why are we so fire-prone? The answer, it seems, lies in ignorance, carelessness, and greed.

With its extensive fire codes, the United States hardly would seem to lack the basic tools for fire prevention. The codes have been written over the past century by a private organization known as the National Fire Protection Association (NFPA), which sells its regulations to local authorities, who adopt them in whole or in part as local law. Failure to enforce the codes is certainly a reason behind many fires, but a growing number of critics charge that the codes themselves are dated and inadequate.

Fire codes originated in response to the massive demographic shift toward American urban areas that occurred in the nineteenth century. As the cities swelled with a flood of new immigrants, so they burned down. Insurance companies were faced with bankruptcy, and premiums skyrocketed. To protect against financial ruin, the insurance industry set up the NFPA, which devised an ingenious solution to the spreading urban infernos: codes and compartmentation. Regu-

In 1980 a fire at the MGM Grand Hotel *(right)* **killed 84 people and injured hundreds of others, despite squads of firefighters and army helicopters.** *(Above)* **People who took refuge on balconies were rescued by means of window washers' equipment.**

Charred slot machines, dripping with melted plastic, line the entrance to the MGM Grand Hotel's casino.

lations were promulgated to encourage the construction of fire walls between buildings and fire floors within structures. These walls and floors were to be made of fire-resistant materials manufactured according to written codes, and fire would be held within compartments formed with these materials until the firefighters arrived to put it out. The codes additionally pushed for the building of stairways and exits, firefighting water supplies, standpipes, hydrants, and other protective apparatus. All this led to a great improvement in fire safety. Even as skyscrapers grew far beyond the reach of the longest fire ladders, compartmentation worked: when a gasoline-loaded bomber hit the seventy-ninth floor of the Empire State Building and burst into flames in 1945, the fire was contained.

In the mid-1960s, though, something went wrong. Fire and deadly smoke began to get out of control in modern high-rises, in both the United States and other countries. These new buildings are riddled with ducts, channels, and gaps above false ceilings, all of which afford passage to flames and lethal smoke. If all the materials inside the buildings were truly fireproof, such pathways would be of little consequence; unfortunately, new synthetic building materials are fraught with unpredictable and dangerous combustion characteristics. Burning plastic, for example, spews forth prodigious amounts of noxious gases and smoke. As fire safety researcher David Lucht comments, "The larger-scale use of synthetic materials perhaps has created smoke and fire problems that codes haven't caught up with yet."

Some fire experts feel that the answer to these new fire hazards is more technology. They claim that fires such as the one at the MGM Grand casino could have been stopped by automatic sprinklers. Indeed, that fire was contained in sprinkled areas adjacent to the casino; yet it is rare to find codes that unequivocally demand sprinklers in large public buildings. Why? Because compartmentation codes are put before sprinklers, says Richard Patton, a fire engineer. "The National Fire

Protection Association has always in essence promoted sprinklers. However, the codes actually made it extremely difficult to put sprinklers into most buildings. They've regulated the fire sprinkler system to the point where sprinklers have not been installed in more than 90 percent of the buildings that need them."

Critics argue that present fire codes have been based, at best, on intuition, guesswork, extrapolation, and the empirical experience of what seemed to go wrong or right at a previous fire. What is needed, they claim, are scientifically based codes written by independent researchers. But our understanding of the physical and chemical dynamics of fire is still in its infancy, so scientists must start with the fundamentals. At the National Bureau of Standards, scientists have begun by studying the flame itself and the particles it creates. The researchers hope to isolate the factors relating soot size to heat radiation—aspects of fundamental fire physics and chemistry. Flame is also being investigated at Harvard University, where scientists study the conditions of fuel and oxygen supply that cause unburned gases to ignite suddenly in a process known as "flashover." This is basic research into one of the many elements of the fire system. The whole system, such as a room burn, is also being investigated at the National Bureau of Standards, and the results of their experiments reveal the devastating speed with which fire can spread:

Ignition. A match is dropped on a newspaper in the corner of a sofa.

A test is conducted in which a smoldering fire—such as a cigarette might cause—ignites a bedspread after 3 minutes (first frame) and the mattress and pillow after 5 minutes (second frame); causes "flashover" of heated gases accumulating at the ceiling after 6 minutes (third frame), and totally engulfs the room a minute later (last frame). The view is through the door of a typically furnished bedroom, with desk, bureau, and closet out of view to the right. The room is equipped with sensors, and at the bottom a plume of smoke shows the direction of air flow. The test was conducted by Harvard University and the Factory Mutual System.

Twenty-five seconds. The flame is 2 feet high. Hot gases are already collecting at the ceiling. The temperature there is 100 degrees Fahrenheit. Oxygen within the room is being burned up. Carbon dioxide is being formed.

One minute, 27 seconds. The superheated soot and gases have raised the ceiling temperature to 800 degrees. The smoke layer is 4½ feet from the floor. Anyone standing upright with his head in this smoke is likely to die.

Two minutes, 11 seconds. Flame is now widespread. The ceiling temperature is 1,200 degrees. The heat radiated from the superheated smoke layer has raised the temperature of the floor to 400 degrees.

Four minutes. The ceiling temperature is now 1,300 degrees, and the smoke layer is within 2½ feet of the ground. All objects in the room ignite in flashover, and the room is an inferno.

To understand precisely what has happened during these four deadly minutes, scientists continue to investigate the qualities of specific materials, alone and in combination, and the nature of combustion itself. As they do so, some of the traditional methods of testing and rating flammability are being called into question. For almost forty years, a device called a Steiner tunnel has been used to determine the fire potential of inte-

rior paneling. The material is ignited at one end of the enclosed tunnel and assigned a rating of 0 to 225 on the basis of how fast combustion spreads. But when the same procedures are applied to plastics, critics charge, the results are erratic.

One step toward an improved fire science might be to re-create those situations by means of computer models. At Harvard University, Dr. Henri Mitler works with programs capable of projecting the hypothetical spread of smoke and flame on the basis of room size, window and door openings, building materials, and furnishings. His colleague Professor Howard Emmons says that the computer can calculate the pattern of destruction of a ten-minute room fire in less than a minute. Emmons calls this a "very favorable" ratio and looks toward the day when advanced computers might swiftly and economically project the spread of combustion in a very large building—the sort of information that is essential to the formulation of updated and effective fire codes.

Revising the codes on the basis of scientific research holds great promise for improved fire management, but some scientists are going one step further in an attempt to prevent fires before they begin. Of the 12,000 deaths caused yearly by fire, 2,300 result from cigarettes dropped carelessly. On the assumption that people will continue

to fall asleep while smoking, many researchers urge the development of "fire-safe" cigarettes, which would stay lit only while being puffed. But the tobacco industry opposes legislation requiring such changes in cigarette manufacture, and the Tobacco Institute has so far been successful in suppressing congressional action. This greatly dismays Massachusetts Representative Joseph Moakley. "There are hundreds and hundreds of children lying in burn hospitals today who don't know what a cigarette is, but they're there as a result of a carelessly discarded cigarette. I think the tobacco industry has the duty to come forward with a safer product." It doesn't, argues Andrew McGuire, executive director of the Burn Council, because the companies "enjoy having the cigarettes left unattended to burn up. It is, I think, the ideal in planned obsolescence. It's probably the classic symbol of twentieth-century America to put out a product that will consume itself whether you're using it or not." Until Congress acts, funds for research on the fire-safe cigarette will remain scant, and pressure on the tobacco industry to improve its product will remain minimal.

As if the prevention of accidental fires weren't difficult enough, America is faced with the growing scourge of arson. It is our most expensive crime, yet the one with the lowest conviction rate—a mere 2 percent.

Arson does not generally originate with the people who live in the buildings that burn; it is, according to David Scondras, a Boston neighborhood activist, a crime most often committed by those who stand to benefit financially from fire. He charges that unscrupulous real estate manipulators burn down overinsured slum properties and gut buildings in promising neighborhoods to pave the way for lucrative redevelopment. His organization uses computers to target arson-prone buildings and cooperates with the state attorney general's office to let owners know that fires on those premises will be regarded with extreme suspicion. The program has shown results, but in the face of the nationwide arson binge, it is like a bucket of water at the Coconut Grove.

The variables in America's deadly fire equation are numerous and complex. Inadequate fire codes, outdated testing apparatus, industrial and legislative foot-dragging, carelessness, and malicious greed all contribute their share to the grim toll exacted by the flames; yet the problem is not insoluble. Fire is most often a preventable tragedy, and if we embrace the full range of scientific, educational, political, and legal measures available to fight it, fire-related deaths might be reduced by as much as 50 percent. The lesson of why America burns is that the death and destruction can be curtailed.

A Is for Atom, B Is for Bomb

(Facing page) **The first explosion of the hydrogen bomb in 1952.**

(Below) **Air-raid drills were regularly practiced in schools in the 1950s when nuclear war was considered imminent.** *(Inset)* **A family fallout shelter.**

Back when TNT rather than plutonium was the instrument of choice for committing global mayhem, the idea of civil defense was hardly controversial. Governments were expected to do all they could to protect civilian populations during wartime, and people routinely cooperated. This attitude persisted through the first decade of the nuclear era, culminating in the great fallout shelter boom of the early 1960s. But as weapons became more lethal, and the policy of "mutually assured destruction" (MAD) prevailed, civil defense was largely discredited. Either we would live, or, if there was war, we would all die. Dust collected in the shelters. Our emergency crackers went stale.

As the 1970s drew to a close, Americans began to think about civil defense once again, and for the first time there was widespread disagreement about whether it was worthwhile to dig in. A deciding factor, it seemed, was our perception of Russian civil defense preparedness.

What were the Russians up to? They seemed to be taking civil defense seriously—100,000 military personnel assigned to CD-related duties, shelters under all new buildings, detailed evacuation schemes, plans for sandbagging vital machinery. Many observers doubted whether any of the Soviet precautions would make much difference if war actually came. But others perceived the Russian preparedness as proof that our adversaries no longer believed in the MAD scenario. To this way of thinking, planning to survive is tantamount to planning to fight, and Americans were urged to reconsider CD.

It might also be worthwhile to consider the threat against which we'd be defending ourselves. In an all-out nuclear war, perhaps 5,000 major cities in the northern hemisphere would be destroyed. Fallout would be a problem for weeks, and subtler forms of radiation would linger far longer. There is even a possibility that the Earth's ozone layer would be depleted, leaving survivors exposed to cancer-causing ultraviolet radiation.

Nevertheless, civil defense advocates point to the past rebuilding of other societies destroyed by war and suggest that America's government and economy would be capable of quick rehabilitation, given the diffusion and protection of data, equipment, and personnel. A reasonable warning period, most likely based upon rapidly intensifying prewar tensions, would of course be vital to this scenario.

Assuming the lead time would be available, the central feature of any survival program is massive evacuation. There already exist plans for the relocation of vulnerable urban populations to small towns beyond the supposed periphery of immediate blast and radiation damage. Since such schemes are largely predicated upon the use of private automobiles, it hasn't taken critics long to postulate gargantuan traffic jams that would reduce evacuation to a cruel joke. The Russians, they say, would be no better off, despite their highly touted plans for getting people out of the cities.

Even if people can be evacuated and do survive in large enough numbers to preserve social organization, they will need an economy and means of production. T. K. Jones, a systems analyst who has worked with

The directors of the Manhattan Project, physicist J. Robert Oppenheimer and General Leslie Groves, stand at the base of the steel tower on which the first atomic bomb was tested. The intense heat of the explosion melted the tower and seared the surrounding sands into jade-green glasslike cinders.

Boeing and the Department of Defense, says that "by protecting our machinery we can get our recovery time down to where we could restart production in something like four to twelve weeks." His optimism is contingent on American adoption of Soviet methods of dispersing plant facilities and burying or sandbagging delicate machines.

The arguments have not abated. "Somebody has to survive," says Allen Zenowitz, director of the New England headquarters of the Defense Civil Preparedness Agency. "Man just is built to survive." Military technology expert Frank Burnaby agrees that some people would indeed survive, but adds that the fortunate ones would be "those in the center of town, who are vaporized."

The power of destruction of the atomic bomb has been evident since the first one was exploded in Alamogordo, New Mexico, on July 16, 1945. It produced an explosion equal to that of 20,000 tons of TNT. Within

weeks the American military had dropped two similar bombs on the Japanese cities of Hiroshima and Nagasaki, helping to draw World War II to a close.

While scientists had deciphered the atom's basic structure in the early 1900s, it wasn't until Einstein's Theory of Relativity came into focus that they understood the vast energy the atom contained. By the 1930s, scientists had succeeded in splitting atoms of uranium by bombarding them with neutrons, and they wanted to go on to pursue the chain reaction such atom-splitting could provide. Physicists recognized the military possibilities of such atomic energy, and in 1939 Albert Einstein wrote to President Franklin D. Roosevelt to apprise him of the situation. Federal funds ensued, and in 1942 the government authorized a united plan to produce the atomic bomb. The Manhattan Engineer District of the Corps of Engineers (dubbed the Manhattan Project) was organized and began work at the University of Chicago and in Los Alamos, New Mexico.

After the Hiroshima and Nagasaki bombs, the United States continued to test atomic bombs on small islands in the Pacific and in Nevada. By 1949, the Russians had tested their first atomic bomb, and the British detonated test bombs in Australia starting in 1952.

By the early 1950s, however, the emphasis began to shift from the atomic, or fission, bomb to the hydrogen, or fusion, bomb. A hydrogen bomb carries thousands of times the power that an atomic bomb does. It is created when the heat from fission sets off the fusion of deutrium and tetrium to form helium, releasing vast amounts of energy. This high energy in turn sets off uranium in a second fission reaction, causing an explosion that can cover an area hundreds of miles in radius, destroying existing life and producing enormous amounts of radioactive fallout.

No single individual can be given all of the credit for so difficult and complex a contrivance as the hydrogen bomb. But if fatherhood is defined by striving, lobbying support, and ambition, as well as by the initial act of

Hiroshima after being devastated by an atomic bomb.

creation, Edward Teller's claim to the title cannot reasonably be challenged. For more than three decades, he has championed the development and deployment of the bomb for reasons heinous, foolhardy, or altruistic, depending on one's point of view. To some he is the embodiment of the mad scientist; to others, he has stood as a bulwark against Soviet domination of the world. Teller's is not a simple case.

Edward Teller is a thickset septuagenarian who is comfortable at a grand piano, a man, according to some, of immense charm, humor, intelligence, and energy. With a heavy accent, Teller spins out his beliefs and his interpretations of the scientific and political events in which he played a part.

He was born in Budapest, Hungary, in 1908, a time in which physics and politics were being reshaped, respectively, by Einstein and by nationalism. Teller was educated in Germany, but in 1935, as the menace of Nazism grew, he left Europe to take a post in the United States. Here he found both political and academic sanctuary, while acquiring a substantial reputation as a physicist. Nevertheless, his career was destined to be dominated by events in the Europe he had left behind: scientists in Germany had split the atom—an event that theoretically made an

atomic bomb possible. Alerted to the possibility by the letter from Einstein, Franklin Roosevelt had ordered the commencement of the intense collaborative effort known as the Manhattan Project, which led in 1945 to the successful testing of the atomic bomb and its use against the Japanese. Hitler, ironically, had never managed to come close to developing an atomic bomb.

Edward Teller was a part of America's atomic bomb research efforts from their inception. He followed the program from Columbia University to the University of Chicago and finally to Los Alamos, New Mexico, where Robert Oppenheimer would lead his team of scientists to the project's awesome conclusion. But almost before the fission bomb research began, Teller fixed upon the notion of translating into destructive force the far more potent energies of nuclear fusion. Years before it could be built, the hydrogen bomb was taking shape in his mind.

Teller the man of science was already well known. But as work on the Manhattan Project progressed, Teller the man of controversy was to emerge. He felt that he had come to Los Alamos not merely to develop the atomic bomb, but to press forward with his own brainchild. Oppenheimer resisted— and Oppenheimer was in charge. For the

Hungarian physicist Edward Teller, the prime mover behind the development of the hydrogen bomb.

The Plutonium Connection

The idea was ingenious, if frightening: assign a twenty-year-old college chemistry student the task of designing an atomic bomb. "NOVA" did it in 1974, and after five weeks' work, using unclassified reference material from libraries and government agencies, the student submitted a plan that, according to a Swedish weapons specialist, outlined a device with a possible yield of one kiloton (about 1,000 tons) of TNT. But the point of the assignment was not to demonstrate the cleverness of one science major, or to show the commonplace physics that underlie this most terrible of modern technologies. Rather, it was to show the practicability and

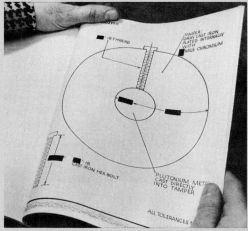

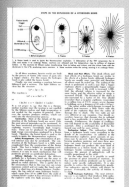

Encyclopedia article telling how an H-bomb works.

likelihood of what would surely be the ultimate act of terrorism—the use of atomic weapons for blackmail.

In the 1980s, the debate over nuclear weapons centers upon reduction of the stockpiles maintained by the United States and the Soviet Union. But even if the two superpowers were to dismantle all of their warheads, the world would not be safe from nuclear terror. The reason is twofold: the basic information on nuclear weapons design is largely declassified and available, and the essential ingredient of fission bombs—plutonium—exists in increasing profusion, as a result of worldwide military and civilian exploitation of atomic power. Naturally, the question arises how and when terrorists or unstable minor powers

might turn these circumstances to their advantage.

It hasn't happened yet. No mad bomber has phoned a newspaper or the police to warn that an atomic device might go off in a major city if demands are not met. But many experts feel that this is a matter of luck rather than logic, that it is merely an indefinite reprieve. Plutonium, a manmade element, is a by-product of the fission of uranium in nuclear reactors. It is processed in such places as Allied General's plant in South Carolina and the Kerr-McGee Company's facility in Oklahoma. In addition to safe handling of this intensely carcinogenic material, a prime concern at these plants is the precise monitoring of inventories. Every milligram of plutonium must be accounted for.

But no systems are foolproof, and each year a tiny amount of the material disappears—whether into the garbage or the wrong hands, no one knows. And even if each processing facility kept track of 100 percent of its plutonium, there would remain a need for elaborate security measures for both the plants and the transportation links that serve them. How effective might these measures be against a planned assault? Foreign energy and weapons development also contribute to the world's stock of plutonium. In the words of Jan Prawitz, the Swedish expert who analyzed the student's bomb plans, "if stealing plutonium is easier in one country than in another, the measures undertaken by the second country will be worthless." The problem of a nation's diverting plu-

tonium from its nuclear power plants is equally unsettling.

Not long ago, a State Department official told the Senate Foreign Relations Committee that, given the equipment and investigative methods available, it would be possible, at least 85 percent of the time, to determine whether nuclear blackmail attempts were bluffs. But the fact that those attempts might even be made is proof that the situation has gotten no better since 1974, when General Edward Giller of the Atomic Energy Commission suggested that we "keep the horse in the barn rather than look for it after it's been taken." The issue, though, isn't whether the horse has been taken. With over 100,000 pounds of plutonium in the world, protecting it is more like keeping flies in a barn.

Reprocessed plutonium.

time being, the disagreement between the two men was overshadowed by the pace and importance of the mission at hand. But it was to reappear once Germany and Japan had been vanquished and the Cold War with the Soviet Union had begun. Teller felt that the centerpiece of our defenses against the Russians should be his still-unbuilt hydrogen bomb, but the Atomic Energy Commission's General Advisory Committee, chaired by Oppenheimer, voted against developing the

new weapon. ("They protested," said Teller, "merely on moral grounds.") Teller pushed successfully for a government reversal of this position. His next objective was the establishment of a second weapons laboratory, independent of Los Alamos. Again he was opposed by Oppenheimer, and again he triumphed through deft yet forceful lobbying of the defense establishment. "He's very eloquent in the way he speaks," former Manhattan Project scientist George Kistiakowsky once said. "He has a passion in his voice which is extremely effective on impressionable politicians." The new laboratory was Lawrence-Livermore, in northern California.

Now the stage was set for one of the saddest and most peculiar episodes of the Cold War—and of Teller's career. Because of his prewar leftist sympathies, his opposition to the policy of massive nuclear retaliation, and his past disapproval of a crash hydrogen bomb program, Robert Oppenheimer ran afoul of Cold War partisans. Teller was called to testify at a 1954 hearing, at which he recalls having said, "I have confidence in the loyalty of Oppenheimer. But Oppenheimer is a complicated individual. I don't understand his actions. I wish that the security of our country was in hands which I understood better and therefore trusted more."

Oppenheimer was stripped of his security clearance, a turn of events that remains controversial to this day. In Teller's own view, the incident harmed him—Teller—as well. "By our old friends," he said years later, "we [he and Mrs. Teller] were practically ostracized." Yet Teller's star continued to rise. He became a champion not only of improved thermonuclear weapons, but of peacetime reactors as well. (To Teller, the Three Mile Island accident was a glass half full, rather than half empty. It proved that even a "stupidly mishandled" reactor killed no one when it malfunctioned.)

Edward Teller's first major political defeat came in 1962, when the nuclear test ban treaty was signed. He had spoken against the agreement, arguing that national security was dependent on continued above-ground testing. Offering a peculiar, lesser-of-two-evils analysis, Teller proclaimed that "not to ratify the treaty would be a mistake. But to ratify the treaty would be an immeasurably bigger mistake." This matter-of-fact acceptance of the instruments of doomsday as a necessary evil and a practical reality remains characteristic of Teller. Not surprisingly, he also favors strong civil defense and claims that "in the wake of a nuclear war, there will be victors and there will be vanquished." Does he

actually expect such a war, and such an outcome? George Kistiakowsky feels Teller believes that "we might just as well get it over with as soon as possible and in the best possible circumstances." But Teller himself professes not to be so resigned: "I hope, the only thing I really hope, that it will never come to that."

The career of Edward Teller, no less than that of Robert Oppenheimer, is emblematic of the dilemma of science in our time. Oppenheimer represented the scientist as warrior—but having helped to let one genie out of the bottle, he hesitated to loose a second, even more lethal force upon the world. Teller appears to have felt no similar compunctions. "The responsibility of a scientist," he argues, "is to make science . . . knowing that it may be used for good or evil purposes." The question of whether there are good purposes for the hydrogen bomb is one that, in the light of his political understanding, Teller has answered for himself. He asks us to answer it as well.

$E = mc^2$

Albert Einstein must certainly rank as one of the greatest geniuses of all time. His contributions to relativity theory, and his efforts to reconcile the physics of particles with the physics of space, ensure his place in the history of civilization.

To the layperson, nothing signifies Einstein's brilliance so much as the equation $E = mc^2$, which reveals an extraordinary relationship between matter and the energy stored within its chemical and nuclear bonds. E, representing energy, is related to m, representing mass, by the square of c, the speed of light. C is an enormous number—299,792,458 me-

ters per second—and its square is suitably vast. Therefore it is apparent that the energy potential of even a small amount of matter is immense. The fission of a handful of uranium on August 6, 1945, grimly demonstrated this counterintuitive proposition when the matter's unleashed fury annihilated the city of Hiroshima and 80,000 of its inhabitants. Einstein was, before all else, a gentle humanitarian; there could not have been a more bitterly ironic confirmation of his genius than this birth of nuclear warfare.

A pacifist, an internationalist, a libertarian, and a Jew, he had been driven from a hostile, fascist Germany to refuge in the United

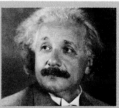

States. Events regrettably led him to help forge the link between science and the government and military, an association now institutionalized in a technocracy of incredible size and power. Yet Einstein remained a philosopher and a humanist to the end. He might have been anticipating his own demise when he wrote, "This death signifies nothing. For us believing physicists, the distinction between past, present, and future is only an illusion, even if a stubborn one."

A Conversation with Bernard Feld

Atom bomb scientist. Theoretical physicist. Arms control expert. Editor of the journal that many see as the voice of the scientific community's conscience. Bernard Feld's credentials would do credit to a quartet of scientific careers. All the more surprising, then, that he expected to major in history when he started at the City College of New York in 1935. His high school had taught science so badly, he explained, "that I thought it was dull."

An inspired instructor, and a large measure of good fortune, put the young Feld on the path to scientific prominence. H. H. Goldsmith, a physicist with an encyclopedic knowledge of science, recruited Feld as a gofer to gather material for a textbook on atomic physics. Feld loved the work. Convinced that he would become a physicist, he enrolled in graduate school at Columbia University in 1939. He quickly landed an assistantship with a professor just arrived from the gathering storms of Europe—Enrico Fermi. That connection inexorably took the young scientist to Chicago, as a group leader of the team that, in 1942, created the first working atomic pile. Two years later, Feld moved west to Los Alamos to become a member of the critical assembly group that worked out the details of the critical mass and purity of uranium for an atom bomb and helped to assemble the parts of the first atomic explosion in the New Mexico desert.

Feld encountered firsthand the roller-coaster emotions of the men and women who built the bomb. Their initial enthusiasm remained even when it became clear that the Germans, whom they had thought they were racing, had no chance of making their own bomb. "We got caught up in a tremendously exciting enterprise," he recalled. "We were close to the point at which we would know whether it would work or not. There was a kind of mass mesmerism." That feeling lasted through the dropping of the first atom bomb on Hiroshima on August 6, 1945. "I remember very well the news of the first bomb," Feld reminisced. "It was an exhilarating experience. It worked. It proba-

bly meant that the war really was over."

Then, eight days later, the United States let loose its second A-bomb, on Nagasaki, and disillusionment was complete. "It was the news of the second bomb that made me sit up and think: was that trip necessary?" recounted Feld four decades later in his brightly lit first-floor office at the Massachusetts Institute of Technology.

He had accepted a post at M.I.T. just after the war, but decided to postpone his move to Massachusetts. Instead of wandering through the marbled corridors of the institute, he prowled the corridors of power in Washington, lobbying against military control of atomic weapons. During that period, he started his association with the *Bulletin of the Atomic Scientists,* a small broadsheet originally published on December 10, 1945, for circulation among the veterans who had turned the atom bomb from theory into reality.

Feld has stayed with the *Bulletin* ever since. In 1950 he joined its board of editors, and a quarter of a century later he became editor in chief.

By then, the magazine had passed through its own crisis. Originally written by and for atomic scientists, by the 1960s it had broadened its appeal, covering issues vital to most of the U.S.—and international—scientific community. But a brief experiment in the early 1970s with professional editors had given the *Bulletin* a shrill, antiscientific tone and had reduced its circulation to fewer than 17,000. "One of my complaints is that the *Bulletin* is one of the most respected and least read journals in the world," commented Feld. "I don't want to reverse that, but I'd like it more widely read."

His stewardship seems to be succeeding. Seven years into his term as chief editor, circulation was up by 50 percent. Undoubtedly his editorial direction—a return to the journal's original emphasis on the dangers of nuclear war and fresh looks at international scientific issues—has contributed to that success.

For readers and nonreaders alike, the true symbol of the *Bulletin*'s influence is its

doomsday clock. Perched in the upper right-hand corner of the cover, the clock ticks off the minutes to midnight—the exact hour at which nuclear holocaust will be inevitable. Over the years, the magazine's editors have moved the minute hand back and forth, between eleven and two minutes to midnight, in response to their changing perceptions of the dangers of nuclear war. In mid-1981, Feld and his editorial colleagues, responding to Soviet adventurism in Afghanistan and Poland and the increasingly hard line of the Reagan administration, advanced the clock by three full minutes, to four minutes to midnight. There it rests, despite pleas to move it in either direction. "We're not giving up on the present administration," explained Feld, "but we're not impressed with their efforts."

Feld is equally unimpressed with the military. Indeed, one of the major functions of the *Bulletin,* as he sees it, is "to keep the military honest." That demands acquiring expertise in military matters such as arms control—a difficult task, given that the military men in Washington rarely acknowledge their critics.

Is it possible to become an arms control expert outside the cozy confines of the Pentagon? Certainly, says Feld. "Essentially, all the information is available, generally in such esteemed journals as *Aviation Week* and *Space Technology,*" he explains, tongue only partly in cheek. "People get interested in specific issues and suddenly find that they're experts." Feld does see one near-essential requirement: "When students say that they want to get involved in arms control, I advise them to get a degree and credentials in physics. In a sense it's like a union card."

Feld's own union card has kept him at the cutting edge of modern physics, first as an experimentalist and more recently as a theoretician. His work on the early particle accelerators in the years after World War II gave him an abiding interest in the structure of elementary particles—a field that he now sees as advancing at a breathtaking rate toward the long-sought goal of understanding nature at its most fundamental level. His opti-

mism stems from two observations: the fact that physicists possess a model that allows them to explain just about every broad experimental measurement made so far at the subatomic level, and the growing likelihood that new, ultrapowerful particle accelerators will allow physicists to create and detect the quarks that are the bases of the model.

What is most exciting, according to Feld, is that experimental efforts to detect the quarks, which are believed to be truly fundamental particles, have united particle physics with an entirely different discipline—cosmology. "We seem to be converging on something important from two ends," explains Feld as he sits in his M.I.T. office surrounded by Einstein posters and modern art. Certainly in this aspect of Feld's life the advancing clock is racing toward an entirely benevolent climax rather than toward doomsday.

Appropriate Technology

Progress is our most important product. The business of America is business. The bigger the better. These are ideas in economics whose time has come and gone, according to the late worldly philosopher E. F. Schumacher, author of the classic *Small Is Beautiful* (1973). Schumacher believed that modern economics, in its effort to be a "true" science, has preferred quantity over quality, never realizing that the consequences of political and business judgments based outside of human values are themselves inhumane.

"If we get rid of the idea 'the bigger the better,' which may well have been a nineteenth-century truth but now has become a twentieth-century myth," Schumacher said, "everything is possible." But in order to create economic policies as if people matter, Schumacher warned, we must start systematically from actual human requirements and longings. Instead of asking people to adapt themselves to machines and technologies, we must create tools that serve human beings.

E. F. Schumacher saw an industrial revolution that had gone amok. Here, in his adopted country of England, smog casts a Dickensian pall over Sheffield.

"One of the deep longings of human nature," Schumacher asserted, "is to work creatively not just with the brain but with the brain and hands, to make something and to take pride in making it. This has been virtually eliminated as a possibility by high-power modern technological development."

Schumacher, a former economic adviser to the British National Coal Board, foretold the current energy crisis a good twenty-five years before it began. Once fuel shortages became a reality, he recommended that we make opportunity out of potential disaster.

"If the high power disappears or is no longer so plentiful," Schumacher believed, "then perhaps this longing for satisfying and dignified work has a better chance, provided it is backed up with real intelligent technology. In other words, the beauty and desirability of creative work can be restored."

Schumacher questioned the wisdom of the current trend in Western nations to mechanize and automate everything around us. Life itself implies energy and depends

on it. To live means to burn it up. But, said Schumacher, there's living within your means and there's living dangerously. Modern societies, by eschewing the simplicity that characterizes many cultures of the Third World, has spent itself into a very precarious situation indeed.

Because we value human energy so highly—and because we maintain an attitude toward work that devalues it and depersonalizes it to the point of drudgery—we get machines to do all the work for us. Since the industrial revolution in the West, automation has always seemed the obvious way out. But it has proved to be a short-sighted policy.

Almost all the energy we use today to keep machines going is, like oil and coal, from fossil fuels, which are nonrenewable. The realization that nonrenewable fuel resources are limited has already sent prices rocketing. It's forced us to think more carefully about how we use them. Cheap fuel gave us cheap transportation, making big industrial centers seem economical. But power stations alone dissipate as much energy as they produce in just keeping cool. We need some alternative to this extravagance.

Schumacher suggested that we reverse the almost irresistible pull toward ever-larger industrial organizations and revert to a plan that will create appropriate technologies, neither too grandiose nor too primitive for the tasks at hand.

Many in the West are now taking Schumacher's ideas very seriously, but it was in the Third World, where the industrial revolution has barely begun, that his thinking has had

its major impact. Despite all their problems, Schumacher predicted that Third World peoples can enjoy a reasonable future, but only if they don't succumb, like the West, to the glossy lure of big technology. Unfortunately, it's difficult to persuade a developing nation that there are any dangers inherent in technological progress.

The noted economist spoke of once meeting a potter in a small village in India who, with marvelous skill, was making beautiful pots on the most primitive equipment.

"Having taken that in," Schumacher recalled, "I then visited an industrial estate which the government of India had sponsored in order to bring industry into rural areas. The first man I encountered was being trained on a very expensive machine-tool. I asked him whether he would be able to use such a tool in the area after he had completed his training. He replied that he could never get hold of such a machine or the money to buy it in all his life. When I asked him what he was going to do, he said that all he could do was go to Bombay and look for a job," Schumacher said. Clearly, the government's policy wasn't working; in fact, its effect was exactly the opposite of what it had intended: instead of bringing industry into

When he visited India, Schumacher *(inset)* encountered radical extremes of technology, represented here by the control room of a nuclear power plant and a man spinning at a portable, hand-powered wheel unchanged for centuries.

the rural area, the program was really drawing people from the rural area to Bombay, a city already crowded with unemployed people.

Schumacher's mind went back to the potter he had seen, whose equipment was worth 5 pounds. Then he thought of the man being trained on equipment worth 5,000 pounds. The 5-pound technology was too primitive to provide a decent living, and the 5,000-pound technology was too expensive to be within the reach of the people whom it ought to benefit.

At that moment Schumacher realized that what was required was a 500-pound technology—an intermediate one that would be much better than what India has, yet much simpler than what Western technologies have produced for themselves.

Indian farmers sort wheat, with the help of traditional beasts of burden. Schumacher felt that abandoning animal power for machines did not necessarily represent "progress."

"Coming back to England," he said, "I started talking about this. Of course, people didn't grasp it immediately. They said, oh, you really want something primitive, not really the best, something second best. But gradually, people saw the point, that the high technology which is so capital-intensive and produces so few workplaces cannot solve the problem of millions of unemployed people in the rural areas of poor countries."

After talking about these ideas for several years, Schumacher yearned for action. And so, with a few friends, he set up a small

organization. "Never mind how small you start," he exhorted. "Start."

The organization eventually had expert teams on a host of subjects. They advise Third World nations on developing their own intermediate technology and on avoiding the Western trend of high-capital, high-energy, and low-manpower intensities. The Intermediate Technology Development Group has produced a catalogue of agricultural implements that don't make their users dependent on fossil fuels. The catalogue's harrows and plows are either animal- or man-powered. Such tools may seem primitive, but their advantage is that they use only renewable resources. What Schumacher wanted was improved versions of smaller-scale machine-tools that incorporate the best modern technologies, but not their worst features.

The almost impossible problem of enabling the Third World to enter the twentieth century and the age of science and technology with as little pain and as much benefit as possible was the focus of much of Schumacher's work. A typical situation encountered in the Third World occurred in Zambia, where the government has a policy of promoting production of the maximum number of eggs to bridge the protein gap in the diet.

Schumacher visited many egg farmers at the request of the Zambians. Repeatedly, he noticed eggs lying about on the floor and was told that they couldn't be brought to market for lack of proper packaging. Nobody knew how to make egg crates in Zambia. Most of the world's egg crates are made by one large multinational company. Inquiries revealed that the smallest production unit would turn out a million egg crates a month. But the whole requirement of a country like Zambia is only a million a year. When Schumacher asked why smaller equipment couldn't be made to match Zambia's need for the product, he was told it would be uneconomical.

Schumacher disagreed strenuously with the assumption that economies of scale can be

Economy of scale, or bigness for its own sake? Schumacher might well have seen these rows of regimented hens as emblematic of human society in the machine age.

achieved only on a large scale. He believed that "we suffer from an almost universal idolatry of giantism" and it is "therefore necessary to insist on the virtues of smallness, where that applies." In keeping with his philosophy, he urged the Zambians to develop their own small-scale egg crate industry.

That such a small-scale alternative to massive technology could either work or pay has shaken many economists—but the Zambians succeeded. Their plant both makes enough egg trays for any farm in Zambia and creates work for the unemployed. Also, unlike mass production, the manufacturing operation requires that workers exercise some skill. People are operating tools rather than babysitting for machines. And the plant is energy-efficient. No fuel is required to dry the trays, because the African sun does it for nothing.

The Third World nations have the unique opportunity to start fresh by establishing their own intermediate technological solutions to production problems. Milling wheat by hand or powering machines with human or animal energy seems primitive to us. We largely rejected such sources of power with our industrial revolution. But by refusing to mimic the West blindly, industrially underdeveloped nations can choose to retain values that have nourished people spiritually for centuries. Gathering together to work and talk fosters camaraderie and friendship among individuals joining as a team to accomplish a goal. Physical labor, in addition to conserving nonrenewable resources, has the added benefit of keeping people physically fit without elaborate or compulsive leisure-time pursuits. And when people work closely with the animals that serve them, they do not forget that our species is just one of many forms of life on Earth.

Some say that our coal and oil reserves are good for another fifty years. Schumacher believed that instead of just exploiting the last coal seams, we should be finding alternative sources of power and better ways of saving it. Most important of all, we need to discover ways of life that don't depend

In return for their work on this dam, these Indian laborers receive food. Actually, they will be compensated again and again, since the CARE-sponsored dam will make possible increased food production.

wholly on burning billions of tons of energy. He not only called for economies in coal and oil, but also questioned the future of big cities. Schumacher pointed out that where 95 percent of a population live in cities, their energy consumption is inevitably high.

Faced with acres of bricks and mortar, it's difficult to see how we can ever decentralize toward smaller communities. But if Schumacher was right and city life eventually becomes insupportable, a massive exodus from the tangled complexity of urban life to the greater self-sufficiency of small towns and villages will come about naturally.

With the promise of new oil reserves and nuclear energy, many people considered Schumacher's analysis of the facts about energy too pessimistic. But was it? To obtain any kind of energy, we must pay a price, and despite our need for power, Schumacher considered today's price too great. This is especially true with respect to nuclear power.

"In order to make nuclear energy a mass phenomenon," Schumacher said, "you have to go to the breeder reactor. You can't produce nuclear power with natural uranium.

The breeder reactor is an awesome proposition," he contended. "There we produce plutonium, a most terrible substance which nature has never produced, and which is not only extremely radioactive but also extremely poisonous. A lump of plutonium the size of a grapefruit, if dispersed, would be enough to kill all creation."

Schumacher also demonstrated that the nuclear power industry itself is uneconomical. Plutonium will remain active for thousands of years, and all that time it will have to be kept safe. But it's not just the safety factor that worries some economists about nuclear power. So far, it's very largely been what's been called an "energy sink," which means we've spent more energy installing nuclear power stations than we've gotten back as electricity. Schumacher expressed fears that we may never break even.

Constructing a nuclear power station uses up a lot of energy, not just in building it, but in many other processes, like mining and enriching the uranium. These tasks take about five years, and for all that time, the plant is using up rather than putting out

Huge combines harvest grain on an American "factory farm." Although they are less labor-intensive, such operations draw heavily upon rapidly depleting energy resources.

power. After construction, it takes a year to get a power plant going, and then at last it produces power for its fairly brief lifetime of about twenty-five years.

Nor is that yet the end of the story. The materials involved will remain radioactive for at least 250,000 years. Throughout that colossal period, just guarding and keeping these by-products safe will involve some energy loss—if this feat can be accomplished at all.

Offshore oil could also prove to be an energy sink, considering the vast energy cost of tapping it and the potential for destroying aquatic and bird life that depend on clean oceans and air. What's worse, we now need this oil not just for transportation, but also for agriculture. As most people live in cities, energy-intensive agriculture was developed to feed them. With so few left to work the land, farming has become a mechanized, chemicalized process.

Farms in Western nations have become factories, with cheap oil providing fertilizers and herbicides to replace the good husbandry, tender loving care, and high manpower needed in the smaller, old-fashioned organic farms. Schumacher worked on a farm for several years. He believed that if we don't go back to organic methods, there soon won't be enough food to go around. He insisted it's the only practical alternative to today's oil-intensive farming.

Now that cheap oil is a thing of the past, Schumacher would want to see us develop a new mentality, recover from travelmania,

and stop devouring energy as we rush ourselves around the world.

"It's necessary to open one's eyes and one's imagination to the unnecessary complexity that is going on today," he said. "For instance, I go into a shop and I want to buy a very straightforward, simple article called yogurt. I find it's been imported from France."

Now, if we visualize the process whereby French yogurt reaches British or American markets, we begin to glimpse the dimension of the problem. Yogurt has been produced on a large scale in a big factory. There is another large factory to make the containers. The containers have to be crated, packed, loaded onto trucks, carted to a shipping port, and shipped or flown to some central distribution point. Then more trucks have to carry the product to stores, where finally the shopper buys it. All this despite the fact that, although it may not be so convenient or so French, anyone can easily make good yogurt at home.

Our tasks, according to Schumacher, are to bring people back into the production process and to reunite the point of production and the point of consumption. Hitherto, it has been easy to ignore the human factor because it has been overshadowed by power-hungry machines. Now, if the power base becomes questionable, so do the machines, which rely on limited resources. And ecologically, machines are mindless. They cause pollution. The human factor is gentle.

"NOVA" Broadcast History

For each "NOVA" program, the following information is provided: title, description, production company, "NOVA" series number, original air date, and principal production staff.

The Making of a Natural History Film
Behind the scenes of nature filming
BBC   #101   3.3.74

BBC Producer: Mick Rhodes
WGBH Exec. Producer: Michael Ambrosino

Where Did the Colorado Go?
Water and electrical power in southwestern U.S.
WGBH/BBC   #102   3.10.74

BBC Writer/Producer: Simon Campbell-Jones
BBC Assoc. Producer: Ben Shedd
WGBH Exec. Producer: Michael Ambrosino

Whales, Dolphins, and Men
The world of marine mammals—research and industry
BBC   #103   3.17.74

BBC Producer: Simon Campbell-Jones
BBC Exec. Producer: Peter Goodchild
WGBH Exec. Producer: Michael Ambrosino

The Search for Life
Research on the origin of life, based on the Mars landers
WGBH   #104   3.24.74

Writer/Producer: John Angier
Assoc. Producer: Cary Lu
Exec. Producer: Michael Ambrosino

The Last of the Cuiva
Primitive nomadic tribes of Amazon
WGBH/Granada   #105   3.31.74

Granada Producer/Director: Brian Moser
WGBH Exec. Producer: Michael Ambrosino

Strange Sleep
Dramatization of discovery of anesthesia
WGBH   #106   4.7.74

Writer/Producer/Director: Francis Gladstone
Assoc. Producer: Elsa Rassbach
Exec. Producer: Michael Ambrosino

The Crab Nebula
The nebula created by the explosion of a star
BBC   #107   4.14.74

BBC Producer: Alec Nesbitt
BBC Exec. Producer: Peter Goodchild
WGBH Exec. Producer: Michael Ambrosino

Bird Brain—The Mystery of Bird Navigation
Bird navigation
BBC   #108   4.21.74

BBC Producer: Alec Nesbitt
BBC Exec. Producer: Peter Goodchild
WGBH Exec. Producer: Michael Ambrosino

Are You Doing This for Me, Doctor, or Am I Doing This for You?
Experimental procedures on humans
BBC   #109   4.28.74

BBC Producer/Writer: Peter Jones
BBC Exec. Producer: Peter Goodchild
WGBH Exec. Producer: Michael Ambrosino

The First Signs of Washoe
Chimp is taught American Sign Language
WGBH   #110   5.5.74

Writer/Producer: Simon Campbell-Jones
Assoc. Producer: Ben Shedd
Exec. Producer: Michael Ambrosino

The Case of the Midwife Toad
Results of experiments into inheritance of acquired traits made suspect by biologist's suicide
BBC   #111   5.12.74

BBC Writer/Producer: Bruce Norman
BBC Exec. Producer: Robert W. Reid
WGBH Exec. Producer: Michael Ambrosino

Fusion—The Energy Promise
Quest for nuclear fusion for use as a fuel
BBC/WGBH   #112   5.19.74

BBC Producers: Stuart Harris, Mike Jackson
BBC Exec. Producer: Peter Goodchild
WGBH Exec. Producer: Michael Ambrosino

The Mystery of the Anasazi
Vanished Indian civilization in southwestern U.S.
WGBH   #113   5.24.74

Writer/Producer: John Irving
Director: Russ Morash
Exec. Producer: Michael Ambrosino

Why Do Birds Sing?
How and why birds sing
WGBH   #201   11.3.74

Writers: Ben Shedd, Graham Chedd
Director: Ben Shedd
Producers: Ben Shedd, Charles Walcott
Assoc. Producer: Terry Rockefeller
Exec. Producer: Michael Ambrosino

How Much Do You Smell?
Humans' sense of smell
BBC   #202   11.10.74

BBC Producer: Mick Rhodes
BBC Exec. Producer: Peter Goodchild
WGBH Exec. Producer: Michael Ambrosino

The Hunting of the Quark
The world of subatomic particles
BBC   #203   11.17.74

BBC Writer/Producer: David Paterson
BBC Exec. Producer: Bruce Norman
WGBH Exec. Producer: Michael Ambrosino

The Secrets of Sleep
A critical look at the function of sleep
BBC/WGBH   #204   11.24.74

BBC Producer: John Mansfield
BBC Exec. Producer: Peter Goodchild
WGBH Exec. Producer: Michael Ambrosino

Inside the Golden Gate
Ecology of San Francisco Bay
WGBH   #205   12.1.74

Producer: John Angier
Assoc. Producer: Cary Lu
Exec. Producer: Michael Ambrosino

The Men Who Painted Caves
Great Ice Age hunters
BBC   #206   12.8.74

BBC Producer: Christopher La Fontaine
WGBH Exec. Producer: Michael Ambrosino

Red Sea Coral and the Crown-of-Thorns
Crown-of-thorns starfish and the ecology of coral reefs
BBC   #207   12.15.74

BBC Producer/Writer: Alec Nesbitt
BBC Exec. Producer: Bruce Norman
WGBH Exec. Producer: Michael Ambrosino

War from the Air
Development of bombs and bombing strategy
WGBH   #208   1.5.75

Writer/Producer: Francis Gladstone
Assoc. Producer: Elsa Rassbach
Exec. Producer: Michael Ambrosino

What Time Is Your Body?
Explanation of biorhythms
BBC   #209   1.12.75

BBC Producer: Dominic Flessati
BBC Exec. Producer: Peter Goodchild
WGBH Exec. Producer: Michael Ambrosino

The Rise and Fall of DDT
Critical look at DDT
BBC #210 1.19.75

BBC Writers/Producers: Alec Nesbitt, Robin
   Bottle
WGBH Exec. Producer: Michael Ambrosino

**Take the World from Another Point of
View**
*Contrasting portraits of two scientists, Richard
Feynman and Richard Lewontin*
WGBH #211 2.2.75

Producer: Francis Gladstone
Assoc. Producer: Elsa Rassbach
R. Lewontin Film Director: Francis Gladstone
R. Feynman Film Director: Duncan Dallas
Exec. Producer: Michael Ambrosino

The Lysenko Affair
*Dramatization of career of T. D. Lysenko, Rus-
sian biologist*
BBC/WGBH #212 2.9.75

BBC Producer/Director: Peter Jones
BBC Writer: John Wiles (based on books by
   Z. Medvedev and D. Joravsky)
BBC Exec. Producer: Peter Goodchild
WGBH Exec. Producer: Michael Ambrosino

The Tuaregs
*Anthropological study of a Sahara Desert
family*
Granada #213 2.16.75

Granada Producer/Director: Charley Nairn
WGBH Exec. Producer: Michael Ambrosino

The Plutonium Connection
A student's attempt to build an atom bomb
WGBH #214 3.9.75

Producer/Writer: John Angier
Assoc. Producer: Barbara Gullahorn-Holecek
Exec. Producer: Michael Ambrosino

The Other Way
*Alternative technology; the philosophy of E. F.
Schumacher*
BBC #215 3.16.75

BBC Producer: John Mansfield
BBC Exec. Producer: Peter Goodchild
WGBH Exec. Producer: Michael Ambrosino

The Lost World of the Maya
Lost Central American Indian culture
BBC #216 3.30.75

With J. Eric, S. Thompson, and Ian Graham
BBC Writer: Magnus Magnusson
BBC Producer: David Collison
BBC Exec. Producer: Paul Johnstone
WGBH Exec. Producer: Michael Ambrosino

Will the Fishing Have to Stop?
Ocean productivity
WGBH #217 4.6.75

Producer/Director: Ben Shedd
Writers: Ben Shedd, Graham Chedd
Assoc. Producer: Terry Rockefeller
Exec. Producer: Michael Ambrosino

Predictable Disaster
*Progress in earthquake
prediction*
WGBH #301 1.4.76

Writer/Producer: John Angier
Assoc. Producers: Barbara Gullahorn-
   Holecek, Paula Apsell
Exec. Producer: Michael Ambrosino

Joey
*Biography of a spastic institutionalized and una-
ble to communicate most of his life*
BBC #302 1.11.76

BBC Producer: Brian Gibson
BBC Exec. Producer: Peter Goodchild
WGBH Exec. Producer: Michael Ambrosino

Meditation and the Mind
Critical look at TM
BBC/WGBH #303 1.18.76

BBC Producer/Writer: Graham Massey
BBC Exec. Producer: Peter Goodchild
WGBH Exec. Producer: Michael Ambrosino

The Planets
*Discoveries about our solar system, learned
from space exploration*
BBC #304 1.25.76

BBC Producer/Writer: Edward Goldwyn
BBC Exec. Producer: Peter Goodchild
WGBH Exec. Producer: Michael Ambrosino

A Desert Place
Desert ecology
WGBH #305 2.1.76

Producers/Directors/Writers: John Borden,
   Neil Goodwin
Exec. Producer: Michael Ambrosino

A Small Imperfection
*Spina bifida—congenital abnormality of the
central nervous system*
BBC #306 2.8.76

BBC Producer: Robert Reid
BBC Exec. Producer: Peter Goodchild
WGBH Exec. Producer: Michael Ambrosino

Ninety Degrees Below
Antarctica
WGBH #307 2.15.76

Producer/Writer: David Kuhn
Exec. Producer: Michael Ambrosino

The Race for the Double Helix
Discovery of DNA
WGBH #308 2.22.76

Featuring Isaac Asimov
VSM Productions Producers: Ronnie Four-
   acre, Peter Shaw
VSM Productions Director: Ronnie Fouracre
WGBH Producer/Writer: Graham Chedd
WGBH Director: David Kuhn
WGBH Exec. Producer: Michael Ambrosino

The Renewable Tree
*Farming techniques to meet demands for
paper*
WGBH #309 3.7.76

Producer/Director/Writer: Ben Shedd
Assoc. Producer: Jane Duderstadt
Exec. Producer: Michael Ambrosino

The Williamsburg File
Archeologist-led tour of Williamsburg
BBC/WGBH #310 3.14.76

BBC Writers: Antonia Benedek, Ivor Noel
   Hume
BBC Director: Antonia Benedek
BBC Exec. Producer: Paul Johnstone
WGBH Exec. Producer: Michael Ambrosino

The Overworked Miracle
Problem of resistance to antibiotics
BBC #311 3.21.76

BBC Producer: Christopher Riley
BBC Exec. Producer: Peter Goodchild
WGBH Exec. Producer: Michael Ambrosino

The Transplant Experience
*Story of Norman Shumway, prominent heart
transplant surgeon*
BBC/WGBH #312 4.11.76

BBC Producer/Writer: Dominic Flessati
BBC Exec. Producer: Peter Goodchild
WGBH Exec. Producer: Michael Ambrosino

The Underground Movement
*Study of life underground: foxes, badgers,
worms*
WGBH/Polytel International #313 4.18.76

Polytel International Producer: Suzanne Gibbs
WGBH Exec. Producer: Michael Ambrosino

Hunters of the Seal
Traditional Netsilik Eskimo life meets 20th century
WGBH #314 5.2.76

Producer/Writer: Barbara Gullahorn-Holecek
Exec. Producer: John Angier

Benjamin
Early child development
BBC #315 5.9.76

BBC Producers: Graham Massey, Robin Bates
BBC Exec. Producer: Peter Goodchild
WGBH Exec. Producer: John Angier

The Woman Rebel
Dramatization of life of Margaret Sanger, birth control advocate
WGBH #316 5.23.76

Producer/Writer: Francis Gladstone
Assoc. Producer: Elsa Rassbach
Exec. Producer: Michael Ambrosino

The Death of a Disease
Smallpox eradication
WGBH #317 6.6.76

Producer/Writer: Paula Apsell
Assoc. Producer: Sari Sapir
Exec. Director: John Angier

Inside the Shark
Life-style and biology of the shark
BBC/WGBH #318 6.13.76

BBC Producer/Writer: Tony Edwards
BBC Exec. Producer: Peter Goodchild
WGBH Exec. Producer: John Angier

The Genetic Chance
Prenatal diagnosis of hemophilia and resultant ethical issues
BBC/WGBH #319 6.20.76

BBC Producer/Writer: Robert W. Reid
BBC Exec. Producer: Peter Goodchild
WGBH Exec. Producer: John Angier

The Case of the Bermuda Triangle
Critical look at supposed supernatural events
BBC #320 6.27.76

BBC Producer: Graham Massey
BBC Exec. Producer: Peter Goodchild
WGBH Exec. Producer: John Angier

Hitler's Secret Weapon
Story of rocket weapon developers in Nazi Germany
WGBH/Long Beach #401 1.5.77

Long Beach Exec. Producer: Don G. Gill
WGBH Producers: Francis Gladstone, Patrick Griffin
WGBH Writer: Francis Gladstone
WGBH Exec. Producer: John Angier

The Hot-Blooded Dinosaurs
New theories from paleontologists about dinosaurs
WGBH/BBC #402 1.12.77

BBC Producers: Robin Bates, Robin Brightwell
WGBH Exec. Producer: John Angier

What Price Coal?
Energy needs vs. environment, health, and safety
WGBH #403 1.19.77

Producer/Writer/Director: Francis Gladstone
Assoc. Producer: Elvida Abella
Exec. Producer: John Angier

The Sunspot Mystery
Possible link between weather patterns and sunspots
WGBH #404 2.2.77

Producers/Writers: Ben Shedd, Graham Chedd
Director: Ben Shedd
Exec. Producer: John Angier

The Plastic Prison
Children born with immune deficiency
WGBH/Yorkshire #405 2.9.77

Yorkshire Producers: Simon Welfare, Kevin Sim
Yorkshire Director: David Green
Yorkshire Exec. Producer: Duncan Dallas
WGBH Producer/Writer/Director: David Kuhn
WGBH Exec. Producer: John Angier

Incident at Brown's Ferry
Investigation of nuclear power plant safety
WGBH #406 2.23.77

Producer/Writer/Director: Robert Richter
Assoc. Producers: Elvida Abella, Marian White
Exec. Producer: John Angier

Bye Bye Blackbird
Habits of common birds and farmers' attempts to control them
WGBH/BBC #407 3.2.77

BBC Producer: Peter Bale
WGBH Exec. Producer: John Angier

The Pill for the People
Why birth control pill doesn't live up to its creators' expectations
WGBH/BBC #408 3.9.77

BBC Producer: Robin Brightwell
BBC Exec. Producer: Simon Campbell-Jones
WGBH Exec. Producer: John Angier

The Gene Engineers
Investigation of recombinant DNA
WGBH #409 3.16.77

Producers/Writers: Graham Chedd, Paula Apsell
Exec. Producer: John Angier

The Human Animal
Arguments for and against sociobiology
WGBH/BBC #410 3.30.77

BBC Producer: Peter Jones
WGBH Exec. Producer: John Angier

The Wolf Equation
Wolf packs in natural and controlled settings
WGBH/Peace River Films #411 4.6.77

Peace River Films Producer: Neil Goodwin
Peace River Films Directors: Neil Goodwin, John Borden
Peace River Films Writer: John Lord
WGBH Producer/Director: Elvida Abella
WGBH Writers: Elvida Abella, Graham Chedd
WGBH Exec. Producer: John Angier

Dawn of the Solar Age
Examination of solar energy
WGBH/BBC #412 4.13.77

BBC Producer: Edward Goldwyn
WGBH Exec. Producer: John Angier

The Business of Extinction
International trade in endangered species
WGBH #413 4.20.77

Producer/Writer/Director: Barbara Gullahorn-Holecek
Assoc. Producer: Sari Sapir
Exec. Producer: John Angier

The Red Planet
History of observation of Mars through 1976
WGBH/BBC   #414   5.4.77

BBC Producer: Tony Edwards
BBC Exec. Producer: Simon Campbell-Jones
WGBH Exec. Producer: John Angier

The Tongues of Men, Part I
The history of language
WGBH/BBC   #415   5.11.77

BBC Writer: George Steiner
BBC Producer: Christopher Martin
WGBH Exec. Producer: John Angier

The Tongues of Men, Part 2
The history of language
WGBH/BBC   #416   5.18.77

BBC Writer: George Steiner
BBC Producer: Christopher Martin
WGBH Exec. Producer: John Angier

Linus Pauling: Crusading Scientist
Life story of double Nobel Prize-winning scientist
WGBH   #417   6.1.77

Producer/Writer/Director: Robert Richter
Assoc. Producer: Marian White
Exec. Producer: John Angier

Across the Silence Barrier
Varieties of deaf experience
WGBH   #418   6.22.77

Producer/Director/Writer: Francis Gladstone
Assoc. Producer: Joan Freeman
Exec. Producer: John Angier

The New Healers
Alternative health care in Tanzania and the U.S.
WGBH   #419   6.29.77

Producer/Writer: Paula Apsell
Exec. Producer: John Angier

The Green Machine
Why and how plants grow
BBC/WGBH   #420   1.11.78

BBC Producer/Writer: Tony Edwards
BBC Exec. Producer: Simon Campbell-Jones
WGBH Exec. Producer: John Angier

In the Event of Catastrophe
Is nuclear war survivable?
WGBH   #501   1.4.78

Producer/Writer/Director: Barbara Gullahorn-Holecek
Assoc. Producer: Elvida Abella
Exec. Producer: John Angier

Blueprints in the Bloodstream
White cell "types" may determine disease susceptibility
BBC/WGBH   #502   1.18.78

BBC Producer/Writer: Vivienne King
BBC Exec. Producer: Simon Campbell-Jones
WGBH Exec. Producer: John Angier

One Small Step
History of humans in space
WGBH   #503   1.25.78

Producer/Writer: Graham Chedd
Assoc. Producer: Marian White
Exec. Producer: John Angier

The Final Frontier
Post-Apollo NASA and the future of humans in space
WGBH   #504   2.1.78

Producer/Writer/Director: Graham Chedd
Assoc. Producer: Marian White
Exec. Producer: John Angier

Bamiki Bandula—Children of the Forest
Anthropology of the pygmies of Zaire
Kevin Duffy Productions/
WGBH   #505   2.15.78

Kevin Duffy Productions Producer/Writer/Director: Kevin Duffy
Kevin Duffy Productions Assoc. Producer: Gerlinde Konig
WGBH Exec. Producer: John Angier

The Trial of Denton Cooley
Dramatization of trial of heart transplant surgeon
WGBH   #506   2.22.78

Producer/Writer/Director: Francis Gladstone
Assoc. Producer: Joan Freeman
Exec. Producer: John Angier

The Great Wine Revolution
Scientific revolutions in wine making
BBC/WGBH   #507   3.1.78

BBC Producer/Writer: Dominic Flessati
BBC Exec. Producer: Simon Campbell-Jones
WGBH Exec. Producer: John Angier

The Case of the Ancient Astronauts
Critical look at von Däniken theories about alien intelligence
BBC/WGBH   #508   3.8.78

BBC Producer: Graham Massey
BBC Exec. Producer: Simon Campbell-Jones
WGBH Assoc. Producer: Betsy Anderson
WGBH Exec. Producer: John Angier

The Mind Machines
Can computers mimic the human mind?
WGBH   #509   3.22.78

Producer/Writer/Director: Paula Apsell
Assoc. Producer: Roy Gould
Exec. Producer: John Angier

Icarus' Children
Winning the Kremer prize for human-powered flight
BBC/WGBH   #510   3.29.78

BBC Producer/Writer: Simon Campbell-Jones
BBC Associate Producers: Robin Bates, Stephen Rose
MacCready footage: Shedd Productions Inc.
Shedd Productions Producer: Jacqueline Phillips-Shedd
Shedd Productions Director: Ben Shedd
WGBH Exec. Producer: John Angier

Still Waters
A year in the life of a beaver pond
WGBH/Peace River Films   #511   4.12.78

Peace River Films Producers/Writers/Directors: Neil Goodwin, John Borden
WGBH Exec. Producer: John Angier

Battle for the Acropolis
Conservation proposals to save Acropolis from pollution
BBC/WGBH   #512   4.19.78

BBC Producer/Writer: Roy Davies
BBC Exec. Producer: Bruce Norman
WGBH Exec. Producer: John Angier

The Road to Happiness
Henry Ford's life and times
WGBH   #513   5.3.78

Producers/Writers: Francis Gladstone, Patrick Griffin
Assoc. Producer: Elvida Abella
Exec. Producer: John Angier

Light of the 21st Century
Lasers
BBC/WGBH   #514   5.10.78

BBC Producer/Writer: Tony Edwards
BBC Exec. Producer: Simon Campbell-Jones
WGBH Exec. Producer: John Angier

The Insect Alternative
Alternative to pesticides
WGBH   #515   5.24.78

Producer/Writer/Director: Graham Chedd
Assoc. Producer: Marian White
Exec. Producer: John Angier

The Desert's Edge
*Process of desertification and possible
countermeasures*
WGBH/BBC   #516   5.31.78

BBC Producer/Director: Richard Taylor
BBC Exec. Producer: Simon Campbell-Jones
WGBH Producer: Sari Sapir
WGBH Exec. Director: John Angier

The Tsetse Trap
Attempts to control tsetse fly
BBC/WGBH   #517   6.7.78

BBC Producer/Writer: Edward Goldwyn
BBC Exec. Producer: Simon Campbell-Jones
WGBH Exec. Producer: John Angier

Memories from Eden
Portrait of the modern zoo
WGBH   #518   6.14.78

Producer/Writer/Director: Barbara
    Gullahorn-Holecek
Assoc. Producer: Joan Freeman
Exec. Producer: John Angier

A Whisper from Space
*History of the universe; implications of 1965
discovery of the 3° cosmic background*
BBC/WGBH   #519   6.21.78

Hosted by Philip Morrison
BBC Producer: Peter Jones
BBC Exec. Producer: Simon Campbell-Jones
WGBH Exec. Producer: John Angier

Alaska: The Closing Frontier
*Controversy between conservationists and
developers*
WGBH   #520   6.28.78

Producer/Writer: Paula Apsell
Assoc. Producer: Linda Harrar
Exec. Producer: John Angier

Black Tide
*Examination of largest
oil spill in history*
WGBH   #601   1.4.79

Producer/Writer/Director: Graham Chedd
Assoc. Producer: Marian White
Exec. Producer: John Angier

The Long Walk of Fred Young
*Life of Frederick Young, nuclear physicist and
Native American*
WGBH/BBC   #602   1.11.79

BBC Producer/Writer: Michael Barnes
BBC Assoc. Producer: Elvida Abella
BBC Exec. Producer: Lesley Megahey
WGBH Exec. Producer: John Angier

A World of Difference
*Portrait of B. F. Skinner, eminent behavioral
psychologist*
WGBH   #603   1.18.79

Producer/Writer/Director: Veronica Young
Assoc. Producer: Patti Polisar
Exec. Producer: John Angier

Cashing in on the Ocean
*International battle over ocean's valuable
manganese nodules*
BBC/WGBH   #604   2.1.79

BBC Producer/Writer: Stuart Harris
BBC Exec. Producer: Simon Campbell-Jones
WGBH Exec. Producer: John Angier

Patterns from the Past
*Anthropology of Q'eros Indians of Peruvian
Andes*
Cohen Films/WGBH   #605   2.8.79

Based on "Q'eros: The Shape of Survival," a
    film by John Cohen
Cohen Films Producers/Writers: John Cohen,
    Sari Sapir
WGBH Exec. Producer: John Angier

The Invisible Flame
Prospects for wide use of hydrogen fuel
BBC/WGBH   #606   2.22.79

BBC Producer/Writer: Alec Nesbitt
BBC Exec. Producer: Simon Campbell-Jones
WGBH Assoc. Producer: Elvida Abella
WGBH Exec. Producer: John Angier

The End of the Rainbow
Research on controlled nuclear fusion for fuel
WGBH   #607   3.1.79

Producer/Writer: Brian Kaufman
Exec. Producer: John Angier

The Beersheva Experiment
*Israeli medical school's attempt to reverse
trend toward specialization*
WGBH/BBC   #608   3.8.79

BBC Producer/Writer: Edward Goldwyn
BBC Exec. Producer: Simon Campbell-Jones
WGBH Producer/Writer: Sari Sapir
WGBH Exec. Producer: John Angier

Einstein
*The scientist and humanist, as seen through
archival footage*
WGBH   #609   3.15.79

Producer/Writer: Patrick Griffin
Assoc. Producer: Alvida Abella
Exec. Producer: John Angier

The Keys to Paradise
Discovery of painkilling endorphins
WGBH/BBC   #610   3.29.79

BBC Producer/Writer: Dick Gilling
BBC Exec. Producer: Edward Goldwyn
WGBH Assoc. Producer: Elvida Abella
WGBH Exec. Producer: John Angier

A Plague on Our Children
*Examination of use and disposal of toxic
substances*
WGBH   #611   10.2.79

Producer/Writer/Director: Robert Richter
Assoc. Producer: Robert Zalisk
Exec. Producer: John Angier

Life on a Silken Thread
*Behavior of spiders, seen with special
photography*
WGBH/German TV   #612   10.9.79

Adapted from a Maran Film Production for
    Suddeutscher Rundfunk
German TV Producer: Jorg Dattler
German TV Writer/Director: Horst Stern
WGBH Writer/Producer: Sari Sapir
WGBH Exec. Producer: John Angier

Sweet Solutions
Sugar as a fuel
BBC/WGBH   #613   10.17.79

BBC Writer/Producer: Vivienne King
BBC Exec. Producer: Simon Campbell-Jones
WGBH Writer/Producer: Theodore
    Bogosian
WGBH Exec. Producer: John Angier

Race for Gold
*Science and technology in the development
of Olympic-caliber athletes*
WGBH   #614   10.30.79

Producer/Writer/Director: Paula Apsell
Assoc. Producer: Joan Freeman
Exec. Producer: John Angier

All Part of the Game
Sports injuries and new medical treatments
WGBH   #615   11.6.79

Producer/Writer/Director: Paula Apsell
Assoc. Producer: Joan Freeman
Exec. Producer: John Angier

India: Machinery of Hope
Development of small-scale, decentralized industry in Third World nations
WGBH #616 11.20.79

Producer/Writer/Director: Barbara Gullahorn-Holecek
Assoc. Producer: Marilyn Hornbeck
Exec. Producer: John Angier

The Bridge That Spanned the World
Study of early iron technology
BBC/WGBH #617 12.4.79

BBC Writer/Producer: Robert Bottle
BBC Assoc. Producer: Robert Symes
BBC Exec. Producer: Karl Sabbagh
WGBH Writer/Producer: Brian Kaufman
WGBH Assoc. Producer: Joan Freeman
WGBH Exec. Producer: Mick Rhodes

Termites to Telescopes
Philip Morrison lectures on nature of civilization
BBC/WGBH #618 12.22.79

2nd BBC Jacob Bronowski Memorial Lecture, given by Philip Morrison at the Massachusetts Institute of Technology
BBC Producer: Peter Jones
BBC Director: David Atwood
WGBH Writer/Producer: Brian Kaufman
WGBH Assoc. Producer: Joan Freeman
WGBH Exec. Producer: Mick Rhodes

Blindness: Five Points of View
Five individuals' experiences with blindness
WGBH #619 12.18.79

Producer/Writer/Director: Veronica Young
Exec. Producer: John Angier

The Elusive Illness: Hepatitis
Hepatitis research
WGBH #701 1.15.80

Producer/Writer: Robin Bates
Assoc. Producer: Theodore Bogosian
Exec. Producer: Mick Rhodes

A Is for Atom, B Is for Bomb
Portrait of Edward Teller, "Father of the H-bomb"
WGBH #702 1.22.80

Producer/Writer/Director: Brian Kaufman
Assoc. Producer: Marilyn Hornbeck
Exec. Producer: Mick Rhodes

Living Machines
How bodies work
BBC/WGBH #703 2.5.80

BBC Producer/Writer: Robin Bates
BBC Exec. Producer: Simon Campbell-Jones
WGBH Assoc. Producer: Theodore Bogosian
WGBH Exec. Producer: John Angier

Portrait of a Killer
Close-up photographic look at hypertension and heart disease
WGBH/German TV #704 2.19.80

German TV Producers: Bo G. Erikson, Carl O. Lofman
WGBH Producer/Writer: Sari Sapir
WGBH Exec. Producer: Mick Rhodes

Umealit: The Whale Hunters
Clash between conservationists' efforts to protect bowhead whales and Eskimo tradition of hunting them
WGBH #705 3.4.80

Producer/Writer/Director: John Angier
Assoc. Producers: Elvida Abella, Sara Jane Neustadtl
Exec. Producer: Mick Rhodes

The Safety Factor
Examination of air travel safety
WGBH/BBC #706 3.11.80

BBC Producer/Director: Alec Nesbitt
BBC Exec. Producer: Simon Campbell-Jones
WGBH Producer/Director/Writer: Paula Apsell
WGBH Assoc. Producer: Linda Harrar
WGBH Exec. Producer: Mick Rhodes

A Mediterranean Prospect
International effort to clean up Mediterranean
BBC #707 3.18.80

BBC Producer/Writer: Peter Jones
BBC Exec. Producer: Simon Campbell-Jones
WGBH Producer: Joan Freeman
WGBH Exec. Producer: Mick Rhodes

Mr. Ludwig's Tropical Dreamland
Billionaire Daniel Ludwig's agricultural experiment in Brazil
BBC/WGBH #708 3.25.80

BBC Producer/Writer: Vivienne King
BBC Executive Producer: Simon Campbell-Jones
WGBH Producer/Writer: Joan Freeman
WGBH Exec. Producer: Mick Rhodes

The Pinks and the Blues
Impact of environment in establishing sex roles
WGBH #709 9.30.80

Producer/Writer/Director: Veronica Young
Exec. Producer: Mick Rhodes

The Cancer Detectives of Lin Xian
Efforts to prevent cancer in China
BBC/WGBH #710 10.7.80

BBC Producer/Writer: Edward Goldwyn
BBC Exec. Producer: Simon Campbell-Jones
WGBH Producer: Robert Zalisk
WGBH Assoc. Producer: Pat Kent
WGBH Exec. Producer: Mick Rhodes

The Sea Behind the Dunes
Coastal ecology
WGBH/Peace River Films #711 10.14.80

Peace River Films Producers/Writers/Directors: John Borden, Neil Goodwin
WGBH Exec. Producer: Mick Rhodes

Do We Really Need the Rockies?
Prospects for extraction of shale oil
WGBH #712 10.28.80

Producer/Writer: Robin Bates
Assoc. Producer: Theodore Bogosian
Exec. Producer: John Mansfield

The Big IF
Interferon, anticancer drug
BBC/WGBH #713 11.4.80

BBC Producer/Writer: Vivienne King
BBC Exec. Producer: Simon Campbell-Jones
WGBH Producer: Theodore Bogosian
WGBH Exec. Producer: John Mansfield

Voyager: Jupiter and Beyond
Mission to Jupiter: first reports from Voyager I
BBC/WGBH #714 11.11.80

BBC Producer/Writer: Fisher Dilke
BBC Exec. Producer: Simon Campbell-Jones
WGBH Producer: Robert Zalisk
WGBH Exec. Producer: John Mansfield

The Wizard Who Spat on the Floor
Biography of Thomas Edison
BBC/WGBH #715 11.18.80

BBC Producer: Robert Vas
BBC Exec. Producer: Peter Goodchild
WGBH Exec. Producer: John Mansfield

The Water Crisis
Pollution and U.S. water resources
WGBH   #716   11.25.80

Producer/Writer/Director: Sari Sapir
Exec. Producer: John Mansfield

Moving Still
History of still and film photography
BBC/WGBH   #717   12.2.80

BBC Producer/Writer: Christopher Haws
BBC Exec. Producer: Simon Campbell-Jones
WGBH Exec. Producer: John Mansfield

A Touch of Sensitivity
Exploration of sense of touch
BBC/WGBH   #718   12.9.80

BBC Producer/Writer: Stephen Rose
BBC Exec. Producer: Simon Campbell-Jones
WGBH Producer: Pat Kent
WGBH Exec. Producer: John Mansfield

The Red Deer of Rhum
Behavior of deer on Scottish island of Rhum
BBC/WGBH   #719   12.23.80

BBC Producer: Peter Jones
BBC Exec. Producer: Simon Campbell-Jones
WGBH Producer: Robert Zalisk
WGBH Assoc. Producer: Pat Kent
WGBH Exec. Producer: John Mansfield

It's about Time
Whimsical investigation into the nature of time
BBC/WGBH   #720   12.30.80

Presented by Dudley Moore; appearance by
    Isaac Asimov
BBC Producer/Writer: Chris Haws
BBC Exec. Producer: Karl Sabbagh
WGBH Exec. Producer: John Mansfield

The Doctors of Nigeria
Traditional and modern medical methods work together
WGBH   #801   1.6.81

Producer/Writer/Director: Barbara
    Gullahorn-Holecek
Assoc. Producer: Linda Harrar
Exec. Producer: John Mansfield

Message in the Rocks
Formation of Earth and beginnings of life according to petrology
BBC/WGBH   #802   1.20.80

BBC Exec. Producer: Simon Campbell-Jones
WGBH Producer: Linda Harrar
WGBH Exec. Producer: John Mansfield

The Dead Sea Lives
Politics and ecology of resources of Dead Sea
BBC/WGBH   #803   1.27.81

BBC Producer/Writer: Christopher Riley
BBC Exec. Producer: Simon Campbell-Jones
WGBH Assoc. Producer: Pat Kent
WGBH Exec. Producer: John Mansfield

Anatomy of a Volcano
1980 eruption of Mt. St. Helens
BBC/WGBH   #804   2.10.81

BBC Producer/Writer: Stuart Harris
BBC Assoc. Producer: John Lynch
BBC Exec. Producer: Simon Campbell-Jones
WGBH Assoc. Producer: Theodore Bogosian
WGBH Exec. Producer: John Mansfield

The Science of Murder
Scientific research in detection of murderers and prevention of homicide
WGBH   #805   2.17.81

Producer/Writer/Director: Brian Kaufman
Assoc. Producer: Linda Harrar
Exec. Producer: John Mansfield

The Malady of Health Care
Comparison of American and British health care systems
WGBH   #806   2.24.81

Producer/Writer/Director: Barbara
    Gullahorn-Holecek
Exec. Producer: John Mansfield

Beyond the Milky Way
What astronomers have learned about galaxies
BBC/WGBH   #807   3.3.81

BBC Producer/Writer: Alec Nesbitt
BBC Exec. Producer: Simon Campbell-Jones
WGBH Assoc. Producer: Robert Zalisk
WGBH Exec. Producer: John Mansfield

The Asteroid and the Dinosaur
New theories on extinction of dinosaurs
WGBH   #808   3.10.81

Producer/Writer/Director: Robin Bates
Exec. Producer: John Mansfield

Animal Olympians
Beauty and power of animals juxtaposed with movement of Olympic athletes
WGBH   #809   3.17.81

Producer/Writer: Jeffrey Boswall
Assoc. Producer: Pat Kent
Exec. Producer: John Mansfield

Resolution on Saturn
First trip to Saturn: Voyager I *continues its journey*
BBC/WGBH   #810   8.25.81

BBC Producer/Writer/Director: Fisher Dilke
BBC Exec. Producer: Simon Campbell-Jones
WGBH Exec. Producer: John Mansfield

Computers, Spies, and Private Lives
Potential for invasion of privacy by computers
WGBH   #811   9.27.81

Producers/Writers/Directors: Theodore Bo-
    gosian, Linda Harrar
Exec. Producer: John Mansfield

Why America Burns
Fire epidemic in the U.S.
WGBH   #812   10.4.81

Producer/Writer/Director: Brian Kaufman
Exec. Producer: John Mansfield

The Great Violin Mystery
Violin building: what makes a violin "great"?
WHA-TV, Madison   #813   10.11.81

WHA-TV Producer/Director: Barry Stoner
WHA-TV Writer: William Jordan
WHA-TV Exec. Producer: Linda Baldwin
WHA-TV Assoc. Producer: Bebe Nixon
WGBH Exec. Producer: John Mansfield

Cosmic Fire
X-ray astronomy
BBC/WGBH   #814   10.18.81

BBC Producer: Dick Gilling
BBC Exec. Producer: Simon Campbell-Jones
WGBH Producer: Robert Zalisk
WGBH Exec. Producer: John Mansfield

Locusts: War without End
Locust control and pesticides
WGBH   #815   10.25.81

Producer/Writer/Director: Mick Rhodes
Exec. Producer: John Mansfield

Did Darwin Get It Wrong?
Controversy between theories of evolution and creation
BBC/WGBH   #816   11.1.81

BBC Producer/Writer/Director: Alec Nesbitt
BBC Exec. Producer: Simon Campbell-Jones
WGBH Assoc. Producer: Theodore Bogosian
WGBH Exec. Producer: John Mansfield

Artists in the Lab
Use of computers and lasers in art
WGBH   #817   11.15.81

Producer/Writer/Director: Barbara
    Gullahorn-Holecek
Exec. Producer: John Mansfield

Notes of a Biology Watcher: A Film with Lewis Thomas
Exploration of two biological principles: connectedness and separateness
WGBH   #818   11.22.81

Producer/Director: Robin Bates
Writers: Robin Bates, Lewis Thomas
Exec. Producer: John Mansfield

City Spaces, Human Places
A view of urban design and spatial relations in city plazas
William Whyte   #819   11.29.81

William Whyte Producers: William Whyte, Pat
    Kent
William Whyte Writer: William Whyte
WGBH Exec. Producer: John Mansfield

Twins
Scientific research on twins—genetics and behavior
CBC/WGBH   #820   12.6.81

CBC Producer/Director: Heather Cook
CBC Writer: Bruce Martin
WGBH Producer/Director: Linda Harrar
WGBH Exec. Producer: John Mansfield

Salmon on the Run
The changing nature of salmon industry in Northwest and disputes over Indian rights
Jim Mayer   #901   1.10.82

Producers: Lynn Adler, Steve Christiansen,
    Jim Mayer
Writer: Jim Mayer
Directors: Steve Christiansen, Jim Mayer
WGBH Exec. Producer: John Mansfield

Test-Tube Babies: A Daughter for Judy
In vitro fertilization and America's first test-tube baby
TVS/WGBH   #902   1.17.82

TVS Producer/Writer: Peter Williams
TVS Director: Gordon Stevens
WGBH Exec. Producer: John Mansfield

A Field Guide to Roger Tory Peterson
Portrait of famous birdwatcher and environmentalist
CBC   #903   1.24.82

CBC Producer/Director: James Murray
CBC Writer: Bruce Martin
CBC Exec. Producer: Nancy Archibald
WGBH Exec. Producer: John Mansfield

The Hunt for the Legion Killer
Medical detective story: Legionnaire's Disease
BBC/WGBH   #904   1.31.82

BBC Producer/Writer: Dominic Flessati
BBC Exec. Producer: Graham Massey
WGBH Producer: Linda Harrar
WGBH Exec. Producer: John Mansfield

Finding a Voice
Communication aids for cerebral palsy victims
BBC/WGBH   #905   2.7.82

BBC Producer: Martin Freeth
BBC Writer: Dick Boydell
BBC Director: Sheila Hayman
BBC Exec. Producer: Graham Massey
WGBH Exec. Producer: John Mansfield

The Television Explosion
New technologies in television
WGBH   #906   2.14.82

Producer/Writer: Thea Chalow
Director: Richard Heller
Assoc. Producers: Mimi Curran, Pat Kent
Exec. Producer: John Mansfield

Life: Patent Pending
The new industry of recombinant DNA
WGBH   #907   2.28.82

Producer/Writer/Director: Robert Zalisk
Assoc. Producer: Eileen Zalisk
Exec. Producer: John Mansfield

Palace of Delights
Frank Oppenheimer's famed science museum
Else, Couturie & Korty, Inc.   #908   3.7.82

Else, Couturie & Korty, Inc., Producers/Writers: Jon Else, Bill Couturie
Else, Couturie & Korty, Inc. Director: Jon Else
Else, Couturie & Korty, Inc. Assoc. Producer:
    Kathryn White
WGBH Exec. Producer: John Mansfield

Animal Imposters
Animals' use of camouflage
Peace River Films   #909   3.14.82

Peace River Films Producers/Writers/Directors: John Borden, Neil Goodwin
WGBH Exec. Producer: John Mansfield

Aging: the Methuselah Syndrome
Medicine's attempt to study and combat the aging process
WGBH   #910   3.28.82

Producer/Writer/Director: Sari Sapir
Exec. Producer: John Mansfield

The Case of the UFOs
BBC/WGBH   #911   10.12.82

BBC Producer: John Groom
WGBH Exec. Producer: John Mansfield

The Fragile Mountain
Independent   #912   10.19.82

Independent Producer: Sandra Nichols
WGBH Exec. Producer: John Mansfield

Young Scientists (working title)
WGBH   #914   11.16.82

Producer: Linda Harrar
Assoc. Producer: Peter Argentine
Exec. Producer: John Mansfield

The Cobalt Blues
WGBH   #915   11.23.82

Producer: Theodore Bogosian
Exec. Producer: John Mansfield

Mississippi (working title)
WGBH   #916   11.30.82

Producer: Barbara Gullahorn-Holecek
Assoc. Producer: Betsy Anderson
Exec. Producer: John Mansfield

Whale Watch
Lazifilm Production   #917   12.7.82

WGBH Exec. Producer: John Mansfield

High-Speed Trains (working title)
WGBH   #918   12.14.82

Producer: Theodore Bogosian
Exec. Producer: John Mansfield

Transcripts of all "NOVA" programs
are available from WGBH,
125 Western Avenue,
Boston, Massachusetts 02134.

Illustration Credits

The line drawings that appear at the beginning of each chapter and in the broadcast history are by Mark Fisher unless otherwise noted.

1, 2, 3, 4 ("NOVA" opening) animation design by Paul Souza, technical direction and animation by David M. Geshwind, additional animation by Becky Allen and Thad Beier, generated at New York Institute of Technology, Computer Graphics Laboratory

6 (left) Dan McCoy, Rainbow; (right) JPL/NASA

7 (top) Kenneth Garrett, Woodfin Camp; (center right) Dan McCoy, Rainbow

8 Paul Souza for WGBH

9 (top, "NOVA" opening) cinematography by Boyd Estus, design by Gene Mackles; (bottom) Michael Lutch

10 (top) Boyd Estus; (bottom) Michael Lutch for WGBH

11 Michael Steinberg for WGBH

12, 13 (all photos) Michael Lutch for WGBH

15 MIT

16 Dan McCoy, Rainbow

18 J. D. Wray, © McDonald Observatory

19 NASA; (inset) courtesy Bell Laboratories

20 courtesy Bell Laboratories

21 (both photos) Princeton University

22 Doug Frueh

23 (both photos) Harvard-Smithsonian Center for Astrophysics

24 (top) AIP Neils Bohr Library; (bottom) Clifford Stoltze

25 (top) Doug Frueh; (bottom) Mount Wilson and Las Companas Observatories, Carnegie Institution of Washington

26 M. Seldner, B. Siebers, E. J. Groth, P. J. E. Peebles, *A. J. 82*, 249, 1977

27 (top) "Horizon," BBC Television; (bottom) *The Churning of the Milky Ocean*, Bikaner Rajasthan, courtesy of Mr. Edward Binney, 3rd

28 (both photos) Lawrence Berkeley Laboratory, University of California

29 (top) Lawrence Berkeley Laboratory, University of California; (bottom and inset) Fermilab

30 (top) Sovfoto; (bottom) The Granger Collection

31 (top) Dan McCoy, Rainbow; (bottom, left to right) Wide World; California Institute of Technology; Stanford University; California Institute of Technology

32 (left) Doug Frueh; (right) Brookhaven National Laboratory

33 Culver Pictures

35 Harvard University

36 Clifford Stoltze

37 John S. Shelton; (inset) James Mason, Black Star

38 "Horizon," BBC Television

39 (top) John Bryson, Sygma; (bottom) World-Wide Standardized Seismic Station WES, Weston Observatory, a part of the World-Wide Standardized Seismic Network

40 (top) California Historical Society, San Francisco; (bottom) Gabriel Moulin Studios, courtesy Chin and Hensolt Engineers, William L. Pereira Associates, Architect

41 John H. Meehan, Photo Researchers

42 © 1980 Gary Braasch; (inset) Ralph Perry, Black Star

43 Ralph Perry, Black Star; (inset) Bill Thompson, Woodfin Camp

44–45 (both photos) © 1980 Gary Braasch

46 courtesy American Museum of Natural History

47 (top) Don Davis; (bottom) courtesy Field Museum of Natural History, Chicago

48 Susan LeVan

49 (top) Helen Michael, Frank Asaro, Walter Alvarez, Luis Alvarez, University of California, Berkeley; (bottom) Mark Fisher

50 (all photos) and 51 (top) Helen Michael, Frank Asaro, Walter Alvarez, Luis Alvarez, University of California, Berkeley

51 (bottom) John S. Shelton

52 (top) Doug Frueh; (bottom) Lawrence Berkeley Laboratory, University of California

53 (top and bottom left) The Granger Collection; (right) Culver Pictures

54 © Walt Disney Productions

55 Harvard College Observatory

56 (top and bottom) NSSDC/NASA; (middle) John S. Shelton

57 Clifford Stoltze

58 Renee Klein

59 Manfred Kage, Peter Arnold

60 Danny Brass, Photo Researchers

61 (top) David Scharf, Peter Arnold; (bottom) Peace River Films

62 (top) Milton Glaser; (bottom) Gardner and Gardner, University of Nevada, Reno

63 (top) WGBH; (bottom) Milton Glaser

64 (top) Gardner and Gardner, University of Nevada, Reno; (bottom) Mark Fisher

65 (top and bottom left) Gardner and Gardner, University of Nevada, Reno; (top right) Mark Fisher; (bottom right) WGBH

67 MIT

68 © 1982 Brian R. Wolff

69 (top) John S. Dunning for the Cornell Laboratory of Ornithology; (middle) Stephen T. Emlen, Cornell University; (bottom) musical notation from *Lullaby from the Great Mother Whale for the Baby Seal Pups*, by the humpback whale, Paul Winter, and Jim Scott, Living Earth Music © 1981, based on whale sounds recorded by Roger Payne and the New York Zoological Society

70 The Granger Collection

71 (top left) The Granger Collection; (top right) courtesy American Museum of Natural History; (bottom left) The Bettmann Archive; (bottom center and right) "Horizon," BBC Television

72 (top) © Sylvia Plachy 1981

72–73 (bottom) Doug Frueh after Niles Eldredge and Ian Tattersall, *The Myths of Human Evolution*, Columbia University Press, New York, 1982

73 (top) Roger Tully, Black Star

74 Douglas Kirkland, Contact

75 (all photos) Fritz Goro

76 The Bettmann Archive

77 (top left and right) Wide World; (bottom) The Bettmann Archive

78 Michael Lutch for WGBH

79 Jean Claude Lejeune, Stock, Boston; (inset) Michael Hayman, Stock, Boston

80 Michael Steinberg for WGBH

81 Jean Claude Lejeune, Stock, Boston

82 JPL/NASA

84 Wide World

85 (top and bottom right) Wide World; (bottom left) The New York *Times*

86 (left) JSC/NASA; (right) Wide World

87 NASA

88 (top) NASA; (bottom) Sovfoto

89 (left) JSC/NASA; (top right) Wayne Thompson; (bottom right) The Granger Collection

90 (all photos) JSC/NASA

91 (center photo) Lick Observatory; (all others) JSC/NASA

92 (inset) JSC/NASA

92–93 NSSDC/NASA

94 (top left) Culver Pictures; (right) Clifford Stoltze

95, 96, 97 (all photos) NASA

98 Pierre Mion © National Geographic Society

99 NASA

101 Jet Propulsion Laboratory

102 (both photos) Mount Wilson and Las Companas Observatories, Carnegie Institution of Washington

103 (top) Lick Observatory; (bottom) NASA

104 (top) Milton Glaser; (bottom) JPL/NASA

105 (top) JPL/NASA; (bottom) NSSDC/NASA

106, 107 (both photos) NSSDC/NASA

108 (top) Sovfoto

108–109 (bottom) Doug Frueh

109 (top) Lick Observatory

110–111 (bottom) Wayne Thompson

111 (top) JPL/NASA

112–120 (all photos) JPL/NASA

121 Dr. James F. Blinn, Jet Propulsion Laboratory

122 (bottom) Arthur Rothstein, Library of Congress

123 (top and bottom left) NASA; (middle left) © Sacramento Peak Observatory, Association of Universities for Research in Astronomy; (top right) by permission of the Houghton Library, Harvard University; (bottom right) The Granger Collection

124 NASA; (inset) Manfred Kage, Peter Arnold

125 (top left) courtesy Museum of Fine Arts Boston, gift of Mr. and Mrs. Edward A. Taft; (top right) Brown Brothers; (bottom right) Smithsonian Institution; (bottom chart and photo) courtesy National Center for Atmospheric Research

126 Kenneth Garrett, Woodfin Camp

128 Manfred Kage, Peter Arnold

129 Doug Frueh

130 Floyd Bloom, Salk Institute

131 Wayne Thompson

132 (top) Martha Stewart, The Picture Cube; (bottom) Mark Fisher

133 Milton Feinberg, The Picture Cube

134, 135 (all photos) Barbara Gullahorn-Holecek for WGBH

136 "Horizon," BBC Television

137 (top) "Horizon," BBC Television; (bottom) Manfred Kage, Peter Arnold

138 "Horizon," BBC Television

139 Peabody Museum of Salem

140 courtesy J. A. Sonnabend

141 Lee D. Simon, Photo Researchers

142 George Lange for WGBH

143 Steve Northup, Black Star

144 Manfred Kage, Peter Arnold

145 (top) K. R. Porter, Photo Researchers; (bottom) © Carol Lee

147 Wide World

148 (left) Wide World; (right) Mousson, Liaison

149 (top) Medical Center Hospitals, Norfolk, VA; (bottom photos) Raymond E. Stephens, Marine Biological Laboratory, from Studies on the Development of the Sea Urchin *Strongylocentrotus Droebachienesis, Biological Bulletin, 142,* Feb. 1972, pp. 132–144

150 (both photos) Television South, England

151 (top) Howard Sochurek, Woodfin Camp; (bottom right) The Bettmann Archive

152 (top) Milton Glaser; (bottom) Doug Frueh

153 (top photos) © 1976 Ted Spagna; (bottom) Mark Fisher

154 (center left) The Granger Collection; (bottom) courtesy Johnny Kelley; (inset) Peter Argentine for WGBH

155 (top) Christopher Brown, Stock, Boston; (bottom) Culver Pictures

156 (top) Peter Argentine for WGBH; (middle) courtesy Jeanne Reaves; (bottom) Peter Argentine for WGBH

157 (top left) Alison Sommers; (top right) Peter Argentine for WGBH; (bottom) Mark Fisher

158 (top) Jeff Foott; (middle) Peter Argentine for WGBH; (bottom) Alison Sommers

159 (top) Alison Sommers; (middle) Peter Argentine for WGBH; (bottom) Mark Fisher

161 Andrus Gerontology Center, University of Southern California

162 Jim Harrison, Stock, Boston

163 (top) Tom McHugh, Photo Researchers; (bottom) Gabor Demjen, Stock, Boston

164 Wide World

165 (both photos) United Press International

166 David Thompson, Liaison

167 Joe Traver, Liaison

168 Fred Ward, Black Star

169 (animation sequence) designed by Paul Souza, illustrated by Lori Snell

170 Mark Fisher

171 The Bettmann Archive

172 John Launois, Black Star

173 Martin Rogers, Woodfin Camp; (top and bottom insets) Laurent Maous, Liaison

174 Ted Spiegel, Black Star

175 Jan Galligan, New York State Department of Health

177 © 1981 Pat Seip

178 (bottom left) FAO; (bottom right) O. Monsen, Liaison

179 FAO

180 (top) Erich Hartmann, Magnum; (illustration) Milton Glaser

181 (top) Photo Researchers; (bottom) David Scharf, Peter Arnold

182 painting by John Bertonccini, ca. 1900, Old Dartmouth Historical Society, The Whaling Museum, New Bedford, MA

183 (top left) Elvida Abella for WGBH; (top right) Hunter's Signal, Pootagook, 1961, reproduced with the permission of the West Baffin Eskimo Cooperative Ltd, Cape Dorset, Northwest Territories; (bottom) Mark Fisher

184 Susan LeVan

185 (top) Brown Brothers; (bottom) Milton Glaser

186 JSC/NASA

187 Wayne Thompson

188 (top) GSFC/NASA; (bottom) Mark Fisher

189 (top) Jen and Des Bartlett, Photo Researchers; (bottom) Susan LeVan

190 Michael Yamashita, Woodfin Camp

191 (top) Alain Nogues, Sygma; (bottom) Mark Fisher

192, 193 (both photos) Loren McIntyre

193 (bottom) Mark Fisher

194 Dan McCoy, Rainbow

196 (center left) The Granger Collection; (right) Anton Higgs

197 courtesy the DuPont Company; (inset) The Granger Collection

198 (all photos) © Don Monroe; (illustration) Doug Frueh

199 (both photos) © Don Monroe; (illustration) Mark Fisher

200 Peter Marlow, Sygma; (inset) © Don Monroe

201 © Martyn Cowley; (inset) © Don Monroe

202 (bottom left) Simone Sacconni, I "Segreti" di Stradivari, Libreria del Convegno, Cremona, 1972; (right) Brown Brothers

203 James Gill, WHA Television, Madison, Wisconsin

204 (top) Simon Sacconni, I "Segreti" di Stradivari, Libreria del Convegno, Cremona, 1972; (bottom) The Great Violin Mystery, produced by WHA Television, Madison, Wisconsin

205 Alexa Pfister, WHA Television, Madison, Wisconsin

206 (center) from the MGM release 2001: A Space Odyssey © 1968 Metro-Goldwyn-Mayer; (bottom) WGBH

207 Manfred Kage, Peter Arnold; (inset) courtesy Sperry Univac

208 (left and right) The Bettmann Archive; (center) United Press International

209 (far left) courtesy SRI International; (center left) WGBH; (center right) J. P. Laffont, Sygma; (far right) Jet Propulsion Laboratory; (bottom) WGBH

210 (both illustrations) Culver Pictures

211 (top and far left bottom) WGBH; (center left bottom) M. Freeman, Woodfin Camp; (right bottom) Tony Karody, Sygma

212 (left) Metropolis, Academy of Motion Picture Arts & Sciences; (right) © Lucasfilm Ltd. (LFL) 1977, All rights reserved.

213 Tadanori Saito, Photo Researchers

215 Ivan Masser, Black Star

216 Ellis Herwig, Stock, Boston

217 courtesy Richard A. Viguerie Company

218 (top) Tyron Hall, Stock, Boston; (bottom) WGBH

219 (top) George Lange for WGBH; (bottom) Doug Frueh

220 (center) John Youssi, reprinted with permission from the 1980 Yearbook of Science and the Future, © 1979 Encyclopaedia Britannica, Inc., Chicago, Illinois; (bottom) Culver Pictures

221 McDonald Observatory; (inset) Lawrence Livermore Laboratory

222 (top) Dan McCoy, Rainbow; (bottom) Wayne Thompson

223 (top) courtesy Spectra-Physics; (bottom) WGBH

224 Michael Lutch for WGBH

225 (top) Doug Frueh; (all photos) Fritz Goro, Life Magazine, © Time Inc

226 Paul Souza

227 (top) NORA, © 1980 by David Em; (bottom) Dr. James F. Blinn, Jet Propulsion Laboratory

228 Harriet Casdin-Silver © 1977

229 (top and center right) WGBH; (bottom left and right) Nelson Max, Computer Graphics Group, Lawrence Livermore Laboratory, University of California

230 (top left) John Mansfield

230, 231 ("NOVA" opening) animation design by Paul Souza, technical direction and animation by David M. Geshwind

232 Harvard Medical Archives

233 Alison Sommers

234 Joe Munroe, Photo Researchers

235 Kenneth Garrett, Woodfin Camp

236 Jeannine Niepce, Photo Researchers; (inset) John Blaustein, Woodfin Camp

237 (top) Joe Munroe, Photo Researchers; (bottom) Alain Nogues, Sygma

238 Nelson Max, Computer Graphics Group, Lawrence Livermore Laboratory, University of California

239 (top left) Dr. Stanley Cohen, Stanford University; (bottom left) David Scharf, Peter Arnold; (right) Renee Klein

240 (top) Andrew A. Stern for NAS; (bottom) Mark Fisher

241 (top) Paul Conklin for the National Academy Forum; (bottom) Doug Frueh

242 (top) B. Ben Bohlool, University of Hawaii at Manoa; (bottom left) A. C. Barrington Brown, from J. D. Watson, The Double Helix, Atheneum, New York, p. 215, © 1968 by J. D. Watson; (bottom right) Rosalind Franklin, from J. D. Watson, The Double Helix, Atheneum, New York, p. 168, © 1968 by J. D. Watson

243 (both photos) Douglas Kirkland, Contact

245 Ralph Crane

246 Patricia Gross, Stock, Boston

247 R. S. Uzzell III, Woodfin Camp

248 (photo) Photo Researchers; (map) Russell Brami

249 (top) Bruno Barbey, Magnum; (bottom) Ghislaine Morel, Liaison

250 (top) Billings Research Energy Corporation; (bottom) Mark Fisher

251 (all photos) Dan McCoy, Rainbow

252 (top) Christa Armstrong, Photo Researchers; (bottom) U.S. Department of Energy

253 (top) Oak Ridge National Laboratory; (bottom) Doug Frueh

254, 255 (all photos) Lawrence Livermore Laboratory, University of California

256 (left) Waite, Sygma; (right) © 1980 Sipa Press, Black Star

257 Wide World

258, 259 (all photos) Richard Land, Home Fire Project, Harvard University

260 Brown Brothers; (inset) Federal Emergency Management Agency

261 U.S. Department of Energy

262 United Press International

263 (top) United Press International; (bottom) Joyce Dopkeen, The New York Times

264 (top) WGBH; (bottom left) reprinted with permission of the Encyclopedia Americana, © 1982, Grolier Inc.; (bottom right) U.S. Atomic Energy Commission

265 Library of Congress

267 (top) MIT; (bottom) reprinted with permission of The Bulletin of the Atomic Scientists, a magazine of science and public affairs, © 1947 by the Educational Foundation for Nuclear Science, Chicago, Illinois

268 The Bettmann Archive

269 (top left) Gianfranco Gorgoni, Contact; (top right) David Newell Smith, Photo Trends; (bottom) Ivan Strasburg

270 Bernard Pierre Wolfe, Magnum

271 George Laycock, Photo Researchers

272 CARE

273 J. P. Laffont, Sygma

274 (both illustrations) Mark Fisher

275–276 Milton Glaser

277–281 (all illustrations) Mark Fisher

288 ("NOVA" opening) animation design by Paul Souza, technical direction and animation by David M. Geshwind

Index

Book Staff

WGBH Design

Designer
Paul Souza

Production Coordinator
Lori Snell

Photo Researchers
Elise Katz
Richard Maurer

Design Assistants
Julie Frankel
Doug Frueh
Ilisha Helfman
Sylvia Peretz

Production Manager
Gail Zimmermann

Manager, Design and Publications
Chris Pullman

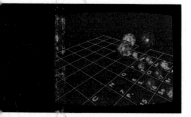

An introductory chapter to this book is by one of the members of the "NOVA" Advisory Panel, Professor Philip Morrison of Massachusetts Institute of Technology. Among the other members of this group of scientific luminaries who have helped the television series for several years are Dr. Daniel Fox of the Health Sciences Center at the State University of New York; Professors Stephen Jay Gould and Gerald Holton of Harvard University; Dr. Arthur Kantrowitz of Dartmouth College; and Dr. Harry Woolf of the Institute of Advanced Study at Princeton.

WGBH Publications

Writers
Joanna Berkman
Elissa Ely
Fred Hapgood
Karen Janszen
Jeff Rothenberg
Bill Scheller
Ron Winslow
And the producers of "NOVA" programs, with thanks to:
John Angier
Robin Bates
Ted Bogosian
Thea Chalow
Graham Chedd
Barbara Gullahorn-Holecek
Brian Kaufman
Pat Kent
John Mansfield
Bebe Nixon
Mick Rhodes
Robert Richter
Sari Sapir
Veronica Young
Robert Zalisk
Conversations
Hannah Benoit
Donald Alan Cooke
Peter Gwynne
Victor McElheny
Dennis Meredith

Editorial Assistants
Carol Hills
Claire Levine

Director of Publications
Karen Johnson

With special thanks to the producers of "Horizon" and the British Broadcasting Corporation, whose productions have contributed to the "NOVA" series and this book

Addison-Wesley Publishing Company

Senior Editor
Doe Coover

Managing Editor
Barbara Wood

Copy Editor
Kendra Crossen

Proofreader
Alan Clark

Editorial Assistant
Jordan Budd

Indexer
Lois Oster

Technical Reviewer
James Cornell

Vice-President of Manufacturing
Jim Georgeu

Promotion and Advertising Manager
Lisa Strawbridge

Director, Marketing and Sales
Joyce Copland

Editor-in-Chief
Christopher Kuppig

Publisher and Corporate Vice-President
Ann Dilworth

The text of this book was set in the Gill Sans family of typefaces on a Linotron 202 typesetter by P & M Typesetting, Inc. in Waterbury, Connecticut. The color separations and electronic page assembly were provided by Color Response, Inc., a Banta Company. W. A. Krueger of New Berlin, Wisconsin, printed and bound the books on 70-pound P.H. Glatfelter Old Forge Offset Enamel paper from Lindenmeyr Paper Company.